Organic Cereal and Pulse Production

A COMPLETE GUIDE

Stephen Briggs

Organic Cereal and Pulse Production

A COMPLETE GUIDE

Stephen Briggs

THE CROWOOD PRESS

First published in 2008 by
The Crowood Press Ltd
Ramsbury, Marlborough
Wiltshire SN8 2HR

www.crowood.com

British Library Cataloguing-in-Publication Data
A catalogue record for this book is available from the British Library.

ISBN 978 1 86126 953 9

Acknowledgements

A big thank-you to Lynn, Harriett and Daisy for allowing me to dedicate time for writing this book.

The author would like to thank the following organizations and individuals for their assistance in providing background information for this book:
Defra, the Home-Grown Cereals Authority (HGCA); A. Biddle at the Processors and Growers Research Organisation (PGRO); the Soil Association; S. Padel at the University of Wales, Aberystwyth; N. Gossett at Organic Grain Link; Dr J. Thomas and Dr D. Kenyon at NIAB; Dr P. Northing at CSL, Dr P Lutman at Rothamstead research, Dr M Shepherd, Dr B Cormack & G Goodlass at ADAS; Dr D. Hatch and S. Cuttle at IGER; Dr E. Stockdale at the University of Newcastle-upon-Tyne; Dr C. Watson and D. Younie at SAC; Dr M. Lennartsson, G. Jones and F. Rayns at HDRA; Dr D. Sparkes at the University of Nottingham; and L Philipps.

Distributed in North, America by:
Diamond Farm Book Publishers
PO Box 537, Alexandria Bay, NY 13607, USA
www.diamondfarm.com

Typeset by S R Nova Pvt Ltd., Bangalore, India

Printed and bound by Replika Press, India

Contents

Preface

This book is aimed at students, researchers, existing organic farmers, those considering conversion, and advisers/consultants providing services in the agricultural sector. The book provides a comprehensive reference text on UK organic cereal and pulse production. It outlines the organic production principles and options, and discusses points to consider when determining management options. It provides guidelines on production options, agronomy, sowing dates and yield expectations, as well as discussion options for management when things do not always go to plan.

The book layout covers the broader issues of organic cereal and pulse production in the context of the wider organic industry and market demands, together with the broad principles underpinning organic production. Topics covered include cropping issues related to cereals and pulses; crop establishment and management; weed management; pest and disease management; appropriate crop storage; the importance of biodiversity management within organic systems; and the production of organic cereals and pulses in the context of the CAP reforms, including how organic production fits into UK government policy, environmental regulations and agri-environment schemes.

The book is written from a UK perspective but draws on information from continental Europe, the USA, Australasia as appropriate, and elsewhere. Information has been drawn together from a wide range of research, published articles, books, market information, and from personal experience.

Published in 2007 during one of the most significant changes to the structure of UK agriculture in nearly three decades, I hope that the book provides a useful reference document to those already involved in organic production, and to others seeking information to help develop an organic production system.

WHAT IS ORGANIC FARMING?

Organic farming has been described as a production system that relies on the management of biological cycles, rather than agrochemical interventions, for nutrient supply and the control of weeds, pests and diseases. The philosophical basis of organic farming can be shown to be distinct from that of other systems of agriculture in that the farm is considered as an integrated whole (or system). This holistic approach recognizes that complex relationships exist between the components (soil, soil organisms, plants, animals, humans and the wider environment) within the farm and with the wider environment, but that the system is capable of responding as a whole to external stimuli. Soil 'life' and 'quality' is central to this philosophy.

Organic farming not only recognizes the need to consider 'whole farm systems' and to understand the complex interactions between components, but also embraces the concepts of local food production, distribution and consumption, and the need to consider the social impact of farming systems. Thus organic farming differs significantly from other modern forms of agriculture, such as integrated crop management (ICM), which involves the modification of individual practices rather than the whole system.

Chapter 1

Introduction

This book is aimed at both existing arable producers who wish to expand their knowledge of organic arable production, as well as those who are considering arable crops as a new part of their rotation. As it is impossible to cover every farm structure and set of circumstances/constraints, and also the variables encountered between growing seasons, which are often out of the control of the producer, this book should be used as a *guide* rather than as a definitive manual for production. It should provoke questions that many farmers ask when considering organic arable cropping, for example, what crop markets are available? Where should the crop fit into the rotation? What problems may be encountered in growing a particular crop? What preventative planning and methods can be used?

At the start of any 'guide', it is always useful to consider the context in which the subject in question has arisen and developed over time. The roots of organic farming have shaped its past, and will as a consequence continue to shape its future development.

THE HISTORICAL ROOTS OF ORGANIC FARMING

The history of organic farming is one of principles, agricultural methods and markets. It is also largely the history of the organic movement, which began as an 'insiders' group of agricultural scientists and farmers, and later expanded to become a grassroots consumer cause.

The organic movement began as a reaction of agricultural scientists and farmers who had major concerns regarding the industrialization of agriculture, the advances in biochemistry, and engineering, which in the early twentieth century led to profound changes in farming. This was leading to a changed landscape, with enlarged field sizes and specialized mono-cropping to make efficient use of machinery, and to reap the benefits of the green revolution. Technological advances during World War II spurred on post-war innovation in all aspects of agriculture, resulting in such advances as large-scale irrigation, fertilization, and the use of pesticides and plant breeding which produced hybrid seeds. Ammonium nitrate fertilizers became an abundantly cheap source of nitrogen. DDT, originally developed by the military to control disease-carrying insects among troops, was applied to crops, launching the era of widespread pesticide use.

Initially organic farming practitioners and advocates focused on the methods, as a definite reaction against the industrialization of agriculture, and remained below the awareness of the food buyer. Only when the contrasts between organics and the new conventional agriculture became overwhelming did organics rise to the attention of the public, creating a distinct organic market. World War II roughly marks the two phases of the organic movement.

Pre-World War II

The first forty years of the twentieth century saw simultaneous advances in biochemistry and engineering that rapidly and profoundly changed farming. The introduction of the internal combustion engine ushered in the era of the tractor, and made possible hundreds of mechanized farm implements. Research into plant breeding led to the commercialization of hybrid seed. And a new manufacturing process made nitrogen fertilizer, first synthesized in the mid-1800s and made affordable and abundant. With increased mechanization and use of fertilizers, fields grew bigger and cropping more specialized to make more efficient use of machinery. But in England in the 1920s, a few individuals in agriculture began to speak out against these agricultural trends.

Consciously organic agriculture (as opposed to the agriculture of indigenous cultures, which always employs only organic means) began more or less simultaneously in Central Europe and India. The British botanist Sir Albert Howard is often referred to as the father of modern organic agriculture. From 1905 to 1924 he worked as an agricultural adviser in Pusa, Bengal, where he documented traditional Indian farming practices, and came to regard them as superior to his conventional agriculture science. His research and the further development of these methods is recorded in his writings, notably his book written in 1940, *An Agricultural Testament*, which influenced many scientists and farmers of the day.

In Germany, Rudolf Steiner's development, biodynamic agriculture, was probably the first comprehensive organic farming system. This began with a lecture series that Steiner presented at a farm in Koberwitz (now in Poland) in 1924. This lecture series, published in English as *Spiritual Foundations for the Renewal of Agriculture*, was the very first publication anywhere on organic agriculture. A number of farmers interested in finding a healthier approach to farming attended the course, and several farms began working with a biodynamic/organic approach. Steiner emphasized the farmer's role in guiding and balancing the interaction of the animals, plants and soil: thus healthy animals depended upon healthy plants (for their food), healthy plants upon healthy soil, healthy soil upon healthy animals (for the manure).

In 1939, influenced by Sir Howard's work, Lady Eve Balfour launched the Haughley Experiment on farmland near Stowmarket, in Suffolk, England. It was the first scientific, side-by-side comparison of organic and conventional farming. Four years later she published *The Living Soil*, based on the initial findings of the Haughley Experiment. Widely read, it led to the formation of a key international organic advocacy group, the Soil Association, which has now become one of the leading organizations in the organic sector.

The coinage of the term 'organic farming' is usually credited to Lord Northbourne, in his book, *Look to the Land* (1940), wherein he described a holistic, ecologically balanced approach to farming.

In Japan, Masanobu Fukuoka, a microbiologist working in soil science and plant pathology, began to doubt the modern agricultural movement. In the early 1940s he quit his job as a research scientist, returned to his family's farm, and devoted the next thirty years to developing a radical no-till organic method for growing grain, now known as 'Fukuoka farming'.

Post-World War II

Technological advances during World War II accelerated post-war innovation in all aspects of agriculture, resulting in big advances in mechanization (including large-scale irrigation), fertilization, and pesticides. In particular, two chemicals that had been produced in quantity for warfare were repurposed to peace-time agricultural uses. Ammonium nitrate, used in munitions, became an abundantly cheap source of nitrogen, and a range of new pesticides appeared: DDT, which had been used to control disease-carrying insects around troops, became a general insecticide, launching the era of widespread pesticide use. At the same time, increasingly powerful and sophisticated farm machinery allowed a single farmer to work ever larger areas of land, and as a result many hedgerows were removed and fields grew bigger in size in the search for efficiency.

In 1944, an international campaign called the Green Revolution was launched in Mexico with private funding from the USA. It encouraged the development of hybrid plants, chemical controls, large-scale irrigation, and heavy mechanization in agriculture around the world.

During the 1950s, sustainable agriculture was a topic of scientific interest, but research tended to concentrate on developing the new chemical approaches. In the US, J. I. Rodale began to popularize the term and methods of organic growing, particularly to consumers, through the promotion of organic gardening.

In 1962 Rachel Carson, a prominent scientist and naturalist, published *Silent Spring*, chronicling the effects of DDT and other pesticides on the environment. A bestseller in many countries, including the UK, and widely read around the world, *Silent Spring* is widely considered as being a key factor in the banning of DDT in many countries. The book and its author are often credited with launching the worldwide environmental movement.

In the 1970s, global movements concerned with pollution and the environment increased their focus on organic farming. As the distinction between organic and conventional food became clearer, one goal of the organic movement was to encourage the consumption of locally grown food and thereby reduce food miles, a theme which is again coming to the forefront of people's consciousness.

In 1972, the International Federation of Organic Agriculture Movements, widely known as IFOAM, was founded in Versailles, France. It was dedicated to the diffusion and exchange of information on the principles and practices of organic agriculture of all schools and across national and linguistic boundaries.

In 1975 Fukuoka released his first book, *One Straw Revolution*, which had a strong impact in certain areas of the agricultural world. His approach to small-scale grain production emphasized the meticulous balance of the local farming ecosystem, with a minimum of human interference and labour.

In the 1980s, various farming and consumer groups around the world began seriously pressuring for government regulation of organic production. This led to legislation and certification standards being enacted through the 1990s and to date.

Since the early 1990s, the retail market for organic farming in developed economies has been growing by about 20 per cent annually, due to increasing consumer demand. While initially it was the small independent producers and consumers who drove the rise in organic farming, at the time of writing the volume and variety of 'organic' products continues to grow, and production is on an increasingly large scale. Concern for the quality and safety of food, and the potential for environmental damage from conventional agriculture, are mainly responsible for this trend.

The Twenty-First Century

Throughout this history, agricultural research, and the majority of publicized scientific findings, has been focused on agrochemical management and fertilizer use, rather than on organic farming. This emphasis has continued to biotechnologies such as genetic engineering (GE), more commonly known as genetic modification (GM). One recent survey of the UK's leading government funding agency for bioscience research and training indicated twenty-six GM crop projects, and only a handful related to organic agriculture. This imbalance is largely driven by agribusiness in general, which, through research funding and government lobbying, continues to have a predominating effect on agriculture-related science and policy.

In Havana, Cuba, a unique situation has made organic food production a necessity. Since the collapse of the Soviet Union in 1989 and the withdrawal of its economic support, Cuba has had to produce food in creative ways, such as instituting the world's only state-supported infrastructure to promote urban food production. Called *organopónicos*, the city is able to provide an ever-increasing amount of its produce organically – although if the USA embargo is lifted, the future of organic urban growing here may be in peril.

Agribusiness is also changing the rules of the organic market. The rise of organic farming was driven by small independent farmers and by consumers. In recent years, explosive organic market growth has encouraged the participation of agribusiness interests. As the volume and variety of 'organic' products increases, the viability of the small-scale organic farm is at risk, and the meaning of organic farming as an agricultural method is ever more easily confused with the related but separate areas of organic food and organic certification.

ORGANIC FARMING TODAY

Organic Conversion and Certification

Organic farming is defined by formal standards regulating production methods, and in some cases, final output. Two types of standard exist: voluntary and legislated. As early as the 1970s, private associations created standards against which organic producers could voluntarily have themselves certified. In the 1980s, governments began to produce organic production guidelines. And beginning in the 1990s, a trend towards the legislation of standards began, most notably the EU-Eco-regulation developed in the European Union.

In 1991, the European Commission formulated the first government system to regulate organic labelling. In one go, the European Regulation (EEC) 2092/91 set the rules in twelve countries, creating a huge market Organic certification, which until then was a voluntary quality control system, became mandatory to all operations and was also to be applied for imports. In the meantime, Europe had become the most prominent market place for organic products, and an increasing number of suppliers all over the world accepted this niche as a new challenge and a rewarding option to export high quality and high priced speciality products. All these supplies, of course, had to comply with the requirements of the European market, and thus the Regulation (EEC) No. 2092/91 became a universal standard for organic production systems.

IFOAM

An international framework for organic farming is provided by the International Federation of Organic Agriculture Movements (IFOAM), the international democratic umbrella organization established in 1972. For IFOAM members, organic agriculture is based upon the Principles of Organic Agriculture and the IFOAM Norms. The IFOAM Norms consist of the IFOAM Basic Standards and IFOAM Accreditation Criteria.

The IFOAM Basic Standards are a set of 'standards for standards'. They are established through a democratic and international process, and reflect the current state of the art for organic production and processing. They are best seen as a work in progress to lead the continued development of organic practices worldwide. They provide a framework for national and regional standard setting, and for certification bodies to develop detailed certification standards that are responsive to local conditions.

The IFOAM principles and standards of organic agriculture were revised in 2006, to more clearly represent the founding principles and desired future direction of the organic sector. These revised principles serve to inspire the organic movement in its full diversity. They are the basic principles and roots from which organic agriculture grows and develops. They express the contribution that organic agriculture can make to the world, and a vision to improve all agriculture in a global context. They also guide IFOAM's development of positions, programmes and standards. Furthermore, they are presented with a vision of their world-wide adoption.

According to IFOAM, the role of organic agriculture, whether in farming, processing, distribution or consumption, is to 'sustain and enhance the health of ecosystems and organisms from the smallest in the soil to human beings'.

The revised IFOAM standards focus on four main principles, and recognize that organic agriculture is based on:

1 The principle of health
2 The principle of ecology
3 The principle of fairness
4 The principle of care

The four IFOAM principles are articulated through a statement followed by an explanation. The principles are to be used as a whole, and are composed as ethical principles to inspire action.

1 Principle of health
Organic agriculture should sustain and enhance the health of soil, plant, animal, human and planet as one and indivisible.

This principle points out that the health of individuals and communities cannot be separated from the health of ecosystems – thus healthy soils produce healthy crops that foster the health of animals and people.

Health is the wholeness and integrity of living systems. It is not simply the absence of illness, but the maintenance of physical, mental, social and ecological well-being. Immunity, resilience and regeneration are key characteristics of health.

As we have seen, according to IFOAM, the role of organic agriculture, whether in farming, processing, distribution or consumption, is to sustain and enhance the health of ecosystems and organisms from the smallest in the soil to human beings. In particular, organic agriculture is intended to produce high quality, nutritious food that contributes to preventive health care and well-being. In view of this it should avoid the use of fertilizers, pesticides, animal drugs and food additives that may have adverse health effects.

2 Principle of ecology
Organic agriculture should be based on living ecological systems and cycles, work with them, emulate them and help sustain them.

This principle roots organic agriculture within living ecological systems. It states that production is to be based on ecological processes and recycling. Nourishment and well-being are achieved through the ecology of the specific production environment. For example, in the case of crops this is the living soil; for animals it is the farm ecosystem; for fish and marine organisms, the aquatic environment.

Organic farming, pastoral and wild harvest systems should fit the cycles and ecological balances in nature. These cycles are universal, but their operation is site-specific. Organic management must be adapted to local conditions, ecology, culture and scale. Inputs should be reduced by reuse, recycling and efficient management of materials and energy in order to maintain and improve environmental quality and to conserve resources.

Organic agriculture should attain ecological balance through the design of farming systems, the establishment of habitats, and the maintenance of genetic and agricultural diversity. Those who produce, process, trade or consume organic products should protect and benefit the common environment including landscapes, climate, habitats, biodiversity, air and water.

3 Principle of fairness

Organic agriculture should build on relationships that ensure fairness with regard to the common environment and life opportunities.

Fairness is characterized by equity, respect, justice and stewardship of the shared world, both among people and in their relations to other living beings.

This principle emphasizes that those involved in organic agriculture should conduct human relationships in a manner that ensures fairness at all levels and to all parties – farmers, workers, processors, distributors, traders and consumers. Organic agriculture should provide everyone involved with a good quality of life, and should contribute to food sovereignty and the reduction of poverty. It aims to produce a sufficient supply of good quality food and other products.

This principle insists that animals should be provided with the conditions and opportunities of life that accord with their physiology, natural behaviour and well-being.

Natural and environmental resources that are used for production and consumption should be managed in a way that is socially and ecologically just, and should be held in trust for future generations. Fairness requires systems of production, distribution and trade that are open and equitable, and that account for real environmental and social costs.

4 Principle of care

Organic agriculture should be managed in a precautionary and responsible manner to protect the health and well-being of current and future generations and the environment.

Organic agriculture is a living and dynamic system that responds to internal and external demands and conditions. Practitioners of organic agriculture can enhance efficiency and increase productivity, but this should not be at the risk of jeopardizing health and well-being. Consequently, new technologies need to be assessed and existing methods reviewed. Given the incomplete understanding of ecosystems and agriculture, care must be taken.

This principle states that precaution and responsibility are the key concerns in management, development and technology choices in organic agriculture. Science is necessary to ensure that organic agriculture is healthy, safe and ecologically sound. However, scientific knowledge alone is not sufficient. Practical experience, accumulated wisdom, and traditional and indigenous knowledge offer valid solutions, tested by time. Organic agriculture should prevent significant risks by adopting appropriate technologies and rejecting unpredictable ones, such as genetic engineering. Decisions should reflect the values and needs of all who might be affected, through transparent and participatory processes.

International Organic Legislation

Legislated standards are established at a national level, and vary from country to country. In recent years, many countries have legislated organic production, including the EU nations (1990s), Japan (2001), and the US (2002). Non-governmental national and international associations also have their own production standards. In countries where production is regulated, these agencies must be accredited by the government.

In India, standards for organic agriculture were announced in May 2001, and the National Programme on Organic Production (NPOP) is administered under the Ministry of Commerce. In the United States, the Department of Agriculture (USDA) established production standards in 2002, under the National Organic Program.

Since 1993 when EU Council Regulation 2092/91 became effective, organic food production has been strictly controlled in the UK, and this regulation also defines the commercial use of the term 'organic'. Farmers and food processors must comply with the legislation in order to use the word organic, and this dictate is policed by trading standards in the UK.

EU and UK Organic Regulations

The principles set out by IFOAM make clear the importance of food quality, natural systems and cycles, the minimization of environmental impact, the maximization of animal welfare, and the need to consider the wider social impacts of agriculture. Organic farming is the only system of sustainable agriculture that is legally defined (EU 2092/91, 1804/99), but it is important to note that this regulation defines a production process and not the product of that process. The framework that regulates organic farming thus protects the integrity of organic production but means that the system is rule based.

Organic production in the EU is defined and governed by two regulations: Council Regulation (EEC) 2092/91, which covers crop production; and 1084/99, which covers livestock production. As such, organic production is the only legally defined method of agriculture in the EU, and each member state has to establish a competent authority to implement the regulation. Within the UK, this authority is part of the Department of Environment Food and Rural Affairs (DEFRA) and is called the Advisory Committee on Organic Standards (ACOS): it provides baseline organic standards for the UK, setting the compliant standards for producers, and it also registers and approves UK organic certification bodies, and monitors their procedures.

There are nine UK-approved organic certification bodies (*see* Chapter 14), which set their own organic standards (based on, and with, the ACOS basic standard as a minimum) and register organic producers and processors. It is the production system that is being certified.

Although regulated and defined by EU regulations, organic standards were originally developed from – and are still underpinned by – a range of underlying principles. These are that organic farming should:

- co-exist with, rather than dominate, natural systems;
- minimize pollution and damage to the environment;
- protect the farm environment with particular regard to wildlife;
- consider the wider social and ecological impact of agricultural systems;
- maintain or develop valuable existing landscape features and adequate habitats for the protection of wildlife, with particular regard to endangered species;
- ensure the ethical treatment of animals.

EUROPEAN ORGANIC AGRICULTURE

The area of land under organic production is growing rapidly, in the EU15 countries. In 1985, certified organic land and land in conversion accounted for less than 0.1 per cent of the total utilizable agricultural area (UAA). By 1998 this had risen to 2.71 million hectares (2.1 per cent of the total UAA), and by 2004 over 6 million ha, or 4 per cent of total farmland, was managed organically on 150,000 holdings across the EU15.

In total there are over 157,000 registered operators (producers, processors and importers) in the EU15; approximately 132,000 of these are farmers. The average organic farm size in Europe (40ha) is larger than the average non-organic farm size (15ha), but still far smaller than the average UK organic holding at 146ha.

In 2003, grassland and fodder crops were grown on 61 per cent (or 3.1 million ha) of the total organic land area in Europe. Arable crops were grown on a quarter of the total land area, and a further 8 per cent (0.4 million ha) was under horticultural production.

In 2003, approximately three million livestock units (LU), or 2.3 per cent of the EU15's total, were certified as organic. Within the EU15, certified dairy cows amounted to 483,000 head, or 2.5 per cent of the total dairy herd, with the UK and Germany representing 40 per cent. Other EU15 certified cattle amounted to one million head or 1.7 per cent of the total cattle (non-dairy) herd, a quarter of which were held in Austria. The number of certified organic pigs was relatively low, with 450,000 head (or 0.4 per cent of EU15's total pig herd). There were six million certified laying hens, the majority of which were in the UK and France. Certified sheep and goats amounted to 2.4 million head (or 2.4 per cent of total sheep and goat herds). Significant proportions of the sheep are located in the UK and Italy, and half of the certified goats are located in Greece.

UK AGRICULTURE IN CONTEXT

The total land area used for agricultural production according to Defra was 18,509,000 hectares in 2005, with 5,777,000 hectares used for arable crop production (Table 1).

Table 1 UK agricultural land use in 2005

	Thousand hectares
Total agricultural area (b)	18,509
Crops	4,443
Bare fallow	140
Total tillage	4,583
All grass under five years old	1,193
Total arable land	5,777
All grass five years old and over (excluding rough grazing)	5,711
Total tillage and grass (c)	11,488
Sole right rough grazing	4,354
Set-aside	559
All other land (d) and woodland	872
Total area on agricultural holdings	17,273
Common rough grazing (estimated)	1,236

The data in this table cover all holdings (including minor holdings) in the UK (a)
(a) Before 2000 Scottish minor holdings were not included; data for earlier years are therefore not directly comparable. From 1997 the Northern Ireland census was based on an improved register of holdings and included all active farms having one or more hectares of farmed land plus any below that size which had significant agricultural output. Data prior to 1997 were revised for comparability. (b) Total area on agricultural holdings plus common rough grazing. (c) Includes bare fallow. (d) In Great Britain other land comprises farm roads, yards, buildings (excluding glasshouses), ponds and derelict land.
Source : Defra statistics

The area of land in the UK used for crop production in 2005 is shown in Table 2. Wheat is by far the most significant combinable crop in the UK, followed in area by barley, oilseed rape, peas and beans, sugar beet, oats, triticale, rye and other crops.

Table 2 UK agricultural crop areas 2005

	Thousand hectares
Total UK agricultural area cropped	**4,443**
Wheat	*1,868*
Barley	*942*
Oats	*91*
Rye and mixed corn	*9*
Triticale	*13*
Total UK cereal area	**2,925**

Table 2 (continued)

Oilseed rape	519
Sugar beet not for stock feeding	148
Hops	1
Peas for harvesting dry and field beans	239
Linseed (c)	45
Other crops	252
Total other arable crops (excluding potatoes)	**1,211**
Potatoes	**137**
Vegetables grown in the open	121
Orchard fruit (d)	23
Soft fruit (e)	9
Plants and flowers (f)	14
Glasshouse crops	2
Total horticulture	**170**

(c) England and Wales only prior to 1992. (d) Includes non-commercial orchards. (e) Includes wine grapes. (f) Hardy nursery stock, bulbs and flowers.
Source : Defra Statistics

UK ORGANIC AGRICULTURE IN CONTEXT

The global market for organic food and drink was worth an estimated £16.7 billion in 2005. The UK has the third largest market for organic food in Europe, after Germany and Italy. More than 50 per cent of Europe's organic land is in Italy, Germany, Spain and the UK.

Organic Food Sales

The UK organic sector has been on a healthy upward trend for the last decade; the latest figures show an annual growth of 30 per cent for UK food and drink sales, amounting to £1.6 billion. As the overall food market in the UK is fairly inelastic, growing by just 2 or 3 per cent annually, the main supermarket buyers are keen to procure organic products. In 2002, only 30 per cent of primary organic produce was UK-sourced; by the end of 2006 nearly 70 per cent of organic foodstuffs sold in supermarkets that can be grown domestically is sourced from the UK.

While nine out of ten people buy organic produce from supermarkets, over 50 per cent would prefer to buy from smaller, local greengrocers, farmers' markets and box schemes. Therefore it seems that the future is local, both for consumer preference as well as to reduce climate-change-causing food miles. This is ably demonstrated by the growth in box schemes (organic vegetable boxes delivered to customers' door steps) and farmers' markets, with sales of organic box schemes and mail order growing by 22 per cent between 2005 and 2006, and with over 550 farmers' markets in the UK, with a turnover of £220 million, since the first was set up in 1997. In 2003, approximately half of consumers were knowingly buying organic food; by 2006 it is estimated that this has risen to nearly two in every three shoppers.

Organic Land Area

Organically managed land now accounts for approximately 4 per cent of the UK's total agricultural land area. According to statistics produced by the Soil Association, the area of agricultural land in the UK under organic management has increased from 8,200ha in 1997 to 631,144ha in January 2006. Within this, the area of in-conversion land increased for the first time in four years by 68 per cent to 87,020ha (215,000 acres). In-conversion land now accounts for 14 per cent of the UK's organically managed land area, with all countries in the UK seeing growth in 2005: Scotland (33 per cent), England (85 per cent), Wales (55 per cent) and Northern Ireland (102 per cent) respectively.

Over the same period, the UK's fully organic land area decreased by 14 per cent to 544,124ha. This can be attributed to a continued decline (–31 per cent) in the area of fully organic land in Scotland, as some extensive hill farms chose to withdraw from organic production at the end of their Organic Aid Scheme agreements. The area of fully organic land increased in England (5 per cent), Wales (6 per cent) and Northern Ireland (24 per cent) in the year to January 2006.

Whilst the area of organic grassland in the UK has slightly declined in recent years (mainly as a result of large upland farms, predominantly in Scotland, withdrawing from organic production at the end of their aid scheme agreements), the numbers of livestock in the UK continues to expand, as does their requirement for organic feed grains.

An estimated 8.9 million organic table birds were slaughtered in 2005, an increase of 55 per cent since 2004. Sales of organic milk exceeded 200 million litres in 2005 and are worth approximately £100 million per year. In 2005, the farm gate value of organic meat and poultry was an estimated £129 million, an increase of 59 per cent since 2004. Hence the market for organic cereals and pulses is expanding rapidly.

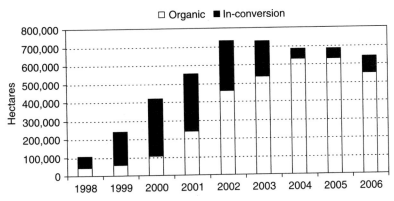

Figure 1 Area of organic and in-conversion land in the UK

Table 3 Area of organic and in-conversion land in the UK by region

	2003	2004	2005	2006
In-conversion		**Hectares**		
North East	15,332	6,812	4,609	6,643
North West	7,708	2,638	2,518	3,236
Yorkshire & Humberside	2,257	1,676	1,279	2,341
East Midlands	2,900	1,611	1,170	2,434
West Midlands	5,977	3,696	2,374	3,218
Eastern	4,140	2,976	2,416	2,649
South West	17,976	10,846	9,089	21,979
South East (inc. London)	11,501	6,530	5,378	10,723
England	*67,791*	*36,786*	*28,832*	*53,223*
Wales	*13,720*	*8,040*	*8,643*	*12,808*
Scotland	*121,283*	*20,375*	*13,666*	*16,724*
Northern Ireland	*1,514*	*825*	*1,604*	*3,196*
United Kingdom	*204,308*	*66,026*	*52,746*	*85,951*
Organic				
North East	12,415	20,470	25,306	29,296
North West	15,096	19,853	19,815	18,858
Yorkshire & Humberside	6,968	8,079	8,560	8,978
East Midlands	11,959	16,107	13,417	13,172
West Midlands	23,423	25,484	26,764	27,011
Eastern	7,753	9,669	10,319	11,782
South West	78,082	86,247	90,500	94,008
South East (inc. London)	28,348	34,288	34,946	35,250
England	*184,045*	*220,197*	*229,626*	*238,355*
Wales	*41,381*	*50,240*	*55,564*	*58,024*
Scotland	*307,325*	*351,888*	*331,600*	*231,206*
Northern Ireland	*4,115*	*6,627*	*4,970*	*6,317*
United Kingdom	*536,866*	*628,953*	*621,760*	*533,902*

Organic Farmers

In January 2006 there were 4,343 organic producers in the UK; organic holdings now represent an estimated 1.4 per cent of all farms in the UK. The average organic farm size in the UK has fallen from 171ha in January 2005 to 146ha in January 2006, but this fall is primarily due to a decline in the size of organic holdings in Scotland, as large hill farms continue to withdraw from organic production. In England, Northern Ireland and Wales, however, the average organic farm size continued to increase.

Organic Land Use

In 2006, 475,885ha or 88 per cent of the UK's fully organic land was under grassland. The area of land under horticultural production is estimated at 8,521ha or 1.35 per cent, and the area of land under arable production is estimated at 47,428ha or 7.5 per cent of the UK's fully organic land. This is a slight decrease over the previous two years, mainly as a result of businesses

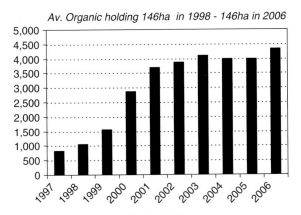

Figure 2 Number of organic producers in the UK

restructuring after the changes bought about by the reform of the Common Agricultural Policy (CAP) and the introduction of the Single Farm Payment (SPS) scheme in the UK.

The majority of organic farming in the UK is on mixed or livestock units, and predominantly in central and western areas, with arable-only farms mainly located in the eastern regions, being reluctant to develop livestock farming systems. The distribution of organic farms in the UK is shown in Figure 3.

There are a number of reasons behind the location and size of organic farms in the UK. EU production subsidies continue to support arable system profitability, and many arable-only farms no longer have livestock management expertise, infrastructure or a willingness to re-introduce livestock enterprises. To contemplate organic conversion, these farms would therefore have to depend on an arable-only or all-arable organic farming system, either initially or permanently. The high level of specialization associated with all-arable farms has limited UK development of organic agriculture, especially in the eastern counties.

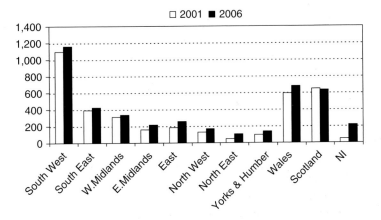

Figure 3 Distribution of organic farmers in the UK

In real terms, there has been a steady decline in the proportion of organic arable production compared to land used for livestock production, which has risen by 72 per cent over the last decade. The lack of arable land under organic management puts increasing strain on the supply of domestically produced organic livestock feed products, resulting in a greater reliance on imported crop, mainly from Europe and the Black Sea region.

UK Organic Cereal and Pulse Production

The current total area of organic combinable crop production is approximately 50,000ha per year (48,495ha in 2004, 51,233ha in 2005 and 47,428ha in 2006). From this, approximately 151,000 tonnes of crop is produced annually, with an estimated farm-gate value in excess of £27 million.

Table 4 Percentage of UK organic arable land area (hectares) under different crops in 2006

Crop	Percentage of UK organic production
Wheat	41.1%
Oats	12.2%
Barley	16.3%
Triticale	6.4%
Rye	0.7%
Peas/beans	13.0%
Maize	0.6%
Oil crops	0.9%
Sugar beet	0.2%
Set-aside	5.0%
Not specified	3.7%
Total	100.0%

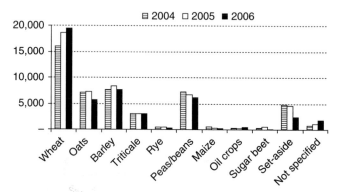

Figure 4 The area of organic arable land area under different crops (hectares)

Wheat is by far the biggest organic crop grown, with over 41 per cent of the UK production area, followed in importance by peas and beans, barley and oats. Surprisingly, the organic area of land under set-aside in 2006 was only 5 per cent; this is a reflection that farms which are 100 per cent organic are not required to set aside land under EU rules, and this small area will represent farms which are part organic, part non-organic, where land is still required to be set aside as part of EU rules on production. The percentage of UK organic arable land area under different crops in 2006 is shown in Table 4. The actual area of organic arable land area under different crops in 2004, 2005 and 2006 is shown in Figure 4.

CONVERSION MOMENTUM

The number of enquiries made to the Organic Conversion Information Service (OCIS) in England and Wales increased by 42 per cent from 473 in 2004 to 895 in 2005. The increase in enquiries can be attributed to factors such as the introduction of the Single Payment Scheme, the introduction of the Organic Entry Level Scheme, and strong market signals. Just over a third (35 per cent) of these enquiries were from beef producers, 31 per cent were from sheep producers, 15 per cent from arable producers, 6 per cent from horticultural producers, 5 per cent from dairy, 5 per cent from poultry and 3 per cent from pig producers.

FUTURE PROSPECTS

The increased demand for grain for livestock feeds, together with an increased demand from the food processing sector, has resulted in an undersupply of domestic organic cereals and pulses and an increased reliance on imports. This increasing demand for organic cereals and pulses was exacerbated under changes to organic standards in 2005, which require organic livestock to be fed on 100 per cent certified organic feed: resulted in a considerable increase in demand for livestock-feeding raw materials both in the UK and right across the EU. Industry forecasts suggest an increased demand in excess of 80 per cent for feed grains, and an over-200-per-cent increased demand for protein crops from organic production as a result of these changes to organic standards in livestock feed allowances.

The effect of de-coupling subsidy payments from production through the SPS has had a significant impact on organic combinable crop production in the UK, many mixed farms having opted to produce only what they need to meet their on-farm feeding requirements. The area and volume of organic arable production in the UK has therefore at best remained static, and in reality has fallen.

The market for organic grain crops has never been stronger, and this is likely to continue for some time with a strong demand from the livestock feed sector and for human consumption products, with at the same time an increased interest in UK sourcing. However, many farmers are still reluctant to convert as they lack sufficient knowledge or confidence to

develop appropriate rotations and nutrient management strategies, especially where farms operate only all-arable enterprises. Hopefully this book
will go some way in helping farmers to bridge the knowledge gap.

THE ORGANIC CHALLENGE

The challenge for all organic farmers is to produce sufficient quantities of
high quality food within the constraints of EU regulation, EEC No.
2092/91. These challenges are exacerbated when the farming system is
predominately based on arable crop production. Organic arable farming
systems will be faced with technical problems such as nitrogen management and weed control, both of which have a considerable effect on the
economic viability of a rotation.

There has been exhaustive research in the non-organic arable sector while
the organic sector has had only limited research dedicated to it. It is hoped
in the future that more research in the organic arable sector will be carried
out. This would reduce the need to transfer research knowledge from the
non-organic sector into organic farm situations, which isn't always ideal.

The outlets for organic cereals are similar to non-organic, for example
livestock feed, soft biscuit, milling for human consumption, and brewing,
along with some small niche markets for cereals such as spelt and rye. For
protein crops such as beans, peas, lupins and soya, the major market outlet is the organic livestock feed sector. The market for organic oil-seed rape
and linseed is small but developing, and is largely constrained by dedicated processing facilities in the UK.

One major difference is that the UK is a net import of organic cereals
and as such there is no export market. As such this requires orderly
marketing throughout the season following harvest to provide crop to end
market users. There is not the ability to deliver large quantities of crop at
harvest without depressing market premiums.

When planning the enterprise balance on the farm, key decisions must
match cropping to market demand and the resources of the farm (soil,
fertility, rotation and so on). This guide aims to take you through the following stages for organic crop production:

- Soil management
- Fertility building options
- Understanding nutrient flows
- Rotation, including crop nutrition
- Organic cereal production
- Organic pulse production
- Choosing varieties
- Crop establishment including cultivation systems, seed rates and
  drilling options
- Weed management before, during and after the growing season
- Managing pests and diseases in the crop
- Harvesting, drying and storage
- Operating organic arable systems

Chapter 2
Soil Management

The soil is regarded as an integral part of the organic production process. In her fundamental text on organic farming, *The Living Soil* (1943), Lady Eve Balfour states:

> A close connection exists between soil fertility and health...the health of humans, beast, plants and soil is one indivisible whole; the health of the soil depends on maintaining its biological balance...

According to this principle, a healthy soil will sustain healthy plants, which in turn produce healthy crops and animals, which lead to a healthy human diet. Organic farmers aim to produce profitable, nutritional food in an environmentally sound manner, whilst aiming to maintain or build soil fertility and to minimize pollution, the use of non-renewal resources and damage to the environment. By operating farming systems that seek to minimize the use of bought-in materials and to optimize the recycling of nutrients within the farm, these objectives can largely be met.

One of the basic principles of soil fertility management in organic systems is that plant nutrition depends on 'biologically derived nutrients'; so instead of readily soluble forms of nutrients, they use less available forms of nutrients such as those in bulky organic materials. This requires the release of nutrients to the plant via the activity of soil microbes and soil animals.

With specific relevance to soil fertility management, the organic standards state that the system must include 'an appropriate multi-annual crop rotation', that the use of readily soluble fertilizers is prohibited, and that manure application and thus stocking rate are limited so as not to allow nitrogen inputs to exceed 170kg N ha^{-1} yr^{-1}, in addition to nitrogen fixation via leguminous plants. Hence organic farming systems aim to achieve a balance between fertility building and soil nitrogen accumulation and exploitative cropping, nutrient supply and release, and farm profit and biodiversity. Thus the foundation of organic production is built upon management systems to create and maintain a living and healthy soil.

But what is a 'living and healthy soil'? How is it achieved, measured, monitored? How does it change over time? How do soils deviate in 'health' status on different soil types and in different locations? What impact do different farming systems have, with or without livestock? How do different soil management practices, cultivations, crops, green manures, and manures and composts affect soil health? And how should soil quality be defined in organic systems?

Given the heterogeneous nature of soils and the large variation in farming systems operated, these questions are not easily answered or quantifiable. But they are fundamental to the success of organic production systems – and more importantly, they are questions that need answering to provide information so that farmers can develop and operate successful and robust organic systems.

Where research has tried to answer these questions, knowledge transfer is often only applied above ground to farming inputs, crop and livestock outputs, and the above-ground physical and living environment. With regard to the below-ground soil environment, these fundamental organic principles are often overlooked or even ignored. This is not surprising, as the below-ground soil environment or 'bio-zone' is often ignored as a precious resource and ecological habitat, it does not command legal protection in most countries, and is not recognized as a distinct 'environment' by any UN convention or environmental body.

To ensure the maintenance of soil fertility and the genuine development and maintenance of sustainable farming systems, any use or sustainable development objectives must move perceptions of soil away from being an inert plant support medium to that of a living and vital resource.

PERCEPTIONS OF SOIL MANAGEMENT

So are the perceptions of soil management different in organic systems? In the latter the emphasis is more on the 'soil processes' of the 'whole' soil environment, rather than on inputs, outputs and the nutrition of plants in isolation from their surroundings. It is not a case of using 'non-chemical' substitutes for the soluble and directly assimilable plant foods used in non-organic systems: rather, it focuses on recycling minerals and soil organic matter in order to maintain a healthy soil environment. Hence the importance of maintaining a good soil microbial habitat – and using management practices to develop the potential of that habitat is critical in order to permit adequate levels of microbial activity and mineralization to ensure adequate crop nutrition. Central to this approach is being able to quantify the level of biological activity below ground, and the impact made by different management strategies, cultivations and crops. This requires the development of improved, cost-effective methods of measuring below-ground biological activity, which currently are not widely available to the farming sector, and/or are prohibitively expensive.

In non-organic production, a strictly analytical approach that concentrates on replacing minerals removed by crops and animals from outside the system, has led to reliance on the application of soluble nutrients. This has in turn stimulated commercial activity to provide soluble nutrient replacements, and research on their use and impact. However, continuing scientific endeavour regarding the manipulation of the soil environment shows that chemical analysis alone can only reveal a very small part of the total picture of soil life and function.

It is not enough to ensure that plants, or indeed any organism, are supplied with nutrients to keep them alive. Plants and the soil microbial biomass coexist with the environment around them, and depend on the maintenance of that environment, reacting to any changes in its balance.

Key Issues, Problems and Questions facing Farmers

For successful organic production, organic principles focus on maintaining a balanced soil ecosystem in a living and healthy soil, rather than on crop nutrition. To achieve this there is a need to 'feed the soil, which in turn will feed the plant'. This relies heavily on recycling crop and soil nutrients while seeking to minimize the use of bought-in materials.

This is a difficult concept for many converting farmers to fully embrace. Most have been used to an output : input culture driven by an analytical approach, which mainly concentrates on replacing with synthetic inputs those minerals removed by crops and animals from outside the system. This leads them to rely on the application of soluble nutrients, which in turn drives commercial activity to find nutrient replacements, and also research activity on their use and impact.

However, continuing scientific endeavour as to the manipulation of the soil environment shows that chemical analysis alone can only reveal a very small part of the total picture of soil life, its function and 'health' status.

THE IMPLICATIONS OF SOIL TYPE

Different farming systems, rotations, and physical and mineral management strategies are required for different soil types under organic management. This may seem obvious, but under organic systems, the differences in a soil's capacity can be more pronounced, whereas under non-organic systems these differences are often reduced by the desire to create uniform soil and crop growing conditions by altering the levels of artificial inputs.

As an example, the natural mineralization of potassium can release 10–20kg K/ha/yr in clay soils with 20+ per cent clay. On sandy soils with less than 20 per cent clay this is not the case. Hence the inherent capacity of the soil to provide crop nutrition without supplementary inputs is directly related to the natural characteristics of the soil type and its management.

As such, the use of 'generic' soil management and crop rotation models is not appropriate in organic farming systems, where soil/site specific choices are more important. However, there is a serious deficiency of appropriate soil management and crop rotation models for different organic farming systems on different soil types. This restricts the system and rotation options that farmers feel confident in operating, especially when considering converting to organic production.

MANAGING SOILS DURING ORGANIC CONVERSION

A healthy and biologically active soil is an essential part of all organic farming systems. During and immediately after the statutory conversion period (typically twenty-four months in the UK), many farmers report reduced productivity levels. But after five to seven years, crop performance and yields increase, and soil structure and ease of soil cultivation improves. These anecdotal reports are common amongst farmers converting to organic production, but as yet there is little quantifiable evidence to support these claims. This needs to be addressed by the scientific community so as to provide farmers with management and economic analyses of soil 'quality' changes that occur during the conversion period.

With organic systems being highly reliant on biological processes, research effort has largely focused on biological indicators when attempting to understand the benefits of organic agricultural production systems. Other areas of soil science – including biological processes such as nitrogen fixation and the role of arbuscular mycorrhizal fungi – have been investigated at a fundamental level, but as yet there is little understanding as to the effects of physical soil management practices, different cropping sequences, and rotation design on these processes. In order for organic farming to develop further, greater research effort is required to understand these biological processes and their interactions.

Long-term phosphorus and potassium monitoring has demonstrated that organic systems can operate with lower levels of 'fertility' and plant-available nutrients than similar non-organic systems. Why is this? Is it due to the lower yields associated with organic systems, or improved nutrient 'scavenging' by plant roots in association with enhanced soil microbial populations and mycorrhizal activity? Or are we mining a long-term phosphorus and potassium build-up from prior to conversion?

In order for organic farming to develop further, a better understanding and greater research effort is required to understand these biological processes and their interactions with different farming systems and management operations. This will then assist in helping determine how to define the changing 'quality' of soils under different management systems.

However, what exactly do we mean by 'soil quality'? On-farm observations of soil quality result in numerous descriptive terms such as healthy, fertile, hungry, degraded. These descriptive terms are often interchangeable, making it difficult to arrive at an agreed definition of 'soil fertility'. This position is further blurred as there is no clear, internationally recognized definition of 'soil quality'.

What is required is a definition of 'soil quality' that encompasses all the attributes of soil fertility by viewing soil as a heterogeneous complex ecosystem, and which describes not only the soil's production capacity, but also its influence on human and animal health. In turn, this has a direct impact on food quality/safety and maintaining environmental quality. This, then, embraces the philosophy of an 'organic farming system'.

ARABLE SOIL MANAGEMENT

Soils are the most important factor in arable production systems, and the fertility of these soils sets the resource constraints to production. In fact the management of soil fertility is one of the major challenges facing the organic farmer, when he needs to consider soil nutrient status, soil physical condition and the biological 'health' of the soil.

Organic farming aims to build up, or at least to maintain, soil nutrient reserves whilst at the same time reducing external inputs. The efficient management of nutrients, soil structure and soil biology should ensure good yields of crops and healthy animals. Poor management can result in poor yields, poor animal health and increased environmental pollution.

To be able to manage the soil well, the organic farmer should have a basic understanding of the underlying scientific principles behind the processes operating in and on the soil, and should be able to apply these principles to practices of managing soil fertility in organic farming systems. In addition the organic farmer should be able to understand the inter-relationships between the effects of operations undertaken, and how they affect individual soil components. In this way practical decisions can be made regarding managing soils in organic systems to maintain soil fertility.

SOIL CHEMICAL COMPONENTS AND THEIR IMPORTANCE

Plants require a number of nutrients to grow, and these can be divided up into major nutrients, minor nutrients and trace elements.

There are six major plant nutrients required by the crop in order for it to function properly: these are nitrogen, phosphorus, potassium, magnesium, sulphur and calcium. These are needed in comparatively large quantities.

There are also six micro-nutrients required by the crop in order for it to function properly, namely manganese, boron, copper, iron, zinc and molybdenum. These nutrients are only needed in very small quantities, and most soils have adequate indigenous reserves.

THE FUNCTION OF MAJOR NUTRIENTS WITHIN THE PLANT

Nitrogen
The plant needs nitrogen more than any other element: it plays a major role in the production of chlorophyll and protein synthesis, and a deficiency causes plants to turn pale yellow, due to the lack of chlorophyll in their leaves. It is the older leaves that turn yellow first as the chlorophyll is transported to the new young leaves in order to maintain growth. Large quantities of nitrogen are present in the soil in complex organic forms that are slowly converted into nitrate once mineralization has taken place.

Mineral soils in the UK may show nitrogen deficiency, although peat soils generally contain far more available reserves.

Under organic farming conditions a good supply of soil nitrogen is important. However, an 'over-supply' of soil nitrogen can create luxury growth in any crop and render it more prone to pest and disease attack. In addition, too much nitrogen will produce greater levels of weed growth.

Phosphorus

This element is used in making up a chemical called adenosine triphosphate (ATP) that is responsible for supplying energy to drive the plants' physiological processes. Visible symptoms of phosphorus deficiency are rare, since much of the lowlands of the country contain adequate amounts. Severely deficient plants show symptoms similar to those of nitrogen deficiency, except that the yellowing is replaced by a purple/bluish discoloration of the leaf, especially at the tips.

Potassium

Potassium regulates the cell water content and therefore cell expansion and plant growth. Deficiency is rare in the UK except on the most sandy and chalky of soils. Deficiency appears as a yellowing of the leaves, and is called marginal chlorosis, or scorch. This spreads in towards the main vein as severity increases. Clay soils generally have adequate potassium reserves. The export of straw residues for sale from cereal crops is a major depletion of potassium from arable fields.

Magnesium

The function of magnesium in the plant is as a component of chlorophyll and a large quantity of enzymes. Deficiency occurs as interveinal yellowing, and tends only to occur in sandy and chalky soils. The crop may occasionally show signs of magnesium deficiency on land that contains sufficient quantities for uptake; these symptoms are often attributed to crop stress, or inadequate quantities of nitrogen available to the plants.

Sulphur

The use of sulphur in the plant is essential in the production of several amino acids. Deficiency in the crop appears as a paling of the younger leaves, rather than the older ones as in nitrogen deficiency. With the ending of the industrial revolution, sulphur deposition has dramatically fallen over farmland, and sulphur deficiency is increasingly found. Adequate amounts of farmyard manure should significantly reduce the chances of the crop being deficient in sulphur. Where a deficiency is found, elemental sulphur may be applied once permission has been granted by the certification body. Sulphur makes up a large constituent of the oil in oilseed rape, and a deficiency is increasing in this crop due to the reduction in sulphur deposition from the atmosphere.

Calcium

Calcium is vital for cell division and cell elongation. Plants showing a deficiency are stunted, and in some cases the leaves may be distorted. Chalk soils or soils on limestone bedrock tend to have large amounts of calcium bicarbonate, and molybdate is reduced. Many areas of grassland have a reducing and acidifying pH trend through the long-term use of ammonium fertilizers where the ammonium has been oxidized to nitrate, unless it is taken up by the crop or used by soil microbes. This nitrate causes a considerable release of hydrogen ions, which then displace the calcium.

TRACE ELEMENTS (MICRO NUTRIENTS)

Manganese

Manganese deficiency is the most commonly found among the trace elements; it is usually seen as a yellowing of the foliage. Severe deficiency can lead to total crop failure. Manganese deficiency is associated with soils with a high pH, and high in organic matter. Other high risk situations are poorly drained clays, sands and ploughed-out pasture as well as peat soils. Deficient crops respond well to a manganese treatment, though repeat applications may be required. This is best achieved in the spring (April to May) as a foliar application. Deficiencies can be determined by soil analysis and confirmed by crop foliar analysis. A derogation for the application of manganese to the crop is required from the certification body on the basis of these tests.

Boron

Boron is involved in the development of the growing points in the plant. It mainly occurs on sands and on light textured soils. Boron deficiency has not been recorded in cereals or grasses, and is confined to root crops. It can occur in peas and beans, where the older leaves develop yellow edges and the youngest leaves wither.

Copper

Copper functions in the enzyme systems as well as being involved in the development of photosynthesis. Copper deficiency tends to be confined to peat soils, organic chalk soils and some sands, and mainly affects a few specific soils; it is best diagnosed with soil analysis, as symptoms in the plant can often be misinterpreted as drought or frost damage.

Iron

The function of iron in the plant is in the manufacture of chlorophyll. Iron deficiency can be seen in plants as severe chlorosis in the younger leaves while the leaf veins remain green. Iron deficiency is rare in the UK, and although symptoms may appear to be iron, it is usually lack of translocation of iron in the plant caused by high levels of other metals such as zinc, copper and nickel.

Zinc

Zinc deficiency is usually confined to nursery stock and apple orchards, and has not been recorded in field crops. Zinc movement in the plant is poor, and its purpose is the production of plant enzyme systems.

Molybdenum

Deficiency of molybdenum is most likely to occur on acidic soils, and it is the only essential trace element that becomes less available at a low pH. It is generally not a problem in arable crops, where its function in the plant is in enzymes. Symptoms of deficiency resemble that of nitrogen.

It is important that when deficiency symptoms occur in the crop, the reasons are identified. For example, a winter wheat crop may show signs of nitrogen deficiency when there are adequate nitrogen reserves in the soil, but the occurrence of the take-all fungus in the soil would be enough to reduce nutrient intake and trigger nitrogen deficiency symptoms.

It should be noted that a deficiency of molybdenum can adversely affect the ability of legumes (including clover) to symbiotically fix nitrogen, as it is a key element required in this process. Growers should therefore ensure that adequate levels exist to ensure optimal performance of fertility building legumes, which are a key element of any organic rotation.

Soil pH

For arable cropping the soil pH should ideally be 6.5 on mineral soils and pH 5.8 on organic (e.g. peat) soils. In order to prevent the soil pH from becoming too acidic there must be adequate calcium on the exchange surfaces on the molecules. These prevent the hydrogen ions from causing low pH and acidity. Soil acidity restricts the growth of most crops by the uptake of toxic amounts of aluminium, manganese or iron. A soil showing consistent manganese deficiency may over time reduce its requirement for manganese where the soil has become more acidic causing the element to become more available. However, as soil acidity increases the availability of calcium decreases making the soil even more acidic.

SOIL TYPES AND THEIR CHARACTERISTICS

The cereal crop is adaptable to many soil types and climatic conditions, and is ideally suited to situations where the soil is workable at the time of crop establishment, whether this is spring or autumn (*see* Chapter 3). Other factors are whether there is sufficient moisture during the growing period, and if the topography lends itself to mechanization. Stone content and depth of soil must be sufficient to allow mechanical cultivation, weeding and root development. There are many potential barriers to full production without selecting sites that unduly restrict output.

Sites prone to flooding or winter kill, and soils prone to frost heave are best not cropped with winter cereals. Spring sowing would be a preferred option on such sites, assuming there will be no damage to soil structure by work at this time of year. Winter wheat is generally tolerant of the winter weather experienced in the UK, as are most winter barley and winter oat varieties, although some do suffer more winter kill than others.

Different cereal types cope better at the extremes of conditions; for example, winter wheat and triticale establish and grow better in colder, heavier soils in wetter climates than other cereals, provided the crop is established relatively early. These soils are often slower to dry and warm up in the spring, and are therefore not well suited to spring sowing.

Spring-sown cereals generally fare better on lighter, warmer and drier soils. Lighter soils, although having greater flexibility in crop establishment, are more prone to nutrient leaching and drying out, and therefore spring crops that mature faster are best suited to these.

The amount and distribution of annual rainfall, in conjunction with soil type, will greatly influence the number of cultivation days available, and when they occur, for crop establishment and weeding. These characteristics will also influence which pests and diseases predominate. The local climate will influence the soil's moisture-holding capacity, and in turn the cultivation characteristics of the soil.

The main considerations of different soil types are how easy they are to work at different times of the year, their cultivation type and timing, soil structure management, weed management, the incorporation of crop residues and manures, seed rates and drilling techniques. When considering cultivation options the soil texture has a major bearing on the options available. The list below shows some of the main characteristics of key soil texture groups.

Light-Textured Soils

These include coarse sands, loamy sands through to sandy silt loams.

Advantages:
- many cropping options (especially if irrigated)
- tilth easily prepared
- wide working opportunity window in spring and autumn
- soon workable after rainfall
- many cultivation opportunities to control weeds

Disadvantages:
- usually weakly structured, therefore require regular loosening
- prone to slumping and surface capping after rainfall
- require consolidation after deep loosening
- low water-holding capacity (unless irrigated)
- low nutrient-holding capacity (need to work hard to maintain this)
- risk of wind erosion when the crop is small (leave surface rough or mulched)

Medium-Textured Soils

These include sandy loams through to clay loams and silt loams; a wide range of soils.

Advantages:
- generally quite readily form a good seedbed
- generally widely acceptable for a range of cropping and cultivation techniques
- seedbeds retain moisture for rapid establishment
- easily consolidated following deep cultivation

Disadvantages:
- the heavier end of medium may require some weathering to form a seedbed
- the optimum conditions for soil working are more restricted than for light soils
- the structure is easily damaged when handled wet
- prone to slumping and surface capping after rainfall

Heavy-Textured Soils

These include sandy clay to clay.

Advantages:
- the structure can be very stable once established
- good water-holding capacity provided drainage is not impeded
- self-structuring clays lend themselves to non-inversion tillage

Disadvantages:
- difficult to create a seedbed unless the weather is dry and settled
- very limited periods when the soil is workable
- a large power requirement for plough-based cultivations
- the structure is easily damaged if handled wet (smear, plastic)
- grass weeds can become difficult to control in rotations favoured on these soils

Peat soils are often easy working and fertile. They are generally well structured, they can be soft to travel on, and with inherently low density, require consolidation to ensure good root/soil contact. Many cultivated peat soils have shrunk, and through partial oxidation have become mixed with the mineral soils that lie underneath. These soils generally take on characteristics closer to the mineral soil type, depending on the relative proportions of each.

Fertility building should be undertaken on all soil types in an organic rotation to maximize residual and mineralizable nitrogen for cereal crops. This often means employing fertility building leguminous crops as an integral part of the rotation.

The type of fertility building crops and the timing of their incorporation, and the choice of cash crops that follow the fertility building crops, is important particularly where higher grain protein is required for milling markets. This is less significant for feed, malt or seed markets.

Sulphur nutrition is becoming increasingly important, with sulphur becoming a limiting factor to plant growth in a number of regions. The most reliable method of assessing sulphur need is either testing grain and determining the nitrogen:sulphur ratio, ideally in an 'old' crop which is still available in store, or new growth can be analysed in the spring to determine the malate : sulphur ratio. This nutrient is available through manures, potassium and calcium sulphate or elemental sulphur may be allowed, but requires some time to oxidize to a form usable by plants.

The atmospheric deposition of sulphur is dropping at such a rate that it makes regional field trials looking at responses almost irrelevant.

PHOSPHORUS (P) AND POTASSIUM (K) AVAILABILITY IN ORGANIC ARABLE SYSTEMS

Soil phosphorus (P) supply relies on both microbial activity to convert ('mineralize') organic P sources, and chemical transformations within the soil. The inter-relationships of biological activity and the availability of P and K from either soil reserves or slow-release fertilizers are complex, with many variables that are not fully understood.

The fraction associated with soil organic matter accounts for 30 to 50 per cent of the total P in most soils, with the remainder present as inorganic forms. Most P compounds in soil are either insoluble or poorly soluble.

The large reserves of P that have accumulated under conventionally managed fields may act as a source of P when farms are first converted to a less intensive, organic management. However, the conversion of poorly soluble P compounds into crop available P is dependent on the maintenance of a neutral to slightly acidic pH (*see* liming section).

Whereas soil phosphorus and soil organic matter are inextricably linked via microbial mineralization, very little soil potassium (K) is associated with soil organic matter – most potassium is present in inorganic forms. Careful management of nutrients to replace the large removals of K from cut grassland as a result of silage or hay cuts is particularly important in organic systems where there is great reliance on the N_2 fixed by clover. Potassium is particularly important in organic systems as it is only possible to have vigorous clover growth if the K supply is adequate. There is also a need to monitor and organize K removal from rotations containing a high proportion of grain legumes (peas, beans) which have a high K demand, and which need to be replaced via mineralization or use of appropriate K fertilizers.

Changes in P availability arising from the activity of mycorrhiza will not be detected by the chemical extraction techniques commonly used for determining plant-available P in the soil. Work to understand how mycorrhizal fungi and plant associations interact with P sources is currently at very early stages, but there could be potential for improved rotational design and nutrient management strategies.

In a ley-arable rotation, P and K offtakes in cash crops are balanced by the recycling of livestock manures. In predominantly arable or all-arable rotations there are not the same opportunities to replace P and K. It is therefore imperative that farmers with little or no livestock manures make best use of manures or green waste composts by optimizing the biological activity within the soil to ensure optimal P and K use from soil organic matter pools and more rapidly releasing sources. If these strategies are not maintaining the P and K supply, then there are restricted fertilizers that a farmer can use upon application to their certification body. These should be seen as a last resort and not a year-on-year solution. It should be possible to maintain positive P and K nutrient balances by applying these strategies. The eleven-year EFRC rotations trial (described in Table 5) showed no significant changes in the available phosphorus over time, although slow release rock phosphate had been applied to the green manure crops when soil analysis revealed a deficiency. These results are supported by the crop off-take data presented in Table 6.

Table 5 Three experimental stockless rotations in EFRC replicated experiment 1987–1998

	Course/year of the rotation			
Rotation	**1**	**2**	**3**	**4**
A	Red Clover	Winter Wheat	Winter Wheat	Spring Oats
B	Red Clover	Potatoes	Winter Wheat	Winter Oats
C	Red Clover	Winter Wheat	Winter Beans	Winter Wheat

(source: Elm Farm Research Centre)

Levels of potassium were not affected over the eleven years of the trial, despite the fact that no application of supplementary fertilizers or livestock residues was made. However, sandy soils with a clay content of less than 20 per cent may need supplementary potassium as there are insufficient clay minerals being broken down to support the K offtakes in the crops. There was no difference in soil potassium levels between rotations A (142.6mg K kg^{-1}) and C (138.0mg K kg^{-1}). However, there was a significant effect between rotation A and rotation B, in which the available K levels were 128.3 mg K kg^{-1}.

SOIL FERTILITY MANAGEMENT

What is Soil Fertility?

'Soil fertility' is the measure of the soil's capacity to sustain satisfactory crop growth, in both the short and the longer term. Organic farming recognizes the soil as being central to a sustainable farming system. Soil fertility is determined by a set of interactions – *see* Figure 5.

Organic farming relies on sound crop rotations to include fertility building and fertility depleting stages, returns of crop residues, nitrogen fixation by legumes/Rhizobium, nutrient retention by green manures, and effective use of manures/composts. Certain other materials, which are essentially slow-release nutrient forms, are also permissible under organic certification. The emphasis is clearly on efficient nutrient cycling, especially as the import of manure on to organic farms is being looked upon less favourably.

Table 6 Cash crop yield and nutrient offtakes (N, P and K kg^{-1} ha^{-1} yr.$^{-1}$) in the EFRC stockless rotations experiment (mean data from 1987–1996) (Source: Elm Farm Research Centre)

Crop yield t ha^{-1}

Rotation	Course of rotation 2	3	4
A	4.29[1] (se ±1.3)	2.64[1] (se ±1.3)	2.03[1] (se ±1.3)
B	29.35[2] (se ±8.8) 14.41[3] (se ±3.6)	4.29[1] (se ±1.3)	3.19[1] (se ±1.3)
C	3.75[1] (se ± 1.3)	4.10[1] (se ± 1.1)	3.99[1] (se ± 1.3)

Off-take of nutrients per course

Rotation	Course of rotation 2	3	4	Mean off-take per course	rotation
Nitrogen					
A	61 (se ±2.65)	36 (se ± 3.57)	25 (se ± 0.18)	41	122
B	87 (se ± 0.02)	64 (se ± 3.14)	39 (se ± 0.14)	63	190
C	56 (se ± 3.07)	150 (se ± 8.59)	57 (se ± 3.12)	88	263
Phosphorus					
A	25 (se ± 1.93)	15 (se ± 1.53)	12 (se ± 3.60)	17	52
B	61 (se ± 4.87)	26 (se ± 2.02)	19 (se ± 2.06)	35	106
C	23 (se ± 1.96)	29 (se ± 1.83)	24 (se ± 1.76)	25	76
Potassium					
A	21 (se ± 1.04)	14 (se ± 1.30)	15 (se ± 1.32)	17	50
B	153 (se ± 4.18)	20 (se ± 1.32)	16 (se ± 1.48)	63	189
C	19 (se ± 1.06)	49 (se ± 3.24)	20 (se ± 1.37)	29	88

[1]Yields adjusted to standard 15% moisture content
[2]Total yield
[3]Ware (marketable) yield

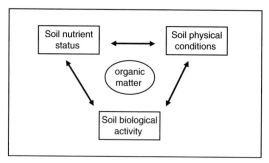

Figure 5 Soil fertility interactions

(Source: M. Shepherd ADAS)

Organic rotations avoid the use of water-soluble ('readily plant-avail-able') nutrients, whilst permitting the use of some slow-release and natu-rally derived inputs (for example, rock phosphate). Organic farming therefore principally relies on the release of nutrients from organic residues or from native soil sources. This is known as mineralization.

Organic matter is essential for soil fertility. It maintains good soil phys-ical conditions (for example soil structure, aeration and water-holding capacity). It also contains most of the soil reserve of nitrogen (N) and large proportions of other nutrients such as phosphorus (P) and sulphur (S). Changes that the management of soil organic matter can bring about include an increase in soil nutrient reserves, an increase or decrease in soil biological activity, a change in the processes of soil nutrient supply, and changes to the soil physical attributes.

Soil life is a critical component in the transformation of minerals from unavailable to available forms. Soil organisms (for example bacteria, fungi, earthworms and a whole range of organisms – the soil microflora, macro- and micro-fauna) form an intricate hierarchy of host predator rela-tionships which compete for food sources and form a web of life in the soil. This has been called the 'soil food web', and it is an essential part of the system, helping to release and recycle nutrients.

The Use of Supplementary Nutrients and Fertilizers

A farm system that is continually exporting cash crops will export nutri-ents in these crops. In time, if a long-term decline in soil fertility is to be avoided, all organic systems will require supplementary nutrients to replace those removed from the farm in crops and animal products, or lost to the environment. This is particularly true of stockless arable systems where no livestock are present on the farm, little crop or residue is retained on the farm as livestock feed, and all cash crops are exported for sale.

In some cases, soil reserves may be able to supply nutrients for hun-dreds of years, but on some soils and in some systems, supplementary nutrients will be required regularly to avoid nutrient run-down. Organic standards recognize this need and allow the use of a limited range of nutrient supplements, such as rock phosphate, potassium sulphate and green waste compost (*see* sections below). Most supplementary inputs are restricted in use, and permission to use needs to be demonstrated, with supporting soil analysis, cropping plans, nutrient budgets and so on.

Declines in measurable soil phosphorus and potassium have been observed on some arable farms after conversion. However, in most cases the combinable crops do not exhibit severe deficiency symptoms. This may be attributed to the fact that the soil analysis system often used measures 'available' P and K, and under organic management, with increased soil biological activity, P and K are being mineralized from an unavailable form to a form available to plants, but which is not being picked up in the analysis method used.

Sulphur deficiency is also likely to become more widespread, particularly on light soil, as low sulphur fossil fuels become the norm and industrial pollution continues to decline in the UK and Europe. Atmospheric deposition of sulphur in the UK fell from more than 70kg/ha/yr in the mid-1970s to less than 10kg/ha/yr by the late 1990s. Responses to sulphur fertilizers have been demonstrated in conventional cereal production and this will also be the case for organic cereals. Deficiencies in crops are manifested in a similar way to nitrogen deficiency, in that not only can it reduce yields, but it can reduce nitrogen fixation by clovers and other legumes.

Even when aiming to maximize the recycling of crop nutrients on the farm, there will always be some net export from the holding as crops and livestock products. The first aim should be to minimize any losses from the system, by avoiding leaching and erosion. Attention should also be paid to the sale of nutrients off the farm, for example as potassium losses from selling straw and conserved forage crops. Secondly, some reliance can be placed on the release of nutrients from soil minerals through mineralization. The purchase of livestock feed on to the farm can also import a substantial amount of nutrients to the farm.

However, if the export of nutrients in crop sales exceeds the purchase/import of nutrients combined with natural regeneration, then it will be necessary to supplement nutrients such as potassium, phosphate, calcium, magnesium and trace elements from external sources. These should preferably be in an organic form, or in a low-solubility form, so that the nutrients are released and made available to plants by the action of soil biological processes. In this way, luxury uptake by plants can be avoided. In some cases, such as potassium, this may not always be possible. Organic standards therefore allow the restricted use of potassium inputs (in various forms) in cases of demonstrable need, and matched to soil type and the crops grown and rotation operated. An up-to-date list of permitted inputs for soil improvement that organic farmers can use, can be obtained from organic certification bodies. The typically permitted inputs are shown in Table 7.

The most suitable supplement will depend on circumstances. Green waste compost is a good source of soil phosphorus, potassium and organic matter, but its availability is limited and cost is high. Where a more specific need is identified, such as correcting low soil phosphorus status, it is probably best to go for a specific product, such as rock phosphate. However, most of these supplementary nutrient sources have relatively low availability and so should be regarded as part of long-term nutrient planning rather than an input for a short-term yield boost.

Table 7 Soil amendments and fertilizers commonly used in organic systems

Product	Price (excl Vat) Bulk (£/t)	Bags (£/t)
Phosphate		
Calcanized aluminium phosphate (Reddzlag)		
(32% P_2O_5)	120–200	120–200
Gafsa rock phosphate (27% P_2O_5)	120–160	180–200
Natural rock phosphate (Tunisian) (27% P_2O_5)	120–160	180–200
Highland slag (15% P_2O_5 3% K_2O)	130–160	180–200
Potassium		
Rock potash (adularian shale) (11% K_2O)	115–160	180–260
Sulphate of potash (50% K_2O)	200–240	180–250
Highland potash (9% K_2O 7% Calcium		
Ca 5% Magnesium MgO)	10–150	150–180
MSL K (Rock Potassium) (8% K_2O 1.7%		
Magnesium MgO)	100–110	110–130
Patentkali (30% K_2O 42% Sulphur SO_3		
10% Magnesium MgO)	185–190	185–190
Kali Sulphate of potash (50% K_2O 45%		
Sulphur SO_3)	200–210	200–210
Magnesia-Kainit (11% K_2O 10%		
Sulphur SO_3 5% Magnesium MgO, 27%		
Sodium Na_2O)	90–100	90–100
Kieserite Mag Sulphate (50% Sulphur		
SO_3 25% Magnesium MgO)	130–140	130–140
Calcium (lime)		
Ground chalk/limestone spread (50% CaO)	13–25	15–85
Calcium sulphate spread (gypsum)	25–40	
Magnesium		
Calcareous magnesium rock spread		
(50% CaO, 15% MgO)	25–50	60–120
Magnesium rock (Kiesertite) (50%		
Sulphur SO_3 25% Magnesium MgO)	130–140	130–140
Compound fertilizers		
Glenphos 75 (0%N 6%P 6% K + trace elements)	125–150	135–160
Pell-N (5%N 4%p 3%K)	150–175	160–185
Complete organic (5%N 5% P_2O_5 5%K_2O)		250–290
Cumulus K (0.75%N 1% P_2O_5 13%K_2O)		230–270
Cumulus NK (5%N 1% P_2O_5 10%K_2O)		250–290
Eco-N (1 litr/ha) £48-£50		
Trace elements		
Dried seaweed meal	580–670	600–700
Seaweed extract	60–200	180–350

NB: Some products are restricted under organic certification and a derogation for use must be obtained from your organic certification body prior to use.

The use of some of the products listed above is restricted, requiring evidence of a need before organic certifying bodies permit their use, usually in the form of crop or soil analysis and cropping plans.

In order to assess if supplementary nutrients are likely to be required, nutrient budgeting should be carried out for macronutrients (*see* Chapter 4), along with regular soil analysis (*see* later in this chapter). Nutrient budgeting is of no use for determining trace element deficiencies. Where deficiencies existed prior to conversion, they are likely to remain an issue in organic production. The advice of an agronomist used in combination with soil and foliar analysis are the main tools for deficiency identification and treatment. Such tests will be required by certification bodies before some macronutrient amendments can be used.

Soil pH and the Use of Lime

The most important soil parameter for guaranteeing nutrient supply to the crop is soil pH. Even when all plant nutrients are present in sufficient quantities in the soil, if the pH is not maintained at the right level (6.0–7.0) the crop will display nutrient deficiency symptoms and will not achieve its full yield potential.

This deficiency is partly due to the fact that in acidic conditions (low pH below 5.5), soil biological activity is reduced, thereby slowing the degradation of organic matter and reducing the release of nutrients. Also, at either end of the pH scale (high pH above 7.8) some major and minor nutrients become insoluble and unavailable to the crop. Other effects of low pH or acidity include:

- deteriorating soil structure
- reduced crop quality
- reduced fertilizer efficiency
- increased nutrient losses
- deterioration of grass swards
- physiological diseases

In soils that have a tendency to become acidic, it is very important to check pH levels every three to four years. Susceptible soils include those that are:

- not formed in chalk or limestone
- sandy
- high in soil organic matter
- in high rainfall areas
- receiving regular inputs of slurry

Some soils are naturally rich in lime (calcareous) and will never need liming. In general, acidic soils should be limed to raise the pH to 6.0 for grassland and 6.5 for arable soils, but it is important not to add too much lime in any one year. Over-liming can result in nutrient deficiency and lost yield due to antagonistic and imbalance effects. As a guide, it is inadvisable to raise the pH by more than half a point in any one application.

How Much Lime should be Used?

The following table (Table 8) gives a guide as to the quantities (tons/ha) of ground limestone required to raise the pH to 6.5. Lighter textured soils need less lime than heavy soils to raise the soil pH to the same degree.

Table 8 Quantities (t/ha) of ground limestone required to raise the pH to 6.5

Soil pH	loamy sand	sandy loam	silt loam	clay loam	heavy clay loam	clay
6.3	0.4	1.1	2.0	2.8	3.8	4.5
6.2	0.7	1.7	3.0	4.2	5.7	6.7
6.1	0.9*	2.2*	4.0*	5.6*	7.6*	9.0*
6.0	1.1	2.8	5.0	7.0	9.5	11.2

*maximum amount of lime to apply in any one application, for given soil texture.

Liming Materials

The liming of agricultural soils is a common practice in the UK. The effectiveness of a liming material depends on its neutralizing value (NV), its fineness, its solubility, and the relative hardness of the parent rock. The NV is a measure of a material's effectiveness relative to pure calcium oxide, expressed as a percentage. Thus, 100kg of ground limestone with an NV of fifty-four will have the same NV as 54kg of pure calcium oxide. Most materials that contain calcium will have some neutralizing value, and one can compare the monetary value of any two liming products by calculating their cost per unit of NV, while taking into account any differences in fineness and hardness.

Not all liming materials are available to organic farmers. The fastest acting materials, slaked lime and quicklime, are not permitted in organic production. Ground chalk, calcium limestone, magnesium limestone and calcified seaweed are the main sources. The finer and softer the product, the higher the solubility and the more rapid the action. Thus, chalk ground to 1mm diameter will have about the same solubility as hard limestone ground to a powder of 0.07mm diameter.

Pay careful attention as to the parent material of the soils on your farm and the liming material used. Often, the least costly material is used, and this is of the same local parent material – for example, a calcium limestone mined locally and applied to soils with a calcium limestone parent base material. However, this has the potential to cause lock-up problems, particularly of trace elements such as copper and selenium, but also of phosphate. Calcium in the soil is often thought to be the same as pH, but it is not, and in fact is very different. Thus soils exhibiting acidifying tendencies with moderate pH levels can have very high calcium levels.

This is true of many of the calcium-rich limestone-based soils in the country that typically stretch in a band from Avon, through Wiltshire, Gloucestershire, the Cotswolds, Oxfordshire, Warwickshire, Northamptonshire, parts of Leicestershire and Rutland, Lincolnshire and up to some parts of Yorkshire, as well as other areas of the country.

Application of a calcium limestone material for 'liming' on these soils can add more calcium to soils that already have an excess, and will further lock up minerals. So using a magnesium limestone in these situations would be far better, even though it is more expensive (because it has to be bought in from other areas). As a rule of thumb, if the old buildings around you are built of a 'Cotswold'-type stone, take note of the calcium level of your soil and think about the type of liming material you use.

Lime Application
The timing of applications will depend on which crops are in the rotation, but it is important to consider that it can take over a year for an effective rise in pH to occur, especially if coarser or harder liming materials are used. A few of the crops to bear in mind are:

Clover: The percentage cover of white clover is rapidly reduced at low pH levels, so it is especially important to maintain a neutral pH in organic systems.
Potatoes: Liming can increase levels of potato scab. Potatoes can tolerate acid soils, so any need for liming can wait until after the potato crop.
Sugar beet: (rarely grown in organic systems) and **barley** are sensitive to acidity. Any required lime should be applied well before these crops.
Fodder beet: Any required lime should be applied well before a fodder beet crop.
Green manures: It may be a good idea to apply lime just before the establishment of lime-tolerant green manures. By the time the following crop is drilled, the soil pH will have returned to the desired level.
Brassica vegetables or green manures (mustard, forage rape): Liming can reduce the risk of club root by improving soil structure.

SOIL PHYSICAL AND STRUCTURAL MANAGEMENT

The management of soil structure is crucial to maintain good crop growth, particularly in organic systems where fertilizers cannot be used to compensate for the effects of bad structure. Careful cultivations, along with regular additions of fresh organic matter, should ensure that good structure is maintained. Particular care should be taken later in the rotation, when the effects of the ley have declined and soils are more vulnerable to damage.

The physical structure of the soil is equally as important as its biological or chemical status. The physical structure is important for a wide range of factors, including water and air movement, seed germination, root development and penetration, nutrient availability and soil erosion. Structure is an important, though often neglected, characteristic of soil. Poor structure can lead to a host of problems, such as increased pest and disease pressure, reduced water and nutrient availability, reduced soil trafficability and soil erosion.

Soil structure affects the size and the distribution of soil pores, which are important for the movement of air and water and for root penetration. Bad structure leads to poor root penetration, reduced access to nutrients, impeded drainage, poor microbial activity, soil erosion and ultimately crop failure. More power will also be required for cultivations, and animal health may suffer in grassland systems.

In order for a crop to grow well it needs a free-draining soil and the ability to explore and exploit the soil profile with roots. A compact layer in the soil can impede rooting and therefore the soil should also have a good granular structure. Platey aggregates in soil impede drainage, with fissures tending to run horizontally rather than vertically; this type of aggregate tends to occur in fields that have had heavy trafficking. A lot of angular, blocky aggregates in a soil can fit tightly together, severely restricting the ability of the roots to explore the soil profile and to extract nutrients. Prismatic structures are usually found at depth, and although they don't impede air, water and root movement through the soil, the roots cannot penetrate the blocks and extract nutrients.

A good soil structure is granular and easy to work. The plants will find it easy to expand their roots and to extract nutrients from the soil, drainage will be free flowing, and the sub-angular blocks that are deeper down will allow roots to penetrate and extract water and nutrients.

Some soils, for example heavy clays, can suffer from compaction, and repeat cultivations at the same depth can create an impeding layer known as a 'plough pan'. In very dry summers heavy clay soils tend to dry out, shrink and crack as water is evaporated, and in the winter they swell as water is absorbed. This shrink and swell characteristic allows for a reasonable level of self-structuring, but mechanical sub-soiling operations may still be needed to relieve severe compaction.

On most farms there is scope to develop a programme for improving the structure of all the soils present on the farm. In many cases it will be important to try and reduce the cultivation depth as much as possible to a maximum of 13–15cm (5–6in) deep for most combinable crops. This limits the disturbance to soil biological life, retains soil organic matter, and helps maintain a good soil structure. Shallow cultivations are also less expensive to undertake than deep cultivations. Depending on the soil types on the farm, shallow cultivations may need to be combined with deeper, loosening cultivations.

On most organic farms the options for reduced or 'min-till' cultivations are very limited; however, there may be opportunities for reduced non-inversion cultivations, reduced numbers of passes, and for running powered cultivators at lower speeds, all of which reduce the damage to soil structure. These measures are particularly important when cultivations are being undertaken in less than ideal conditions.

Once soil has been cultivated, it is more prone to compaction, so trafficking, especially with heavy machinery, should be avoided or kept to an absolute minimum. Using low ground pressure tyres, dual wheels or caged wheels can reduce compaction problems, but their effectiveness declines as the soil becomes wetter.

Crop harvesting is also another danger point with regard to soil compaction. Combine harvesters may be fitted with low ground pressure tyres to help reduce compaction, but often tractors and trailers containing 10 tons or even 15 tons of grain can be seen driving indiscriminately over the field, rather than their route being limited to headland areas, and combines unloaded at the end of field runs. Indiscriminate driving over the field can cause widespread soil damage, particularly in late harvested crops or in wet summers (*see* Table 9). This is especially important when harvesting cereals that have been undersown for the following fertility building phase, as there will be limited opportunities to remedy soil damage. Subsequent fertility building leys with poor structure will have reduced growth and nitrogen fixation, and later cash crops will have a potentially reduced yield.

If livestock are grazed on stubbles or during the ley phase, poaching should be avoided by controlling stocking rates and limiting damage in vulnerable areas by regularly moving fencing, feeding and watering points. It is often overlooked that the ground pressure exerted under the hooves of a sheep is greater than the ground pressure from a combine harvester!

Soil structural problems can be difficult and expensive to correct, particularly on heavy soils, and so they are best avoided in the first place. Cultivation or travelling with machinery on land when the soil is too wet, particularly on heavier soils, is the easiest way to damage soil structure. Under these conditions soils smear and clay particles are disrupted, blocking air spaces and forming hard impermeable clods and layers in the soil. Over-cultivating a soil, particularly with powered cultivators when the soil is dry, can also cause problems, as the dust produced can be washed deep into the soil profile, clogging soil pores and blocking drainage channels.

Table 9 The effect of the number of tractor passes on the degree of soil compaction

Number of passes	Proportion of compaction caused
1	50%
2	10% more
3	6% more
4	3% more

Source: 'A guide to better soil structure' Cranfield University

Sub-soiling

Sub-soilers typically consist of a number of heavy steel legs on a heavy duty frame, which are used at a depth of 25 to 45cm in dry soils. The wings on the bottom of the legs produce a sort of 'bow wave' effect in the soil as they move, and create a fracturing of the soil around them. This improves the soil structure at depth and allows air, water and plant roots to move more freely. It is important to note that a soil pit should be dug prior to sub-soiling, as the cost of sub-soiling can be high and the soil needs to be in a brittle state in order to be effectively fractured by the machine. The moisture state of the soil is of paramount importance, because sub-soiling in damp or wet soils that have a 'plasticine' state is a waste of energy and time, as all it will do is strike channels through the soil, without any relief from compaction; indeed, the operation itself can result in more compaction damage.

Good soil structure.   *Poor soil structure.*

Managing Soil Structure

To some extent, soil structure is determined by soil texture (the proportions of sand, silt and clay particles). These individual soil particles, along with organic matter, form clusters known as aggregates. It is the type and arrangement of these aggregates that determine soil structure.

Because soil texture has such a strong influence on structure, the type of soil may limit what is achievable; thus:

- Weakly interacting sand grains cannot form aggregates.
- Strongly interacting clay minerals can form stable aggregates, resistant to trampling by animals, cultivation, the weight of machinery, and rain impact.
- Many silt soils have a tendency to 'slump' and are structurally weak.

Good biological activity is essential to maintain soil structure. Plant roots and fungal filaments cover soil aggregates, while gums produced by bacteria feeding on organic matter help stick sand, silt and clay particles together. Poor biological activity and low organic matter levels tend as a result to lead to poor aggregate stability and poor structure. It is, however, possible to manage improvements in soil structure by using a combination of organic matter additions and careful and timely cultivations, combined with a little understanding and some simple observation.

Soil Examination

Soils should be regularly examined in order to assess structure, identify problems, and decide on restorative action. Know your soil texture, because this determines how the soil should be managed. Structure can be improved by:

- using timely cultivations;
- organic matter additions;
- avoiding livestock poaching.

Recognizing structural problems early on before they become difficult and expensive to correct is an important skill. Indications of structural problems include:

- fields slow to dry and quick to waterlog;

- poor seed germination and or emergence due to capping;
- patches of poor crop growth;
- increased pests, diseases and weeds;
- increased problems with producing a good seedbed.

Assessing the structural condition of a soil is easily done in the field and should help spot problems before they begin to affect the crop.

SOIL PHYSICAL ASSESSMENT METHODS

The best time to assess structure is when the soil is moist with a growing crop in place; assessment in dry soils is very difficult. Also, soil structure tends to deteriorate during the season, which should be taken into account when doing assessments. Repeat at points across the field – areas such as gateways and around feeding troughs are especially prone to poor structure and may provide an early warning of more general problems.

- Dig and remove a block of soil about 30cm deep.
- Bang the spit of soil on the ground to see how it breaks up.
- Assess the soil visually, and compare with Table 10.
- As well as assessing the plough layer, a further assessment should ideally be made at deeper levels with fewer pits to determine if deep compaction is occurring.

Table 10 Visual indicators of soil structure

Indicators of good structure	Indicators of bad structure (very bad structure in italics)	
	Medium to heavy soils	**Light soils**
small to medium aggregates easily broken when moist	mainly large clods	lack of cohesion
loose friable overall residues	clods resist breaking clods have smooth surfaces	surface capping persistent crop residues
lots of pore spaces	poor root penetration	compacted layers
good root penetration	persistent crop residues	poor root penetration
plentiful earthworms	few earthworms compacted layers *flattened clods* *horizontal cracking* *mottled orange grey colours* *sulphurous smell*	

Good management of soil structure broadly involves a combination of avoiding operations that damage the soil, such as excessive trafficking and overstocking, and maximizing operations that benefit structure, such as appropriate cultivations and the addition of organic matter.

CULTIVATIONS

Cultivation can have positive effects on soil structure – breaking up hard clods, or compacted layers – but also negative effects – breaking up aggregates, and smearing and moulding the soil into clods, particularly if done when the soil is too wet. If the soil is ploughed to the same depth every year, compacted plough pans can develop. Therefore the most appropriate method and degree of cultivation will depend on a number of factors: the purpose of the cultivation, the machinery available and the soil type. The influence of soil type can be summarized as follows:

- Light-textured soils do not form strong aggregates and are prone to compaction, therefore they should be ploughed.
- Clay soils can form strong, stable aggregates and so cultivations can be kept to a minimum, particularly if the soil contains free lime, which tends to result in a more robust structure.
- Keep secondary cultivations for seedbed preparation to a minimum to avoid pulverizing aggregates, particularly on weaker soils.
- Excessive cultivation should be avoided. Reduced tillage has become increasingly common because of the many benefits, though in organic systems, the need for cultivation to control weeds means that there is often limited scope for this.
- Do not cultivate when the soil is too wet (including harvesting operations), particularly on heavier soils. Damage can take considerable time and effort to remedy. Any short-term benefits of working in wet conditions must be weighed against the long-term costs of yield reduction, and the time and effort required to remedy the damage.

If a new ley is established by undersowing a cereal, care must be taken during harvesting operations to avoid trafficking over the field more than is necessary. Very poor soil structure has been recorded in leys established by undersowing, which will impact on their productivity. Other types of organic matter, such as FYM and crop residues, will also help improve soil structure, though stabilized organic matter such as compost will have a smaller (though more long-term) effect.

Adding any form of organic matter will help to improve structure in the soil. Organic matter stimulates soil bacteria and fungi, which help to bind soil aggregates together. There is also a smaller but longer-term effect resulting from the humus produced by this microbial activity.

The main influence of adding organic matter is relatively short-lived. After ploughing a grass/clover ley, soil structure is generally very good as fungi and bacteria multiply as they break down the organic matter. However, there is a measurable decline in soil structure over the next three to four years, as the organic matter is used up and the activity of fungi and bacteria declines. This means that more care must be taken with cultivations and trafficking later in the rotation.

SOIL BIOLOGICAL MANAGEMENT: 'THE LIVING SOIL'

Non-organic farming has placed greater emphasis on soil chemistry than soil biology. Organic farming, however, places a far greater importance on soil biological processes. In organic farming systems the emphasis is more on soil nutrition, and in developing and maintaining a healthy 'living' soil full of a diverse range of living organisms.

These living organisms in the soil are at the heart of organic farming. Though earthworms are the most visible soil organisms and perform an important role in organic matter breakdown, it is the microscopic organisms that are the key players. Because of the greater reliance on biological soil life to process soil nutrients and make them available to plants, the key is to feed the soil, and thus allow the soil to feed plants and animals.

In agricultural soils, the largest group of soil organisms, both in terms of numbers and biomass, are the bacteria, followed by fungi, followed by protozoa and nematodes. On a typical grassland, their total biomass exceeds that of the animals grazing on it several times over. It is said that there are more living animals in a teaspoonful of healthy soil than walk the earth above ground; the main challenge is that only about 6 per cent of the soil dwellers have been categorized!

The practical difficulty in studying microscopic organisms in the soil means that there are large gaps in our knowledge about soil micro-organisms. What we do know is that, as well as huge numbers, there is huge diversity within all the different groups. The result is that there is a highly complex *'food web'* within the soil, with different organisms feeding on soil organic matter and other soil organisms.

- Good biological activity is central to organic farming (and benefits all farming systems).
- This ranges from earthworms down to bacteria – micro-organisms play a vital role in nutrient cycling, soil structural development, and pest and disease control.
- Encourage biological diversity/activity in soils by:
  o providing good soil structure;
  o providing fresh organic matter.
- 'Specialized' micro-organisms such as N-fixing bacteria and mycorrhizal fungi provide large benefits to organic systems – plan rotations and management to encourage them.

IMPROVING BIOLOGICAL ACTIVITY

Most of the activities of soil organisms are beneficial, and it should be one of the central aims of the organic farmer to stimulate soil biological activity. The options available to stimulate the activity of soil micro-organisms, which will in turn enhance the benefits they provide, are simple:

- Provide comfortable living conditions = good soil structure.
- Provide a food (energy) source = fresh organic matter.

The principal option is the addition of organic matter in the form of FYM, crop residues, green manures, compost and so on. Organic matter supplies carbon, which is the food source for the soil micro-organisms.

Fresh organic matter, such as crop residues, provides carbon in a form more easily processed by micro-organisms than composted organic matter, and so results in a larger though short-term stimulatory effect. As the carbon is used as a food source, nutrient elements such as N are excreted, just as with larger animals. These nutrients are then available to plants for uptake.

A very large addition of organic matter can result in a large flush of nutrients, such as that which occurs after ploughing in a ley. In the absence of fresh organic matter, soil micro-organisms break down native soil organic matter at a very slow rate, such that a large proportion of the humus in soils is many hundreds or even thousands of years old. As well as providing a flush of nutrients, the addition of organic matter also stimulates many of the other services provided by soil micro-organisms, such as pest and disease control.

More generally, crops that return a large amount of high nutrient residue to the soil will cause a large (but short-lived) rise in microbial activity, while those which produce low nutrient residues, such as cereal straw, will produce less of a rise in soil microbial activity.

The other main management practice that can influence soil biological activity is cultivation. In general, a small amount of cultivation is beneficial, as it breaks up soil clods exposing organic matter, and aerates the soil. More vigorous cultivation can have a negative impact. Deep ploughing, in particular, kills earthworms and disrupts soil microbial activity by exposing organisms at the surface. Mycorrhizae are also disrupted by vigorous cultivations.

Mycorrhizal fungi are amongst a group of specialized soil micro-organisms that perform specific roles, such as nitrogen-fixing bacteria or, in the case of mycorrhizal fungi, increasing the capture of phosphate through a symbiotic relationship between the host plant and the fungi. These require more specific consideration if they are to be encouraged.

Specialized Micro-Organisms

Adding organic matter high in available nitrogen can suppress nitrogen fixation, but adding phosphorus and potassium will stimulate nitrogen fixers. In contrast, adding phosphorus will suppress mycorrhizal fungi (which form a symbiotic relationship with many crop species, helping them take up nutrients), while adding nitrogen can stimulate their activity. Specific amendments intended to stimulate more specialized groups of soil micro-organisms, such as compost teas, are available, but as yet there has been little research into their effectiveness.

Other factors to consider when addressing soil microbiology include crop type and cultivations. For instance, though many crops form a symbiotic relationship with mycorrhizal fungi, some, such as the brassicas and sugar beet, do not. Planting a non-mycorrhizal crop reduces the subsequent mycorrhizal infectivity of the soil, and can have a significant impact on the yield of highly dependent mycorrhizal crops.

Therefore, it is inadvisable to plant a crop that is highly dependent on mycorrhizae (such as maize) after a non-mycorrhizal crop (such as a mustard cover crop or oil-seed rape). The table shows temperate crops that are highly dependent on mycorrhizae, and those that are non-mycorrhizal.

Table 11 Crop mycorrhizal dependency

Non-mycorrhizal crops	High mycorrhizal dependency crops
Brassicas	Maize
OSR/Fodder rape	Alliums
Sugar beet	Linseed
Leaf beets	Soybean
Lupins	Sunflower

'Subterranean' Livestock Farming

Many arable farmers have little or no interest in farming livestock, but what many do not realize is that the arable farming systems that they operate are totally dependent on the livestock they unknowingly farm. The soil organisms – from the large earthworms to the tiny microscopic organisms such as soil nematodes, bacteria, fungi and protozoa – are vital players in making the soil function, in processing soil organic matter and creating plant foods, in forming and manipulating soil structure and in maintaining a healthy living soil. If organic arable farmers do nothing else, they should learn to farm the 'subterranean' livestock they already have on their farm.

SOIL SAMPLING AND ANALYSIS

The accepted method of soil analysis in the UK uses an 'index'-based system to quantify the amount of plant-available nutrients in the soil. The soil index

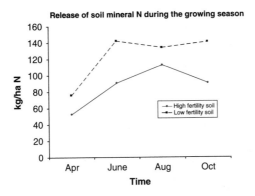

Figure 6 Release of soil mineral – during the growing season

system is used as a guide for the farmer to measure the level of soluble nutrients in the soil. The index runs from one to nine and is used to measure phosphorus (P), potassium (K) and magnesium (Mg) in milligrams per litre.

This system has largely found favour with non-organic input-based cropping systems, where it is used to judge the level of additions and the replenishment of soluble synthetic fertilizers. It is also used as a crude measure of whether the levels of major soil nutrients are increasing, decreasing or remaining static.

The soil index system is not a measure of the 'total' level of major nutrients in the soil, rather the level of soluble nutrients immediately available to the plant at that time. The chemical extraction methods used therefore only provide a snapshot of the nutrient content at a particular time. They are less suitable for determining plant-available nitrogen (N). Most soil N is in organic compounds, a small amount of which is gradually released as plant-available mineral N over the year (*see* Figure 6). The amount of mineral N present at any one time is a poor indicator of the amount that will be released during the rest of the season.

This information is useful to organic farmers and growers, but only to a point, as its primary function is for planning the replenishment of soluble synthetic fertilizers, which of course is not permitted in organic farming systems. For organic farmers and growers an understanding of the 'total' level of soil nutrients is important, together with the level of soluble nutrients immediately available to the plant and the interrelationships of soil organic matter and microbial function. From this information decisions can be made regarding the programming of fertility building or cash cropping, and additions of appropriate levels of compost, farmyard manures and permissible fertilizers.

If relying only on the index-based system, farmers and growers should aim to keep soils around index two. Indexes of around one will provide enough nutrients for growth, though if the soil becomes too depleted it can fall to index zero. This would indicate that there are insufficient levels of soluble nutrients in the soil, even though in some circumstances the 'total' level of nutrients in the soil may be sufficient. This may be due to an imbalance of nutrients, creating a 'lock-up' situation of a particular nutrient, or poor soil structural conditions, waterlogging or poor microbial function levels, all of which would impair the processing of nutrients into a soluble form available to plants. It should also be noted that the indexing system is not the 'be-all and end-all', mainly because each index number covers a large window of nutrients. Therefore they should only ever be used as a guide.

Growers should aim to maintain soil nutrient levels to index 2 for P, K and Mg, so as to ensure adequate nutrient availability for optimal production. Levels of soluble nutrients can be influenced by adding manures and composts, and if required, permitted slow-release nutrient compounds.

Many commercial laboratories offer services for determining the contents of P, K, Mg, S and trace elements in soils, and also soil pH (acidity). These simple chemical extractants cannot, however, duplicate the processes operating around the plant root, and the results are therefore only a guide to availability. Not all laboratories use the same methods.

Interpretation of the results must take account of the analytical method used. The most commonly used extractants in England and Wales are the Olsen reagent for plant-available P, and ammonium nitrate solution for K and Mg. Nutrient contents can be expressed as availability indices ranging from 0 (deficient) to 9 (excessive). Different extractants are used in the SAC system in Scotland, but otherwise the approach is similar.

There has also been recent interest in the basic cation saturation ratio (BCSR) and soil audit system introduced from the USA. This more exhaustive series of analyses differs from the other methods in providing recommendations that seek to achieve an 'ideal' ratio of cations (Ca, K, Mg and Na) in the soil. This soil audit also provides measures of available P and N and of 'active humus'. There is no evidence that this approach is superior to the more commonly used methods, nor has there been the extensive field testing to validate the concept under UK conditions. However, with the right interpretation, this approach does allow a more holistic review of soil health, mineral supply and structure to be undertaken, as compared to the simpler 'index'-based soil analysis whose primary function is to provide guidance as to the replenishment of soluble fertilizers.

Assessing Soil Fertility

Because of the fertility-building and fertility-depleting stages of organic rotations, it is difficult to define the overall fertility of an organically farmed soil from measurements at a single stage of the rotation. It is also more important to include measurements of the reserves of less readily available nutrients – for example, organic P and non-exchangeable K – in assessing fertility than with non-organically farmed soils.

Careful attention to nutrient movements on and off the farm and around the farm will allow maximum benefit from nutrients to be gained, and will avoid potential damage to the environment through pollution. Combined with using simple nutrient budgets (*see* Chapter 4), soil analysis can indicate whether the system is in balance, or losing or gaining nutrients.

Whatever soil analyses are used, it is important that the correct procedure for collecting the soil samples is used. Samples should be taken in sufficient sample numbers to be representative of areas and soil types on the farm. Unrepresentative areas (including gateways, around water troughs, in-fill areas and suchlike) should not be sampled. Sample locations should be recorded, so wherever possible they can be repeated in later years to provide a level of continuity in sampling and assessment. The physical condition of the soil is best assessed by visual inspection of the soil in the field.

Taking Soil Samples

- Sample soils with a coring tool that takes the same diameter core through the soil. Do not bias the sample by using a trowel, for example.
- Samples should be representative of the field and taken from the correct depth for long-term grass or arable land as specified by the laboratory carrying out the analyses.

- Samples should be collected at the same time of year, and should not be taken within two months of applying manure.
- Care should be taken to avoid contamination of the samples once they have been collected.
- It is generally unnecessary to sample fields more frequently than once every four years. For the examination of long-term trends, samples should be collected at the same stage of the rotation.

Soil Analysis for Organic Conversion

It is advisable that a programme of soil analysis should be initiated prior to converting a field to organic production. The soil analysis should include a measure of the organic matter content and a textural analysis, as well as available and reserve levels of major nutrients and trace elements. This should be carried out for each field prior to entering conversion, to establish the soil nutrient status so that any mineral imbalances or structural problems can be corrected prior to entering conversion.

Subsequent soil analysis should be performed on a regular basis to monitor any mineral or structural disorders. Maintenance of pH levels and the mineral balance through soil amendments and the use of fertilizers appropriate to an organic system, and sanctioned under the certification body standards, should be given a high priority. Routine inputs of phosphate fertilizers may be required. Potash fertilizers may only be added to the lighter soils under derogation where analysis and soil type dictates.

Soil Fertility Measurement and Monitoring

The mainstream perspective of soil fertility is largely based on an output/input-based system, seeing fertility as a means of achieving maximum agronomic production. Where low productivity exists, additions of synthetic N, P, and K are used as a means to improve fertility and hence productivity. This produces tangible measurements of inputs and outputs, but largely ignores mineral availability as a result of soil processes.

Whilst this approach may be valid for a non-organic system that is reliant on replenishing plant-available minerals with synthetic inputs, relying solely on a limited few chemical analyses is not appropriate for organic production systems, which depend more on biological processes. An approach is therefore required for organic production whereby soil analysis should be used to determine both plant-available nutrients and a measure of soil nutrient levels that are in 'reserve' (which can potentially become plant-available by microbial mineralization), in combination with quantifying soil biological activity and function.

More importantly, soil analysis should be used to monitor long-term soil trends in conjunction with changes in management or cropping practice, rather than to determine short-term crop-production requirement needs, so as to provide a more accurate and inclusive picture as to the implications of system operation, management practice and rotation influence on the overall quality of the soil resource.

Typical use of the analyses of pH, P, K, Mg reported as indices, whilst useful as a measure of plant needs, tells us very little about the 'soil fertility' – rather, they are a measure of plant-available nutrient levels and deficiencies, and an indication of what synthetic inputs need to be added for crop production in current and immediately following seasons. Within this approach there is also no consideration of the biological or physical components, or their influence on 'soil fertility'.

By determining the solubility of minerals and microelements under various pH conditions, analytical results and their relation to each other can be used to indicate if minerals or trace elements are available to the plants, or are in an unavailable form in the soil. Moreover, if different extraction strengths are used in a series of laboratory tests, a gradient of 'unavailability' can be determined, whereby the measure of the minerals that are transferable from a plant-unavailable to plant-available form by microbial mineralization can be established. The German worker, Dr Balser and his contemporary Dr Albrecht, have both used similar methods to describe not only the chemical status of the soil but also its biological activity. These methods are very useful to the organic farmer in understanding the chemical and biological balance in the soil, and what plant foods are available and what are 'potentially' available through careful soil management. However, as with all soil analysis, the interpretation of the results requires a measure of experience and on-farm observation. This requires a 'holistic' approach rather than a reductionist view, so as to concentrate on the whole soil environment rather than its constituent parts.

By focusing on the total health of the soil and the 'processes' which determine the environment for soil life, rather than the plant and its nutrition in isolation from its surroundings, plants can be provided with the means to develop their roots in well structured soils that have a balanced supply of nutrients, whilst avoiding nutrient excesses or imbalances that can lead to a high susceptibility to pest and disease damage.

ORGANIC SOIL MANAGEMENT IN PRACTICE

Assuming that organic systems focus on improving 'soil health' and optimizing soil biological and recycling processes, what are the practical implications for farmers?

If organic practitioners are only applying these principles above ground to farming inputs, crops, livestock outputs, physical operations and the living environment, and are overlooking or even ignoring the below-ground 'soil environment', then best practice is being compromised in terms of rotations, crop choices and management operations. If the focus is solely on above-ground production, rather than on maintaining and improving soil chemical and physical habitats capable of supporting an active, diverse, balanced and well functioning below-ground soil ecosystem – and if this focus is combined with inappropriate above-ground management, cultivation timing and methods of crop and soil management – then there is potential to seriously undermine the robust nature of any organic system.

Where organic production systems can be frequently undermined is where rotations do not balance in terms of soil mineral improvement, retention, release and disturbance to the soil microbial biomass. Physical soil management, cultivations, cropping rotation and choice have a major influence on this. However, the influence of different organic management strategies on soil mineral retention and release is not well understood, especially over a range of soil types. Many organic mixed and all-arable rotations are based on building soil fertility and then exploiting it. This often leads to a highly 'cyclic' system, building soil fertility with legumes over a period of, typically, two to five years, and then exploiting the fertility with a sequence of cash crops, typically over two to four years.

Whilst overall nutrient accumulation and utilization by crops may be in relative balance, these types of rotation result in large soil nutrient peaks and troughs throughout the rotation, and different levels of disturbance to the soil microbial biomass, especially when fertility building green manures are incorporated, with a subsequent large flux of soil N, P, K and so on.

If managed appropriately so as to minimize microbial disturbance and nutrient losses via leaching, volatilization and suchlike, these 'cyclic' systems can be very robust. However, when best practice is not followed, there is potential for this robust system to become 'leaky' in terms of nutrient retention and use, and for significant damage to be induced on soil microbial populations and activity. Many organic systems are based on large 'cyclic' nutrient flows, but this has potential environmental pollution implications, as well as resulting in sub-optimal use of soil nutrients for crop production.

To move organic systems forwards, there is a need to move away from land-use systems that are highly cyclic and that can undermine soil microbial activity or populations, and to operate with smaller nutrient peaks and troughs. This can be achieved by changing the system in a number of ways, such as shorter fertility building and cropping sequences; the use of reduced or minimum cultivations; the development of bi-cropping systems where soil N is fixed at the same time as it is utilized; improved crop sequencing; and the use of green manures. However, a number of technical and operational considerations impact upon the development of such changes, in particular weed management.

One of the main problems in ensuring 'best practice' is that many farmers are not fully aware of either the location or magnitude of such nutrient peaks and troughs in the rotation (other than by crop performance indications), or the implications of altering cropping or soil management practice.

The adoption of using techniques such as nutrient budgeting to examine changes in rotation design, crop choice and sequence, and the strategic use of inputs in the form of compost, FYM and supplementary inputs, is fundamental to understanding the cyclic nature of the rotations used, and the impact of management and operational changes. However, the data that is available on plant leguminous nitrogen fixation, and crop nutrient demand and removal, has largely been generated from non-organic farming systems, in soils with different inherent capacities to retain or release nutrients, cropping situations and management. This needs to be urgently addressed by the scientific community via appropriate investigation and research.

Whilst it is more difficult to make fundamental changes in organic systems – such as the development of bi-cropping and minimum tillage systems – it is easier to integrate shorter fertility-building and cropping sequences, improved crop sequencing, improved use of green manures, and best practice with regard to cultivation timing, so as to minimize losses.

By ensuring that appropriate crops are grown in a sequence that makes optimal use of available nutrients, the likelihood of excessive mineral loss can be reduced and even avoided. Cultivations should be carried out to minimize disturbance to the soil microbial biomass; this often results in shallower surface cultivations, combined with deeper soil loosening.

When incorporating fertility-building green manures from a highly 'cyclic' system, avoiding early cultivation when soils are warm and biological populations are very active may help to reduce mineralization 'peaks', large subsequent nutrient fluxes and potential leaching losses.

The incorporation of legume-based, fertility-building green manures in early autumn, followed by the establishment of winter-sown cereals in the late autumn, results in large levels of soil N mineralization which is available to the newly established crop. However, large volumes of available soil N at an early crop development stage is not ideal, and high levels of leaching and loss may occur over the winter period. Crops make better use of soil N in early and late spring.

There is a need to delay the release of soil N from the incorporation of the green manure to the spring. By incorporating the green manure and establishing a quick-growing, non-leguminous over-winter green manure, soil N can be retained and released after incorporation of the over-winter green manure in the spring.

Further enhancement of this system can be made by matching over-winter green manures with different C:N ratios to the release profile of the soil N to the demand profile of the crop being grown. Using green manures with a higher C:N ratio will delay soil N release to later in the crop development stage. This may prove important in releasing soil N later in the growing season of milling quality cereals, so as to improve grain protein contents.

As a practical example, grain protein contents in organic bread-making wheat are often sub-optimal. Using this principle, it may be possible to use green manures with higher C:N ratios to delay the release of soil N to later in the growing season for these cereals: later seasonal release of soil N should assist in improving grain protein contents.

To be able to adopt such practices, information is required as to the decomposition characteristics of different green manures in relation to different soil types and soil fertility conditions. However, information is lacking relating to decomposition rates in different soils, and whether or not the rate of decomposition of green manures is faster or the same as in non-organic systems.

It is assumed that the decomposition characteristics of green manures are inextricably linked to soil biological processes and soil microbial populations and activity. However, research is required to determine if larger microbial populations or increased levels of microbial activity have an impact on the rate of green manure decomposition.

Central to this is being able to quantify the size of the below-ground microbial population, its level of biological activity, and the impact that different management strategies, cultivations, crops and so on, have. This does require the development of improved, cost-effective methods of below-ground biological activity measurement, which currently are not widely available to the farming sector and/or are prohibitively expensive.

Fundamental to these assumptions is the question 'Do organically managed soils have a larger microbial biomass, or higher levels of microbial activity'? With organic systems being highly reliant on biological processes, research is required to determine if it is the size or the activity level of the microbial biomass that is the important factor in the decomposition rate of green manures, and in determining levels of microbial mineralization of soil nutrients. This question is explored in later chapters.

Chapter 3
Fertility Management

BUILDING SOIL FERTILITY

The emphasis in non-organic farming is often on crop nutrition. In organic farming systems the emphasis is more on soil nutrition and in developing and maintaining a healthy 'living' soil. Because of the greater reliance on biological soil life to process soil nutrients and make them available to plants, the key is to feed the soil, thus allowing the soil to feed the plant.

The soil on the farm and the parent material of the soil determine the baseline soil fertility. This can be modified by cropping, land management and the addition of livestock and manures. Nutrients can also be imported on to organic farms by several routes: these include the use of imported farmyard manure and slurries, some insoluble rock-based fertilizers, some soluble fertilizers, and plant foods for major and trace elements and nutrients contained in livestock feeds. The importation and use of nitrogen other than contained in farmyard manures and livestock feeds is prohibited in organic systems, with reliance placed upon 'building' soil fertility and in particular nitrogen by utilizing legumes. The cornerstone of the organic philosophy is the use of alternating fertility-building and fertility-depleting phases.

N release & utilization

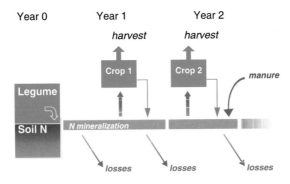

Figure 7 The soil fertility cycle

Source: M. Shepherd ADAS

It is especially important to minimize losses of nutrients from the soil to the wider environment. This helps to maintain the efficiency of the organic system.

Nitrogen

Organic farming aims to be self-sufficient in nitrogen through the fixation of atmospheric nitrogen (N_2), the recycling of crop residues, and the application of manure. As well as legume-based leys, organic rotations also often include an extra nitrogen boost by growing a grain legume (for example, field beans or peas) during the fertility-depleting phase. However, it is the legume-based ley or green masure that is the most important element of fertility building.

Crop Response

Cereal types vary in their responsiveness to site fertility. The most responsive are winter wheat and then winter barley, and the least responsive are triticale, rye and spelt. Grain legumes are more sensitive to low levels of phosphorus and potassium than cereals, and most are sensitive to very low or very high calcium levels; indeed lupins are intolerant of high calcium levels, which significantly impair development. It is important that a soil test is carried out at least every four years in each field so that soil nutrient levels can be monitored.

FERTILITY-BUILDING CROPS

Typical crops under UK conditions include red clover, white clover, vetch, lucerne, sanfoin, grass/clover leys (white clover/perennial ryegrass or red clover/Italian ryegrass) and grain legumes such as peas and beans. Lupins and soya are also being used in southern Britain with more limited success.

Nitrogen supply to the cash crops following the nitrogen-fixing legume in the rotation relies on mineralization of the residues that have been accumulated during the fertility-building phase. Captured nitrogen can also be returned to the soil as manure, either directly by grazing animals or indirectly in manure produced by animals that have been fed on leguminous forage or grains exported from the field. The key is to optimize the level of nitrogen fixation, and minimize any losses from the system. Inputs and losses from the system can be summarized as in Table 12.

Table 12 Typical system nutrient sources and losses

Nutrient sources	Nutrient losses
• Fixation of atmospheric N	• Nitrate leaching
• Purchased feedstuffs	• Ammonia volatilization
• (Cover crops)	• N2 and (NO_x) emissions
• Imported manure/compost	• Crop/animal produce
• Rainfall	• Exported manures

Other aspects of fertility-building crops also influence the design and performance of rotations, particularly the effect of pests and diseases during the fertility-building phase, and its implications for the following crops. These effects may be detrimental (for example, clover and lucerne could increase the risks of *Sclerotinia trifoliorum* for field beans) or beneficial (a green cover may reduce the incidence of common scab of potatoes to the rotation).

Soil Organic Matter

Additions of organic matter occur from the ley phase of rotations and from green manures, crop residues, animal manures and composts. The benefit of these additions depends on the amount and quality of the organic material added. Grazed or mulched crops will add more organic matter to the soil than where crops have been harvested and much of the growth removed from the field.

Legumes will generally accumulate more organic matter than non-fixing crops, whose yields are limited by the small amounts of soil nitrogen that are likely to be present at the start of the fertility-building phase. In grass/clover leys, most of the build-up of organic matter occurs in the first three years, and there is relatively little additional benefit from extending the ley period much beyond this.

Periodic inputs of crop residues and manures at other stages of the rotation, outside the fertility-building phase, are also important as sources of fresh organic matter.

MANAGING NUTRIENT SUPPLY

Nutrients are a valuable resource to all farmers, and organic farming is no different. Nutrient management is one of the main challenges facing the organic farmer. In the short term, the challenge is to supply sufficient nutrients at the correct point in the rotation and to the crop during its development to achieve economically viable yields. At the same time, avoiding the loss of nutrients is of paramount importance.

In the long term, the challenge is to balance inputs and offtakes of nutrients to avoid nutrient rundown or environmental pollution.

Within most organic arable systems, there are five main aspects to nutrient management:

- The fertility-building ley, containing legumes to add nitrogen to the system.
- Leguminous green manures grown between cash crops – cut and mulched.
- Grain legumes in the rotation as cash crops.
- Manures used to redistribute nutrients around the farm.
- Imported manures or acceptable fertilizers.

Nutrients other than nitrogen are imported on to the farm mainly in bought-in feed and animal bedding, though other sources such as green waste compost and permitted fertilisers may be important in some systems.

Fertility-Building Leys

The fertility-building ley is the cornerstone of most organic rotations. A well managed ley will provide nitrogen to cash crops and to forage for animals, and will help in the control of weeds, pests and diseases.

Despite the importance of the ley in fixing atmospheric nitrogen, there is a remarkable degree of ignorance within both the farming and research communities about how much nitrogen a ley will fix, and how this nitrogen is released after incorporation ('mineralization'). Being able to predict these two aspects more accurately would prevent nitrogen losses due to excess nitrogen, and crop failures due to too little nitrogen.

The challenge in trying to determine the level of nitrogen fixation and mineralization is that many factors are involved, such as legume species, soil type, climate, pests and disease. However, for a particular legume species, there is generally a close relationship between total nitrogen content and the volume of biomass produced by the legume, measured as yield. Careful management of legumes to encourage growth can therefore maximize the amount of this nitrogen that is fixed, as opposed to taken up from the soil – that is, maximize the nitrogen input to the farm.

This is an important consideration when deciding how to manage the ley. For instance, it is common practice to add manure to the ley: ground conditions make spreading easy, and it provides phosphorus and potassium, especially important if the ley is cut for silage. However, the nitrogen in manure applied can reduce nitrogen fixation by the legume, and repeated applications could reduce the proportion of legume in the ley over time. Legumes only fix significant amounts of nitrogen if they cannot obtain it from the soil. Thus, anything that adds nitrogen to the ley, such as farmyard manure, potentially reduces fixation. Conversely, anything that removes nitrogen, such as silage removal, increases nitrogen fixation by the legume. Recent

Ploughing of a red clover and grass ley.

work as part of a DEFRA-funded research project (OF0316: *see* www.organ-icsoilfertility.co.uk) demonstrated that repeated cutting and mulching clover and grass (a common practice on stockless arable farms) can reduce nitrogen fixation by between 50–60kg/N/ha/yr, as compared to cutting and removing the biomass produced by the legume grass mix. The same function is operating in that returning a nitrogen-rich material to the soil in the form of the mulch reduces nitrogen fixation by the legume, as it can derive some nitrogen being released by the mulched material.

An alternative strategy is to add the manure to other parts of the rotation. The nitrogen will boost cash-crop yield, while the phosphorus and potassium will still be available to the ley. The use of manure containing a high proportion of nitrogen in readily available form, such as slurry and poultry manure, is particularly detrimental to nitrogen fixation. The least detrimental is composted farmyard manure, which has the smallest proportion of nitrogen in an available form, but which retains good phosphorus and potassium availability.

Cutting and removing plant material (as silage, for instance) also promotes nitrogen fixation, when compared with grazing or cutting and mulching, as it removes nitrogen from the field. Although this means there is less nitrogen left for following cash crops, the nitrogen removed can be recycled back into the field in manure.

Though good ley management can increase nitrogen fixation, soil nitrogen eventually builds up and the rate of fixation falls. For a grass/white clover ley the optimum length is probably three to four years, after which there is little net fixation and incorporation is appropriate (*see* Figure 8).

For a grass/red clover ley, which has more rapid early development, the optimum length is probably closer to eighteen months to two years, after which there is little net fixation and incorporation is appropriate.

Even more difficult than predicting fixation, is predicting nitrogen mineralization and subsequent utilization by crops. This microbial process is influenced by temperature, moisture and soil type, as well as the nature

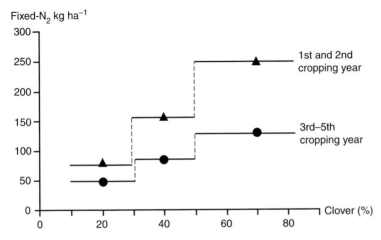

Figure 8 Nitrogen fixation by white clover over time at different percentage sward clover contents

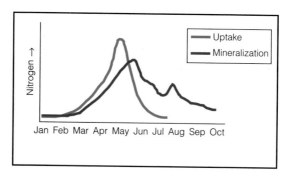

Figure 9 Nitrogen mineralization by the soil and uptake by a winter wheat crop

of the incorporated material and how it breaks down and releases nitrogen. However, it is possible to say that there will be a large peak in nitrogen availability soon after the ley has been incorporated, with increasingly small amounts released in subsequent years (*see* Figure 9).

The mismatch of nitrogen mineralization in the soil and crop uptake requires careful management to avoid nitrogen losses and optimize utilization. A good example of this is that winter wheat develops slowly during the autumn, and significant levels of nitrate may be lost by leaching before the spring, when the main demand from cereals occurs.

The best way to minimize this risk is with spring incorporation and spring cropping, which also more closely matches release of nitrogen for the following crop. This may be impractical on many soils, in which case late autumn incorporation is preferred, when soil temperatures reduce the action of microbes and slows the rate of mineralization of the incorporated residues.

Alternatively, adopting systems of autumn cultivations followed by catch crops and then spring planting, or potentially catch crops intercropped with winter wheat, could improve the utilization of nitrogen for crop performance.

A combination of using catch crops and spring wheat is more ideal for nitrogen resource utilization, and subsequently has the potential to improve the baking quality characteristics in wheat.

FERTILITY BUILDING OPTIONS

Building a Nitrogen Supply

Farmers have exploited the nitrogen-fixing role of legumes for centuries, but they now require more information about the management of legumes and the amounts of nitrogen they supply. In 1996 Stopes *et al* reported on a trial that investigated approaches to managing leguminous green manures in a situation that did not rely upon a livestock enterprise. The objective of the research was to provide information as to which green manures accumulated most biomass (dry matter (DM)) and nitrogen under differing lengths of green manuring, six to twenty-four months. The influence on subsequent cash crops was also recorded. The results from the trial showed that one year's green manuring with pure red or white clover swards could increase both cereal yield and grain nitrogen. In this instance the stand of pure trefoil was not a successful green manure crop; the trial did not, however, investigate the benefits of mixing the legume species.

The work of Schmidt *et al* (1999) confirmed that nitrogen accumulation from green manure crops is capable of supporting up to three years of cash cropping, provided that the green manure crop establishes well. The establishment and performance of green manure crops has not always been reliable, and so in order for all-arable systems to be recognized, it is important that establishment using direct sowing and undersowing techniques, and the subsequent management of green manure crops, is improved.

Green manures may be cut and mulched as many as three to six times during the season, depending upon growing conditions. Data from the EFRC all-arable research programme reported by Philipps *et al* in 1999 produced an average dry matter production of 11.0t ha^{-1}, and the above-ground nitrogen accumulation was 240kg ha^{-1}. The rotations that leave the green manure *in situ* longer accumulate the highest amount of dry matter and the largest quantities of nitrogen.

The EFRC rotation trial was established in 1987 as a result of changes in the Common Agricultural Policy (CAP) and with the introduction of the set-aside programme (EU regulation No. 1765/92). Set-aside, promoted with subsidies and a facility for organic farmers to use more than 5 per cent of legumes in set-aside mixtures, has increased the possibilities for fertility-building periods in all-arable systems to be longer than one year. This facility has played a major role in the conversion strategies of all arable organic farms and many mixed organic farming systems in the UK.

Recent work has evaluated a much wider range of legumes under UK conditions, and their potential for use as fertility-building green manures in organic farming systems. The legumes can be grouped into three main types: (i) forage legumes, suitable for nitrogen fixation and grazing and/or cutting as forage; (ii) grain legumes, suitable for nitrogen fixation and potential grain harvest; and (iii) 'other' legumes, suitable for nitrogen fixation, but not capable of being grazed or taken for grain harvest in the UK. These are listed in Table 13.

Table 13 Legume species (nitrogen fixers) with potential use in organic farming systems

Forage legumes

Trifolium repens	White clover
Trifolium pratense	Red clover
Trifolium incarnatum	Crimson clover
Trifolium subteranneum	Subterranean clover
Trifolium hybridum	Alsike clover
Trifolium resupinatum	Persian clover
Trifolium alexandrinum	Egyptian clover
Medicago sativa	Lucerne/alfalfa

Table 13 (continued)

Forage legumes

Vicia sativa	Winter vetch, common vetch
Vicia lathyroides	Summer vetch, goat
Vicia hirsuta/villosa	Hairy vetch
Anthyllis spp	Kidney vetch
Onobrychis viciifolia	Sainfoin
Lotus pedunculatus (L. uliginosus)	Large bird's-foot trefoil
Lotus corniculatus	Bird's-foot trefoil
Medicago lupulina	Black medick
Lupinus alba	White-flowering lupin
Lupinus luteus	Yellow-flowering lupin
Lupinus angustifolius	Narrow-leafed/ blue lupin
Trigonella foenum graecum	Fenugreek

Grain legumes

Pisum sativum	Field peas
Pisum arvense	Fodder peas, grey peas
Vicia faba	Field beans, fababeans
Phaseolus vulgaris	Green/dwarf/French/pinto beans
Glycine max	Soya beans
Lens culinaris	Lentil

Other

Galega orientalis	Galega, goat's rue
Lathyrus spp.	Vetchling, sweet pea
Melilotus alba	White sweet clover
Cicer arietinum	Chickpea
Castanospemum australe	Black bean

Table 14 Non-legume (nitrogen-holding) green manure/cover crops, with grazing or forage potential commonly used in the UK

Sinapsis alba	White mustard
Brassica campestris	Stubble turnips
Secale cereale	Forage rye
Lolium multiflorum	Italian ryegrass
Lolium perenne	Perennial ryegrass
Raphanus sativus	Fodder radish
Brassica oleracea	Kale
Brassica napus var. napus	Forage rape
Beta vulgaris subsp	Fodder beet
Tanacetifolia	Phacelia
Helianthus annuus	Sunflower
Fagopyrum esculentum	Buckwheat

This work has confirmed the suitability of a range of well known legumes such as red clover, white clover and vetch: these are the green manures most widely used in the UK. The work has indicated that a number of other legumes not currently utilized in organic systems may also be suitable as fertility-building green manures and covers.

The advantages and disadvantages of the different legumes are considered in Table 15.

Table 15 Advantages/disadvantages of different green manures

Crop Forage Legumes	Advantages	Disadvantages
Red clover	Good N fixation Deep-rooting Vigorous Large biomass for forage High protein content forage Can be grazed Suited to most soil types	Can disappear in grazed swards Difficult to establish in autumn in north Typical productive life of 3 years Stem eelworm susceptible Can be very competitive when undersown Erect growth habit Relatively slow to establish
White clover	Good N fixation Viable for 5–9 years Good regeneration after drought Suitable for undersow/ bi-cropping Prostrate growth has good ground cover Good biomass for forage High protein content forage Can be grazed Suited to most soil types	Lower biomass productivity than red clover Withstands heavy grazing Relatively slow to establish
Crimson clover	Good N fixation Resistant to clover rot Attracts pollinating insects Stunning red flowers	Forage can be stemmy and less palatable Erect growth habit Does not like competition in a bi-crop Autumn establishment difficult Does not like heavy soils
Subterranean clover	Rhizome regenerates below ground and survives cultivation Very prostrate growth (50–75mm) Survives grazing well	Rhyzome regenerates below ground and survives cultivation Slow spring growth Does not like wet soils
Alsike clover	Deep rooting Vigorous Large biomass for forage High protein content forage	Tolerant of wet soils Tolerant of acid soils Not drought resistant

Table 15 (continued)

Crop Forage Legumes	Advantages	Disadvantages
Persian clover	Not frost hardy Attracts pollinating insects High protein content forage Can be grazed Suited to most soil types	Lower biomass productivity than red clover Not frost hardy Slow to establish Spindly erect growth
Egyptian clover	Not frost hardy Attracts pollinating insects High protein content forage Can be grazed Suited to most soil types	Lower biomass productivity than red clover Not frost hardy Slow to establish Spindly erect growth
Lucerne/alfalfa	Good N fixation Deep rooting, High fodder potential Can be undersown Productive for 5+ years	Slow to establish in year 1 Careful management to avoid overwinter loss/ dieback Has higher bloat risk as fresh forage Requires high pH (pH6+)
Winter vetch, common vetch	Good N fixation Late sowing possible Highly productive biomass Very competitive/ smothers weeds Useful cover for brassica cash crops	Viable seeds can become weeds in subsequent cash crops Avoid green bridge if peas in rotation Does not like repeated cutting or grazing Potential grain production
Summer vetch	High-yielding as whole-crop silage when grown in mix with cereal Long flowering period Relatively frost hardy Exhibits allopathic effects	
Hairy vetch Kidney vetch	Good N fixation Lower productivity than other vetches Slower development	Viable seeds can become weeds in subsequent cash crops Avoid green bridge if peas in rotation Does not like repeated cutting or grazing Potential grain production
Sainfoin	Good N fixation High protein forage Highly productive Bloat free High palatability	Not suited to autumn sowing

Table 15 (continued)

Crop Forage Legumes	Advantages	Disadvantages
Large bird's-foot trefoil	Good N fixation Tolerant of anaerobic/waterlogging	Marsh trefoil (*L.uligi nosus*) is intolerant of high pH
Bird's-foot trefoil	Native plant Prostrate growth habit Can be undersown Tolerant of shade Useful cover for brassica cash crops – can reduce black spot Low bloat risk when grazed	Lower biomass production Does not withstand heavy grazing Relatively slow to establish
Trefoil, black medick	Good N fixation Cold tolerant Prostrate growth habit Can be undersown Tolerant of shade Low bloat risk when grazed Different species than clover for rotation use	Lower biomass production Withstands moderate grazing Relatively slow to establish
White-flowering lupin	Very good N fixation Fodder/grain legume potential	Poor biomass yield Not very competitive
Yellow-flowering lupin	Very good N fixation Fodder/grain legume potential	Poor biomass yield Not very competitive
Narrow-leafed/blue lupin	Very good N fixation Fodder/grain legume potential Prefers light sandy soils Slow early development	Poor biomass yield Not very competitive
Fenugreek	Good N fixation Quick-growing and very competitive Good forage potential Has pharmaceutically important chemicals Long productive season (into late autumn)	Not totally frost hardy Strong flavour, can be unpalatable Strong flavour, can taint milk
Grain Legumes		
Peas (fresh or dried)	Well known and understood Good N fixation Maple varieties have better leaf cover Potential as cash crop, and animal feed Can be used as mulched green manure	Not very competitive Semi-leafless varieties have reduced cover Predators – birds etc. – must be controlled Cannot be grazed

Table 15 (continued)

Crop Forage Legumes	Advantages	Disadvantages
Field beans	Well known and understood Good N fixation Very competitive Winter or spring sown Potential as cash crop, and animal feed Can be used as mulched green manure	Suited to many soils Predators – birds etc. – must be controlled Cannot be grazed
Green/dwarf/French beans	Good N fixation Potential as cash crop Can be used as mulched green manure Potential for high revenue (human food)	Not very competitive Predators – birds etc. – must be controlled Cannot be grazed
Pinto beans	Good N fixation Can be used as mulched green manure Potential for high revenue (health food premium)	Not very competitive Predators – birds etc. – must be controlled Slow to establish Grain harvest not possible in UK Cannot be grazed
Soya beans	Good N fixation Limited potential as cash crop Can be used as mulched green manure Potential for high revenue (health food premium)	Not very competitive Predators – birds etc. – must be controlled Grain harvest very difficult in UK Slow to establish Cannot be grazed
Lentils	Good N fixation Can be used as mulched green manure Potential for high revenue (health food premium) Competitive ground cover Rapid establishment Drought tolerant	Predators – birds etc. – must be controlled Grain harvest not possible in UK Cannot be grazed Does not like wet soils Little known about management
Other Legumes		
Galega, goat's rue	Moderate N fixation Can be used as mulched green manure	Very slow to establish Predators – birds etc. – must be controlled Grain harvest not possible in UK Cannot be grazed Does not like wet soils Little known about management
Vetchling/sweet pea	Very competitive Good N fixation	Viable seeds can become weeds in subsequent cash crops

Table 15 (continued)

Crop Forage Legumes	Advantages	Disadvantages
	Can be used as mulched green manure Competitive ground cover Rapid establishment Drought tolerant	Does not like repeated cutting or grazing Potential grain production
White sweet clover	High N fixation Very fast-growing and very competitive Tall-growing but straggly in nature Biennial	Bitter at maturity and not palatable to livestock Can cause internal bleeding if fed as forage Too vigorous for under sowing in cereals
Chickpea	Good N fixation Can be used as mulched green manure Competitive ground cover Rapid establishment Drought tolerant	Predators – birds etc. – must be controlled Grain harvest not possible in UK Cannot be grazed Does not like wet soils Little known about management
Non-legume cover crops **Mustard**	Good potential for N uptake Competitive Quick-growing Large rooting system Useful to suppress wireworm populations	Need to establish early for good growth, disease implications for cash brassica crops Moderately winter hardy Club-root host Can be grazed
Stubble turnips	Good potential for N uptake Good winter forage	Need to establish early for good growth, disease implications for cash brassica crops in rotation
Grazing rye	Rapid establishment Best for N uptake Can be grazed or cut for forage Can be taken to grain harvest Exhibits allopathic effects	Can be a green bridge for pests and diseases in cereal rotation
Ryegrass (perennial ryegrass/Italian ryegrass)	Inexpensive Good establishment Good for N uptake Can be grazed or cut for forage	Can be a green bridge for pests and diseases in cereal rotation Can become a weed May be difficult to kill in spring
Phacelia	Rapid establishment Good potential for N uptake	Cannot be grazed Will taint forage if included

Table 15 (continued)

Crop Forage Legumes	Advantages	Disadvantages
	Deep-rooting Winter hardy Non-brassica cover, attractive to bees Attracts pollinating and beneficial insects Pest-control advantages	
Fodder radish	Good potential for N uptake Competitive Quick-growing	Need to establish early for good growth Disease implications for cash brassica crops
Kale	Good potential for N uptake Competitive Quick-growing	Need to establish early for good growth, disease implications for cash brassica crops Club-root host
Forage rape	Good potential for N uptake Competitive Quick-growing Large rooting system	Need to establish early for good growth, disease implications for cash brassica crops Club-root host
Chicory (usually as part of a mixture)	Good potential for N uptake Competitive Quick-growing Deep-rooting system Large biomass produced Drought resistant Tolerant of heavy grazing	Intolerant of high soil N conditions
Buckwheat	Tolerates poor soils Annual Rapid development	Frost sensitive Cannot be grazed or cut for forage
Fodder beet	Good potential for N uptake Good winter forage grazed Can be lifted as root crop	Need to establish early for good growth, disease implications for cash brassica crops in rotation Soil damage if lifted
Sunflower	Exhibits allopathic effects Tolerates poor soils Rapid development	Frost sensitive Cannot be grazed or cut for forage Should be undersown as bare ground beneath

The work on legumes has identified some green manures that have the potential for wider use in organic systems, or whose use could be adapted for specific needs; these are as follows:

Alsike clover (*Trifolium hybridum*), **Persian clover** (*Trifolium resupinatum*) and **Egyptian clover** (*Trifolium alexandrinum*): these clovers are not frost hardy. They can be planted as a green manure in the summer/autumn, but will be naturally killed back by frost prior to spring planting. They can also be planted in combination with grass for grass seed production, with the clovers dying back from frost, leaving the grass as a single stand for grass seed harvest.

Subterranean clover (*Trifolium subteranneum*) is commonly used in Australia and New Zealand, and has the potential for use in the UK. It could be grazed and semi-destroyed with cultivations, and planted with a spring cereal cash crop. Limited work has shown that the subterranean clover will then re-grow and fill up the ground below the cereal to create a carpet of clover – thus establishing a bi-crop. A suitable innoculant should be used when planting for the first time, as the correct rhizobia are unlikely to be present in the soil.

Lucerne/alfalfa (*Medicago sativa*) is commonly used by dairy and beef farmers in the UK on alkaline soils, as a cut forage for three to five years. Organic arable farmers can use it to fix nitrogen over shorter periods of time, one to two years. It can be undersown into cereals for establishment. A suitable innoculant should be used when planting for the first time, as the correct rhizobia are unlikely to be present in the soil.

Sainfoin (*Onobrychis viciifolia*) was once commonly used by dairy and beef farmers in the UK and throughout Europe before there was widespread reliance on fertilizers. It can be sown as a green manure to fix nitrogen for one to three years, and cut as forage. It also has the benefit of natural anthelmintic properties, which can help combat internal parasitic worms in livestock. It can be easily undersown into cereals for establishment.

Vetch (*Vicia sativa, Vicia lathyroides, Vicia hirsuta/villosa*) is used by horti-culturalists more than arable farmers. It can be grown as a summer or twelve-month green manure. It fixes large amounts of nitrogen, and smothers weeds well. It can be left to set pods, but should be mulched and killed prior to setting viable seed, or volunteers will become a major weed problem in following crops.

Vetch can be planted in combination with a cereal in the autumn and harvested as an arable silage in early spring, prior to undertaking summer fallow operations.

Vetch exhibits allopathic effects. When incorporated with cultivations the exudates from the roots inhibit the germination of seed. Always delay planting seed by three to four weeks after incorporation (vegetable growers use this to their advantage by planting pre-germinated modules which get a head start against weeds).

Lupins (*Lupinus alba, Lupinus luteus, Lupinus angustifolius*) have been grown mainly as a high protein feed crop, either for grain harvest or as an inclusion in an arable silage. Whilst not very competitive as a single stand, they should not be ignored as a fertility-building green manure in combination with other green manures. However, growth is severely limited in soils with a high calcium content. A suitable innoculant should be used when planting for the first time, as the correct rhizobia are unlikely to be present in the soil.

Black medick (*Medicago lupulina*) is commonly sold as 'yellow trefoil', as it has yellow flowers similar to trefoil but without the bird's-foot-shaped leaves. In fact the plant is not of the *Trifolium* family, and as such can be used in rotation with clovers to help provide an adequate break between 'like' families.

Black medick is gaining popularity with organic arable farms for under-sowing in cereals and field beans, to act as a green manure understorey which develops late in the season and is useful to out-compete weeds as the crop canopy dies back in the summer. It can be left post harvest over the winter to out-compete weeds, and to retain and fix soil nitrogen. It is very shade tolerant but not very drought tolerant, so should not be sown too late in the season. It can be easily undersown into cereals for establishment.

It also has the benefit of natural anthelmintic properties, which can help combat internal parasitic worms in livestock.

Fenugreek (*Trigonella foenum graecum*) is mostly known for its strong 'curry' taste. It is used by horticulturalists mainly. It has great potential to be used as a nitrogen-fixing green manure. It develops rapidly and produces a dense cover to suppress weeds. The main drawback is that if grazed it is rather unpalatable, and could taint the milk of dairy animals.

White sweet clover (*Melilotus alba*) is widely used in North America as a green manure. Some varieties are white flowered and some are yellow flowered. It produces phenomenal growth in a short space of time with a fantastic competitive ability and has a long tap root which helps with improving soil structure. It has the potential to fix more nitrogen than just about any other green manure. Its main drawback is that it is a true green manure and cannot be grazed. It should not be cut and used as a dried forage for livestock as there is a high risk of causing internal bleeding lesions in ruminants. A suitable innoculant should be used when planting for the first time, as the correct rhizobia are unlikely to be present in the soil.

Lentil (*Lens culinaris*), **chickpea** (*Cicer arietinum*) and **Pinto bean** (*Phaseolus vulgaris*) are not commonly grown in the UK. They have great potential to be used as nitrogen-fixing green manure. They develop rapidly and produce a dense but low growing cover to suppress weeds. Their ability to be grazed and/or cut for silage in UK conditions is not known. A suitable innoculant should be used when planting for the first time, as the correct rhizobia are unlikely to be present in the soil.

Field peas (*Pisum sativum*) are commonly grown as a cash crop in the UK. They can also be grown as a nitrogen-fixing green manure, in isolation or in combination with other species. They fix large amounts of nitrogen. This is normally taken off with the resultant grain crop, leaving only a slight surplus for following crops. This would be best achieved as a spring-sown pea in combination with grazing rye or ryegrass.

Field beans (*Vicia faba*) are commonly grown as a cash crop in the UK. They can also be grown as a nitrogen-fixing green manure, in isolation or in combination with other species. They fix very large amounts of nitrogen. This is normally taken off with the resultant grain crop, leaving only a slight surplus for following crops. Beans can be winter or spring sown, and when incorporated will leave nitrogen for the following crop. Their main benefit is that they are very frost hardy, and are one of the few leguminous plants that will fix nitrogen over the winter.

Fertility Management for Bread-Making Cereals

Grain protein content is a limiting factor in the production of bread-making cereals. Even when sequenced as the first cash crop after the fertility-building phase, soil nitrogen release occurs from the point where the preceding green manure or legume-based ley is disturbed and cultivated. This results in a flux of soil nitrogen at germination and emergence. To improve grain protein contents, a flux of soil nitrogen release at grain development would be more appropriate. This would require the retention of soil nitrogen at establishment, and the release of soil nitrogen later in the crop production cycle.

For winter cereals this may prove difficult. On heavier soils it is often impossible to cultivate and plant in the spring, which results in soil nitrogen being lost over the winter period via leaching. On land that can support spring-sown crops, an option may be to cultivate in late summer/early autumn and establish a quick-growing green manure such as mustard or forage rape, which could then be incorporated prior to cereal drilling in the spring.

Using Green Manure as Nitrogen Holders and Fixers

Green manures and cover crops form an important part of some organic systems. There is significant scope for many farmers, including organic producers, to use green manures more and to use them more effectively. There are two principal types:

- those that **capture nitrogen**, preventing it from being leached;
- those that **fix nitrogen**, providing a boost to soil fertility.

Where they are adopted in organic farming systems, green manures are used to reduce soil-nitrogen losses, to out-compete weeds, and to improve soil structure and soil organic matter levels. The deep-rooting systems of many green manures also have the benefit of drawing up minerals such as potassium and phosphate from deeper in the soil profile and relocating them to the rooting zone of grain crops.

Research has shown that green manures and cover crops used regularly in the rotation for short periods of less than six months between cash crops can be as effective at maintaining soil nitrogen concentrations and yields as long-term (two-year) leys.

A green manure also adds fresh organic matter to the soil, which increases soil microbial activity and protects the soil from erosion, and some can be used for forage. Choice depends on a range of factors, including utilization for forage and/or grazing; length of time in occupation; weed competition; other crops in the rotation of the same family; and so on.

Grazing rye, for instance, is good at capturing nitrogen, vetch is a good nitrogen fixer, and clover, forage rape and stubble turnips provide good forage. Other factors to consider are the time of year involved (as some species are not frost tolerant), soil type, climate and seed cost. There are also implications for disease. Growing mustard in a rotation that includes a brassica crop can increase the risk of clubroot, though in contrast, trefoil can reduce the incidence of take-all in wheat.

The release of nitrogen is, again, an important consideration. The more mature the green manure is, the more slowly it will break down after incorporation, and the slower the subsequent nitrogen release. In some cases there may be an initial period of nitrogen immobilization or 'lock-up' by incorporating very mature green manure residues, as microbes normally associated with nitrogen mineralization and release divert their energies to the decomposition of the mature green manure material.

Some green manures, such as rye and vetch, have also been shown to have an allopathic effect, inhibiting seed germination for several weeks (typically three to four weeks, depending on soil and climate) after they are destroyed and incorporated into the soil. The effect is much less of a problem for large-seeded crops than for those with small seeds, but still needs to be considered where these green manures are used. On a number of occasions the author has looked at crops that have germinated poorly after being drilled straight after the incorporation of vetch green manures.

Horticultural growers use the allopathic effect to their advantage, incorporating green manures such as rye and vetch which then inhibit weed germination, and planting transplants or modules (which are already germinated) so that the transplants have a two- to three-week head start over the weeds. Even in arable cropping situations, with good planning this effect can be put to use, because it reduces weed seed germination.

Green manures planted between crops over the winter period are used to retain soil nitrogen and reduce leaching. When these green manures

Vetch used as a nitrogen-fixing cover crop.

are subsequently incorporated, their decomposition stimulates microbial activity and soil nitrogen release, which is available to the following crop. The ratio of the amount of carbon to the amount of nitrogen in the green manure crop, or C:N ratio, influences the rate of decomposition of the green manure and nutrient availability. C:N ratios vary depending on the composition of different materials and their growth stages. Young green material with C:N ratios of 15 will break down rapidly and release nitrogen. Older, more 'woody' material with a C:N ratio of 80:1 will break down more slowly and release nitrogen over a longer period. Material with a high C:N ratio has a low percentage of nitrogen, and conversely a low C:N ratio has a high percentage of nitrogen. For example, cereal straw contains only around 35kg N/ha and has a wide C:N ratio, compared with more than 150kg N/ha for some vegetable residues, with a narrow C:N ratio.

The narrow C:N ratio of green leafy residues means that nitrogen is released much more rapidly than from cereal straw. As a result, different types of residue will require different management if maximum benefit is to be obtained.

After incorporating green leafy residues, there is liable to be a rapid increase in soil N. Establishing the next crop quickly will enable this to be exploited. If that is not possible, a cover crop could be used to retain the nitrogen.

Residues with a C:N ratio in the mid-twenties will make soil nitrogen readily available as they decompose. However, mature plant residues with a C:N ratio of over 40:1 (Table 16) may cause temporary problems in the supply of nitrogen to plants, as micro-organisms may immobilize surrounding soil nitrogen to aid their growth and reproduction, thus diminishing the amount of nitrate and ammonium available for crop development. Incorporating a low nitrogen residue, such as cereal straw, can thus have the opposite effect, immobilizing nitrogen from the soil and in the short term reducing the nitrogen available to cash crops.

Table 16 Carbon: nitrogen (C:N) ratios and moisture contents of selected organic materials

Material	C:N ratio	Moisture
Soil micro-organisms	7:1	n.a.
Soil	10:1–12:1	20–30%
Poultry manure	3:1–10:1	65–75%
Cattle manure	10:1–30:1	65–90%
Vegetable crop residues	13:1–30:1	75%
Clover	13:1	75%
Compost	15:1	70%
Grass clippings	17:1	82%
Grazing rye	36:1	70%
Maize stems	60:1	30%
Wheat straw	80:1–150:1	30%
Fresh sawdust	400:1–750:1	20–25%

There is potential to use different green manures alone or in combination, which when incorporated decompose at different rates, so as to release soil nitrogen at different stages to the growing crop. This can be used to limit the size of the soil 'nitrogen flush' after leguminous green manures are incorporated, and release soil nitrogen later in the growth of the grain crop.

THE CATCHES OF GREEN MANURES?

Green manures are good in theory, but how practical are they as cover crops? Well-grown cover crops and/or green manures can decrease leaching, and can supply a welcome boost of fresh organic matter; but they are not always easy to accommodate in rotations, and can be especially difficult to manage on heavier soils where the timeliness of cultivation is critical. However, the benefits they can provide means they are worth persevering with.

Timeliness of cultivation: The timeliness of cultivations is important to avoid soil damage, and is always a compromise in some way or another. Beware of the long-term damage that cultivating in wet conditions might cause, and take steps to minimize these occurrences.

Fertility on light-textured soils: Maintaining nitrogen for the rotation on light-textured, less nutrient-retentive soils will always be a challenge. Increasing organic matter levels and adopting practices to minimize nitrate loss will help.

Pest and disease carry-over: Maintaining green cover can increase the risk of some pest and disease carry-over if the same or similar host species are used. Choice of species and careful planning of the rotation will minimize carry-over.

Weed management: Good cover crops and/or green manures are useful at out-competing weeds; however, their presence does mean that there is less opportunity for mechanical weed control between crops.

Stockless systems: Managing systems that do not benefit from animals to increase and recycle soil fertility (feed import and manure production) provides special challenges. Good rotation planning and nutrient budgeting are essential.

MANURE AND COMPOST MANAGEMENT

Manure Management

The main route of entry for nutrients brought on to the farm is usually via animal feed and bedding. Animal manure provides an important method for redistributing nutrients around the farm, particularly nitrogen, phosphorus, potassium, sulphur and magnesium. Manure also supplies valuable organic matter.

Fresh manure, especially slurry and poultry manure, contains a considerable proportion of nitrogen as ammonium, which is easily and rapidly lost to the atmosphere. Older farmyard manure may contain considerable nitrate, which can be washed out by rainwater. Both nitrate and ammonia losses cause environmental pollution as well as representing a loss of nitrogen that could be used by the crop. Manure also contains significant potassium in an available form, which is also easily washed from manure heaps by rainwater if the heaps are not well managed or covered.

Nutrients in manure are all too easily lost, thereby losing a valuable resource and potentially resulting in environmental pollution. Despite the obvious importance of manure to the organic farmer, there is plenty of scope for improved management on many organic farms.

The effective management of manures can minimize nutrient losses and maximize the benefits to the crop. There are two main treatment options available, namely composting and stacking (or aerating or not for slurry). For slurries there is the additional option of mechanical separation, which can reduce the volume of material to be transported and enable irrigation of the separated liquid.

Organic standards encourage the active composting of manure and aeration of slurry; however, on many farms this does not occur. Solid manure is usually stacked until spread, and slurry is left undisturbed in large tanks or lagoons. Both methods of dealing with manure have their merits and problems as far as nutrient management is concerned.

Composting

Composting is the biological and aerobic process of converting organic material, including manures, into a relatively homogeneous and stable product that can be used to build long-term soil fertility. Significant quantities of compost are now being produced from municipal and garden waste (2.5 million tonnes in 2004) by local authorities and commercial composting companies. Municipal composts are often available at a relatively low cost and can provide a valuable source of nutrients and soil organic matter, although their use is restricted and organic standards impose strict limits on the heavy metal content of municipal manures, and prohibit the use of any containing GMO (genetically modified organisms) contamination.

Farmyard manure heaps can lose valuable nutrients.

The effective composting of solid manures and aeration of slurry has a number of benefits:

- reduced odours;
- weed seeds and pathogens killed;
- reduced volume of material to spread;
- material with reduced bulk density (resulting in less soil compaction when spreading);
- production of a more uniform product;
- nitrogen stabilized in an organic form (solid manure).

However, the turning process in composting solid manure can cause a large loss of nitrogen as ammonia, while the aeration of slurry can also increase nitrogen losses, particularly if the aeration is either too vigorous or not thorough. Carbon is also lost as carbon dioxide. More than half the carbon and nitrogen can be lost to the atmosphere during the composting process. Nitrogen, potassium and phosphorus can also be lost to leaching during composting if rainfall ingress is not managed well.

When applied, the nitrogen in composted solid manures is also less available in the short term to the crop following application; rather, it is available over a longer period of time as the more stable medium decomposes. As a rule of thumb, less than 10 per cent of the total nitrogen will become available in the first year after application, as compared with 10 per cent for autumn application and 25 per cent for spring application of fresh farmyard manure.

These problems can be overcome, to some degree, by a number of methods. Increasing the amount of organic carbon in manures can reduce nitrogen losses, so using more straw per animal will be beneficial. However, adding too much straw increases ammonia losses due to the extra aeration of the stack caused by the more open structure. Covers are often recommended to reduce nitrogen losses from leaching and rainfall (see notes below). Covering slurry stores is also the most effective method of reducing losses of nitrogen from a liquid manure system. The decision to compost or not, or to aerate slurry or not, will depend on your objectives.

Effective composting can be a time-consuming and expensive operation, which can result in significant nitrogen losses. It also reduces the amount of nitrogen available to crops immediately after spreading. The advantages are a reduced volume of material to spread, with a lower bulk density (and thus weight); a reduction in weed seed numbers in the manure/compost; and a more uniform sterile medium, which contributes to long-term soil fertility. Effective composting systems are described below.

Slurry aeration is also expensive and difficult to achieve effectively; ineffective aeration especially can significantly increase nitrogen losses. However, when done effectively, it reduces weed seed numbers, reduces odours (which can be a particular problem in liquid manure systems); and produces a more homogenous material. The best system for a particular farm will depend on what the farmer is trying to achieve, and what resources are available.

On-Farm Composting

Many farms believe that they 'compost' when in reality they undertake manure management operations with a much reduced effect. Clearing out a livestock shed and heaping manure in a field, perhaps moving it once, is *not* composting!

There is significant scope for farmers to improve the management of their manures and slurries, and to undertake composting operations that will result in all the benefits described above. The main difference between manure management (handling and moving manure) and composting is that composting involves the 'aerobic treatment of manures'. Compost is only formed if air is introduced into the heap as the peak of heat obtained by the initial heap formation has passed. This requires regular movement and heating and cooling of the developing compost.

Fresh manure is dominated by bacteria and is anaerobic in nature. These bacteria create alcoholic compounds, and the strong foul smells associated with manure. By contrast, true 'compost' is dominated by fungi and has a sweet, almost pleasant smell similar to silage or soil. The fungi involved are almost all beneficial fungi which, when incorporated into the soil, help repel pathogenic fungi and assist with nutrient cycling and availability. The change from a bacteria-dominated manure to a fungae-dominated compost is achieved via careful management of an aerobic environment within the compost-making process, which is often likened to cooking.

The composting process can be broken down into a number of relatively well defined phases. The heap or windrow creation is followed by an initial rapid heating to around 60–65°C, as micro-organisms (mainly bacteria) break down readily decomposable compounds such as starches, sugars, fats and proteins. This phase lasts two to three weeks, and it also serves to kill weed seeds and pathogens. As the readily compostable components are used up, the decomposition rate slows and the temperature of the heap begins to fall. Different organisms (including fungi) take over the degradation of the more resistant compounds, such as cellulose. Finally,

Tractor-mounted compost windrow turner in operation.

Fertility Management

Farmyard manure spreader with doors can be used to create a windrow.

fungi come to dominate the heap and begin to break down the most resistant compounds such as lignin, though this is a very slow process and at this point the composting process is effectively over, with the heap temperature similar to the ambient air temperature.

Compost can be made by moving heaps of manure on a regular basis to develop an aerobic environment. Good compost can be undertaken on farm by creating a windrow of FYM (farmyard manure) and turning it regularly with a specialist compost windrow turner (*see* photo on page 85). The windrow is best formed by creating a windrow of no more than 4m wide, by 2m high (6.5 by 13ft); any larger than this can result in pockets of anaerobic manure which do not compost well. Many organic farms have adapted FYM spreaders to create windrows by adding rear doors to the spreader which can be partially closed when discharging the spreader to create a windrow (*see* above). Many FYM spreader manufacturers are now offering the rear doors as an optional extra. Passing FYM through a spreader at least once does significantly help to break it up, and to induce oxygen into the process.

Compost Creation and Turning

Compost can be formed by turning the windrow approximately two to three weeks after creation, turning again after four to five weeks, and again after eight weeks, and if possible, utilizing or applying within ten to twelve weeks. Failure to turn the heap leads to anaerobic conditions, in which the whole decay process slows down or stops. The heat profile generated should be monitored carefully (with a temperature probe), and ideally the developing compost should be allowed to generate 60–65°C and should maintain this temperature for seven to ten days.

Farms should consider purchasing a specialist compost-turning machine (*see* photo on page 85), which could be shared between farms. Whilst this may seem a luxury, it vastly improves the quality of the compost produced, significantly reduces the spread of weeds around the farm, and can make substantial savings in time compared to turning with a tractor foreloader. The cost of windrow-turning machines varies depending on size and capacity, but at the time of writing a small tractor-mounted machine typically costs £12,000–£15,000, with larger, self-propelled machines costing in excess of £100,000. Specialist information and machines can be obtained from Westcon Equipment in Wimborne, Dorset, and Morawetz Composting in Barnsley, South Yorkshire (*see* 'Other useful addresses' in Chapter 14 Sources of Information).

The main factors to consider in the preparation of materials for composting are outlined in Table 17.

Table 17 Main factors to control in preparing material for composting

Factor	Acceptable range	
Carbon : nitrogen ratio	20 : 1–40 : 1	↓
Moisture content	50–60%	Increasing importance
Particle size	1–1.5cm	↓
pH	6.0–9.0	↓

When implementing a practical composting system on farm, after clearing livestock sheds (or receiving manure on to the farm), heap the 4m (13ft) wide, 2m (6.5ft) high windrow in a dedicated sacrificial area, and turn on a regular basis as described above. The use of a dedicated area will help limit the spread of weeds around the farm, and will limit compaction to a small area. Avoid heaping manures in fields and undertaking composting in fields on a movable basis, as this will result in soil compaction and weed seed spread.

During the composting process, much of the organic carbon is respired by the micro-organisms. The remaining material is a mixture of resistant compounds including organic matter and highly complex molecules formed by the continuous re-combination of organic matter during the decomposition process. These are similar in nature to soil humus and contain stable organic nutrients, including most of the nitrogen and sulphur, and a significant proportion of the phosphorus.

Applying Composts

The application of composts should be undertaken on the basis of total nitrogen content, as with manures. Up to 250kg/ha/yr total nitrogen may be applied per year under the code of good agricultural practice for the protection of water, or 500kg/ha total nitrogen every two years in catchments sensitive to nitrate leaching. Organic standards limit nitrogen input to 170kg/N/ha/yr over the entire farm unit. Organic standards do allow an application of up to 250kg/ha/yr total nitrogen provided that the 170kg/N/ha/yr over the entire farm unit is not exceeded.

At this rate, one-off applications of compost have little direct impact on crop yields, as scarcely any of the nitrogen is available. However, a significant proportion of the phosphorus and potassium is available, and can contribute to crop nutrition. The relatively high proportions of phosphorous and potassium content in composts means that they should not be used principally as a nitrogen source, but rather programmed for the maintenance of phosphorous and potassium.

As well as supplying valuable nutrients, composts add useful organic matter to the soil. Typically composts contain 20–30 per cent organic matter. However, unlike manures, the organic matter is stabilized, only breaking down slowly in the soil, and as a result regular applications can quickly build up soil organic matter which improves moisture retention, soil structure and the traficability of the soil.

Compost spreading is best undertaken when the soil is warm enough to bind the nutrients in, and the plants are able to make use of them; this tends to be between March and September. Ideally applications should be targeted on to arable land, or silage or hay aftermaths or stubbles, and a rear-exit, vertical beater machine with spreading discs should be used to achieve an even application. Side-slinging farmyard manure spreaders are not suited to compost application, as they are designed to deal with farmyard manure, which has a far higher bulk density and water content.

Within the rotation, aim to target applications to silage aftermath cuts (first and second cut), clover leys, and prior to spring cultivations for spring-planted crops, root crops (potatoes), to stubble turnips after being grazed off, to undersown clovers after harvesting cereals, or on to green manures (mustard) if not being grazed off. Spread typically at 10t/acre (25t/ha) to ensure that you remain within the 170kg/ha/yr maximum nitrogen input, as specified in organic standards and NVZ rules.

Compost and Farmyard Manure Storage

If the compost has to be stored over winter, ideally it should be transferred to the area of hardstanding. Bunding hardstanding is a good measure to minimize the chances of runoff/pollution. This is necessary, as the nutrients contained in composts and farmyard manure will be a significant source of fertility for many organic farms. If an area of hardstanding is not available, a sacrificial field area should be used, rather than storage within different fields each year.

A cover for compost is not necessary during the active composting process. However, when stored over the winter, covering the farmyard manure/compost will be essential to reduce leaching and loss of nutrients, as well as improving the breakdown process. Covering will reduce losses of nitrogen (and phosphorous and potassium) by stopping rainwater leaching through manure and compost heaps, and can reduce gaseous losses by 60 per cent. Composting within a building is the ideal solution so that the heap is protected from water and rainfall, and leaching and potential pollution avoided.

If a building is not available, breathable covers such as 'toptex' are the preferred option over and above plastic sheets, since there is always a risk of combustion if a manure heap heats up too much under a plastic sheet. Another simple system is to use a straw chopper to blow on a bale of straw on top of the manure heap to create a thatch. This will help shed rainwater and the loss of nutrients via leaching.

Farmyard Manure, Compost and Slurry Application

As well as good quality handling, management and storage of manures and slurries is important, avoiding nitrogen loss when spreading is essential. Uncomposted manure still contains a significant amount of nitrogen in an available form, which can be lost by leaching or volatilization after spreading.

To avoid leaching losses, manure should be applied in spring, after field drains have stopped running. This is also the period of maximum nutrient demand by crops. The worst time to apply is between September and November. To prevent gaseous losses, rapid incorporation is the most effective method. Avoiding spreading in hot and/or windy weather will also reduce gaseous losses. Table 18 below shows the effect of delaying the incorporation of manures on nitrogen losses.

Table 18 Nitrogen losses following delayed incorporation of manures

| Manure type | Proportion of available N retained | |
	90%	50%
Slurry	Immediate	6 hours
FYM	1 hour	24 hours
Poultry	6 hours	48 hours

(Source : Shepherd.M ADAS)

Where best to use manure in the rotation depends on the objective. Manure is often spread on to leys, but this can have a detrimental effect on nitrogen fixation. Use of manure on grazing land can cause nutrient imbalances in grass in spring, reduced herbage intake, and disease problems, and is best avoided.

Nitrogen is the main factor limiting production in organic systems, so it makes more sense to use manure on cash crops. Top-dressing cereals with slurry in spring has been shown to be particularly effective at increasing yield and protein content, for instance.

Though nitrogen losses through volatilization can be large, this can be reduced using a band spreader. Solid manure should be added before cultivations, though because of the danger of leaching of nitrogen during winter, it is best applied to spring crops.

Composted manure supplies little nitrogen to the crop in the short term, and so it is more appropriate for spreading on leys.

Chapter 4

Understanding Nutrient Flows in Organic Systems

Achieving a balance between the nutrients exported from the farm in products sold (grain, forage, meat, milk) and nutrients imported to the farm (from nitrogen-fixing legumes, rock fertilizers, animal feeds, bedding materials, manures) is important, both for short-term productivity and long-term sustainability. Also the sequence of crops in the rotation has an important influence on soil nitrogen use and for weed management, as does the use of green manures to retain and release soil nitrogen over different time scales.

Understanding where nutrients flow within the farming system is an important aspect of this process, as is optimizing the use of nutrients whilst minimizing any losses. Developing a 'nutrient budget' is a useful tool in understanding nutrient flows. This can be undertaken on a 'rotational' basis (to evaluate the sustainability of a rotation or cropping change) or at a 'farm scale' basis (to evaluate the long-term sustainability of the farming system).

HOW TO USE NUTRIENT BUDGETS

Nutrient budgets are important tools for organic farmers. They can be used to assess how well nutrient inputs and outputs are balanced with current management, or to examine the implication of alternative management systems. They are also required by organic certification bodies to demonstrate the sustainability of the planned cropping rotation.

At a 'farm scale', if the input of nutrients to the farm is insufficient to replace the nutrients leaving in agricultural products, there will be a gradual depletion of the nutrient reserves. In contrast, where inputs exceed outputs, nutrients will either accumulate within the farm or be lost to the environment. Significant surpluses of nitrogen and phosphorus, though less likely on organic farms, are indicative of potentially damaging losses to the wider environment.

The simplest form of budget is the farm gate budget, which compares the nutrient input in purchased materials (feed, bedding, seed, manure, fertilizer) entering the farm, with the quantities removed in products (crop, milk, livestock sales, and so on). The necessary information can be readily obtained from farm records and from appropriate standard values for the nutrient content of the various materials entering and leaving the farm. Surpluses may indicate an excessive input of manures or livestock feed, whereas a deficit, of potassium for example, may indicate a shortage that could be remedied by increasing the amount of straw bedding brought in to the farm.

Although adequate for phosphorus and potassium, these simple budgets are unsuitable for nitrogen as they do not include the input of fixed nitrogen from the atmosphere or from nitrogen-fixing legumes. Nitrogen budgets require a separate estimate of the quantity of nitrogen fixed by individual legume crops on the farm, and therefore involve greater uncertainties. They should also include an estimate of the input of nitrogen in rainfall. These simple budgets do not include any estimate of the quantities of nutrients lost by leaching or to the atmosphere.

However well inputs and outputs are balanced, some loss is inevitable. Ideally, therefore, budgets should show a small surplus to balance these unavoidable losses. Typically, leaching losses from well managed organic farms are likely to be less than 5kg P/ha and 10–20kg K/ha per year. Losses of nitrogen are far more variable and often considerably larger. There is less information available for determining budgets for other nutrients.

Although nutrients may be balanced at the farm scale, it is equally important to balance inputs and outputs for rotations and individual fields, so that inputs are sufficient to replace the nutrients removed in harvested crops. Determining simple nutrient budgets for rotations and individual fields can assist in developing an effective manure management plan and directing manures where they are most needed. For example, animal manures should be preferentially applied to those fields with the greatest nutrient offtakes (for example, silage fields). Nutrient budgets can also be useful when planning rotation modifications. Substituting different crops will produce different removal and addition levels: for example, more cereals and fewer legumes in a rotation will result in a higher nitrogen and potassium demand, and a larger requirement for fertility-building phases. Substituting more grain legume crops will reduce the nitrogen-fixation requirement, but will increase potassium requirements.

One limitation of the nutrient budget is that the calculations describe total quantities of nutrients, rather than plant-available forms. Because of the uncertainties in these calculations, budgets should be used with care and as indicators of deficits and surpluses, rather than relying on them for absolute gains and losses from any system.

Budgets may be particularly misleading where insoluble phosphorus fertilizers have been applied. Because of the high total content of phosphorus in the fertilizer, the budget may indicate a relatively large surplus; in reality, however, only a small proportion of the fertilizer phosphorus

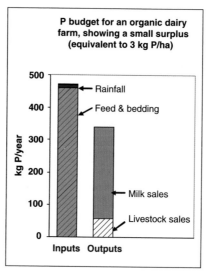

Figure 10 A phosphate budget for an organic dairy farm

(Source: Shepherd M. ADAS)

will be available to crops in any one year. Most will remain inactive in the soil so that its contribution to the budget should be spread across the full rotation. A phosphate budget for an organic dairy farm (Figure 10) demonstrates that large amounts of phosphorus are imported in the form of feed and bedding, and that this exceeds the phosphorous exported in livestock, meat and milk sales.

Budgets should be used to provide an indication of likely long-term changes in nutrient status, but should be regarded as a supplement to soil analysis rather than as a substitute for direct measurement of the soil nutrient status.

Example of a Nutrient Budget

It is normal to produce nutrient budgets for nitrogen, phosphorus and potassium, as these elements are likely to see the largest fluxes and are most likely to limit crop production. Where livestock will form part of the system, it is also useful to include magnesium and calcium in a nutrient budget, as conserved forage and protein crops associated with livestock production can remove large amounts of these nutrients.

Determining nutrient inputs and outputs is relatively straightforward in some cases – for example with supplementary nutrients – but very difficult in others – for example, the amount of nitrogen fixed by a clover/grass fertility-building ley. For better accuracy, the nutrient content of crops harvested and manures added should be determined. In practice, many farmers use standard values as a guide.

Nutrient budgets for stockless or predominately arable rotations have shown some degree of variation in nutrient balance, dependent upon soil type and rotation design. Most organic systems show a surplus of nitrogen and phosphorus, with potassium sometimes in surplus, sometimes in deficit, depending on the rotation and crops grown.

Nitrogen in a well balanced rotation should be adequately supplied; phosphorus can be maintained through soil analysis and the use of some supplementary fertilizers when appropriate; and potassium can be sustained through the mineralization of clay minerals, providing the soil type has a reasonable clay content – more than 20 per cent – although a greater understanding of how to manage these processes requires more research. Table 19 shows a 'farm scale' nutrient budget from a stockless arable rotation as part of the ADAS Terrington stockless research programme, while Table 20 shows a 'rotation' nutrient budget for a commercial organic farm in the Midlands operating a stockless rotation.

The 'farm scale' nutrient budget for the ADAS Terrington stockless research programme (Table 19) shows that the rotation has a small deficit of nitrogen, phosphorus and potassium. However, the soils on the farm are deep, fertile silt loams with inherent potassium reserves and a good ability to supply nitrogen from the mineralization of organic materials. Provided phosphorus levels are maintained through the importation of approved phosphate sources, the rotation has proved to be nutrient sustainable for over fifteen years. On a different soil type and in a different location, the rotation may not be so sustainable.

Table 19 Farm scale nutrient budget for ADAS Terrington stockless organic research trial (Source E. Stockdale et al)

	N	P	K		N	P	K
Inputs				**Outputs**			
Fixation	31.9			crop sales	88.8	13.1	45.2
Deposition	30.0		5	volatilization (from cut and mulching)	5.0		
Seed	4.3	1.2	2.6				
P fertilizer (*redzlagg*)		9.3					
Total	**66.2**	**10.5**	**7.6**		**83.8**	**13.1**	**45.2**
Balance					**−17.6**	**−2.6**	**−37.6**

The 'field scale' nutrient budget for the commercial organic farm in the Midlands (Table 20) shows that the winter wheat, spring beans and spring cereal all have deficits of phosphorus and potassium at some level, whereas there is a slight balance of phosphorus and potassium in the clover phase (when not cut and removed as forage). The resultant averages show that an importation of phosphorus and potassium may be required, depending on soil type. The nutrient budget also identifies the 'peaks' and 'troughs' of phosphorus and potassium demand in association with different crops in the rotation, winter wheat and winter beans being the most demanding crops in terms of phosphorus and potassium. Provided levels of phosphorus and potassium are maintained through the importation of approved phosphate sources (and potassium sources where permitted under organic standards), the rotation can be operated sustainably.

Table 20 Field scale nutrient budget for a stockless rotation on a commercial Midlands farm (Source E. Stockdale et al).

Rotational Nutrient budget		W Wheat		S Beans		S Cereal		R Clover		Average	
(kg/ha)		P	K	P	K	P	K	P	K	P	K
Deposition		0.3	2.5	0.3	2.5	0.3	2.5	0.3	2.5	0.3	2.5
Seed		0.6	1.0	1.1	2.5	0.6	0.8	0	0.1	0.6	1.1
Manure		0	0	0	0	0	0	0	0	0	0
Fertilizers		0	0	0	0	0	0	0	0	0	0
Crop sales		10.6	16.8	19.0	38.0	7.7	12.0	0	0	9.3	16.7
Straw		2.1	16.8	0	0	2.0	16.0	0	0	1	8.2
Silage		0	0	0	0	0	0	0	0	0	0
Animal products		0	0	0	0	0	0	0	0	0	0
Balance		**−11.8**	**−30.1**	**−17.7**	**−33.0**	**−8.8**	**−24.7**	**0.3**	**2.6**	**−9.5**	**−21.3**

A 'rotation' nutrient budget is shown for a commercial organic farm in the Midlands (Table 21). This rotation has an apparent surplus of nitrogen, but a deficit of phosphorus, potassium, calcium and magnesium. However, nearly 30 per cent of the nitrogen input is from farmyard manure (FYM) application, and some of this will not be immediately available. Without the FYM input the rotation would not be nitrogen sustainable.

The phosphate deficit will need to be addressed by importing rock phosphorus or manures to balance phosphorus offtake in crop sales. The crop deficit may be a problem on lighter, sandier soils unless FYM or appropriate potassium amendments are used to balance the offtake. The calcium deficit should not present a problem on soils rich in calcium, but would need to be addressed with inputs on an acid soil or a soil deficient in calcium. The magnesium deficit is not significant.

Different cropping and management will have an effect on the budget – for example, incorporating straw rather than removing it will significantly reduce potassium offtake. Increasing the proportion of grain legumes will remove more phosphorus and potassium from the system, as will taking forage cuts from clovers. The removal of livestock from the clover will also alter the balance of nitrogen, phosphorus and potassium inputs and outputs from the rotation.

Table 21 Rotation nutrient budget for a stockless system on a commercial farm

YEAR	CROP CROPS	Yield (t/ha)	GAINS (kg/ha)					LOSSES (kg/ha)				
			N	P	K	CaCo	MgO	N	P	K	CaCo	MgO
1	Red clover grazed/forage	10	300	0	0	0	0	-300	-70	-250	-220	-50
2	Red clover grazed/forage	10	300	0	0	0	0	-300	-70	-250	-220	-50
3	Winter Wheat	4.3	0	0	0	0	0	-64	-34	-22	-26	-13
	straw removal	3.5	0	0	0	0	0	-18	-10	-46	-18	-4
4	Winter Beans	3.7	295	0	0	0	0	-240	-74	-166	-130	-30
5	Winter Wheat	4.3	0	0	0	0	0	-64	-34	-22	-26	-13
	straw removal	3.5	0	0	0	0	0	-18	-10	-46	-18	-4
6	Spring barley straw incorp	3.5	0	0	0	0	0	-58	-31	-20	-23	-12
	6 years leaching	0	0	0	0	0	-180	-18	-60	0	0	0
	LIVESTOCK											
	none	0	0	0	0	0	0	0	0	0	0	0
	INPUTS											
1	Cattle FYM @ 25t/ha	0	130	59	165	50	50	0	0	0	0	0
2	none	0	0	0	0	0	0	0	0	0	0	0
3	Cattle FYM @ 25t/ha	0	130	59	165	50	50	0	0	0	0	0
4	none	0	0	0	0	0	0	0	0	0	0	0
5	none	0	0	0	0	0	0	0	0	0	0	0
6	none	0	0	0	0	0	0	0	0	0	0	0
7	none	0	0	0	0	0	0	0	0	0	0	0
1>6	Atmospheric gains		180	0	0	0	0	0	0	0	0	0
	TOTAL GAINS & LOSSES		1335	118	330	100	100	-1242	-351	-882	-681	-176
			N	P	K	CaCo	MgO					
	NUTRIENT BALANCE (kg/ha)		93	-233	-552	-581	-76					
	ANNUAL AVERAGE (kg/ha)		12	-29	-69	-73	-10					

These nutrient budgets illustrate the complexities and challenges of managing nutrients in organic farming systems, where the sources available, such as nitrogen from legumes, and phosphorus and potassium from manures, do not always contain nutrients in the required ratios.

Nutrient budgets are not a substitute for soil analysis (*see* Chapter 2), but they can provide a useful tool for judging what is happening in a particular organic farming system, rotation or field context. One significant drawback of nutrient budgets is that they do not show when the nutrients will be made available. In the case of nitrogen, for example, it is important that this happens at the times when crops have a demand. To be able to achieve this, a more sophisticated approach is required. Limited model development work has been started in this area, but at the time of writing this is not available in a form useful to most farmers.

Chapter 5

Rotations

The rotation is the cornerstone of most organic farms. Organic arable farming systems depend to a greater or lesser extent on a diverse rotation, working on the principle that, as with natural ecosystems, diversity and complexity provide greater stability and less risk in agriculture. In contrast, non-organic arable systems often operate systems based on only two or three crops, often repeating the same crops over a number of years, or even in successive years, being driven more by market forces than good agronomic practice. This in turn leads to a higher level of inherent risks, which necessitate the use of inputs.

Organic farms depend on well planned and well designed rotations to ensure their biological and financial viability. Key biological factors in organic rotations include soil fertility management, maintenance of soil structure, weed management, and pest and disease control. All of these factors are regulated to some extent by organic standards generally requiring rotations that:

- provide sufficient crop nutrients, and aim to minimize losses;
- provide a self-sustaining supply of nitrogen through the inclusion of legumes in the rotation;
- minimize the build-up of weeds, and pest and disease problems;
- maintain or improve soil structure and soil organic matter;
- provide sufficient livestock feed (where appropriate);
- maintain a profitable output of cash crops and/or livestock.

Farms vary widely between locations, with different soils, climate and resource availability. Even on a single farm, the variation in the productive potential of different fields can be enormous. As such there can never be one single 'blueprint' organic rotation.

Research in the years 2000 to 2004 at the University of Nottingham in the UK examined alternatives to the common two-year clover-ryegrass ley conversion strategy, investigating the balance between soil nitrogen fixation, crop sale revenue, and variable and fixed costs. The influence of soil nitrogen on yield, and the importance of organic premiums to maintain a viable income, was identified as a key driver. Alternative conversion strategies and the subsequent organic rotations were highly influenced by the farm enterprise balance, resource constraints and expertise, which have been shown to change during and after organic conversion and can vary at a farm level or even at a field or part-field level. The work undertaken did emphasize the importance of designing conversion strategies and rotations that are farm and site specific rather than generic.

In some systems, such as all-grass farms or in perennial cropping, there may be no rotation at all. Instead, diversity is achieved through species and varietal mixtures in space rather than over time. In arable systems, the rotation should form the basis for fertility building and exploitation phase(s) with cash crops as well as the main management tool for weed, pest and disease control. However, invariably the market for crops changes more quickly than most rotations, which are likely to be four, five or even seven years in length. This, then, requires a level of flexibility to respond to market demands without endangering the basis for successful production.

MAINTENANCE OF SOIL PRODUCTIVITY

The starting point for the design of any rotation should always be the resource capability of the farm in terms of climatic conditions, soil type and structure. Good soil management is essential for successful organic production. Without the quick-fix input-based solutions of non-organic mineral fertilizers, available to non-organic farmers to make good the shortcomings of poor soil structure and biotic activity, the maintenance of fertility, good structure and the avoidance of nutrient loss should be of paramount importance in any rotation. The principles of organic soil management are discussed at length in Chapter 2, along with fertility management in Chapter 3. The emphasis must be on maintaining a good soil structure in order to allow free movement of air and water so as to allow plant roots to penetrate all areas of the soil in search of nutrients and water.

Weed Management

Weed management is one of the main challenges facing organic arable farmers. A diverse rotation with a range of crops and cultivation methods can help prevent the build-up of weeds in the system. When matched to a mixture of spring and winter cropping, there is a potential to target cultivations and weeding activities to different weeds or the same weeds at different developmental stages, and avoid the same weed problems developing in successive years. This can also help reduce the return of weed seeds to the weed seed bank. Many of the implications of rotations for weed control are discussed in Chapter 10.

Pest and Disease Management

The main approaches to pest and disease control in organic arable systems are discussed in Chapter 11. Without the quick-fix pesticide input solution, the focus in organic production is more on problem avoidance than treatment. This can be achieved in the rotation by reducing pest and disease carryover between crops by having a diverse cropping rotation with no two 'like' crops next to each other, thus denying a host opportunity to any pests present. This is particularly true for soil-borne pests and diseases.

MARKETING AND CROPPING BALANCE

The rotation is not only important for agronomic reasons, but also for continuity of supply to market, and in reducing the potentially very cyclical nature of crop production. Operating the rotation over the farmed area so as to provide a roughly equal area of each crop and land use, helps ensure that a similar tonnage of a particular crop is available to market each year. In Figure 11, a rotation for a 100ha farm is split into five equal blocks, so that 20ha of wheat, beans and triticale can be grown and marketed each year alongside 40ha of fertility-building clover; this will help spread the workload and farming operations over a wider period, and it also has good environmental benefits in that a patchwork of cropping and land use provides a greater diversity of habitats for animals, insects and birds.

The rotation for a 100ha farm shown in Figure 12 operates a more block cropping approach, more akin to a non-organic system. Whilst it may simplify the cropping and field operations, it will result in bottlenecking the workload for each crop and will therefore reduce the environmental benefits associated with diversity. This type of approach will also result in a 'famine versus feast' marketing situation, with plenty of crop to market for three years, but then no crop to market for two years. This approach is not ideal for operating a good, sound rotation in an organic system.

Enterprise Balance in Rotations

The rotation should aim to create a balance between, on the one hand, building fertility (with leguminous crops and green manures), and on the other, exploiting the fertility that has been built up with non-leguminous cash crops (cereals, roots, oil-seed crops), whilst at the same time balancing competitive and non-competitive crops and the need for intervention weed, pest and disease control. This then requires a balance of crop and land cover types. As a guide the proportions of different crop types in arable and mixed rotations are shown in Table 22.

Area	Year	1	2	3	4	5
Block 1	20ha	Clover	Clover	Wheat	Beans	Triticale
Block 2	20ha	Triticale	Clover	Clover	Wheat	Beans
Block 3	20ha	Beans	Triticale	Clover	Clover	Wheat
Block 4	20ha	Wheat	Beans	Triticale	Clover	Clover
Block 5	20ha	Clover	Wheat	Beans	Triticale	Clover

Figure 11 A well balanced arable rotation

Area	Year	1	2	3	4	5
Block 1	20ha	Wheat	Beans	Triticale	Clover	Clover
Block 2	20ha	Wheat	Beans	Triticale	Clover	Clover
Block 3	20ha	Wheat	Beans	Triticale	Clover	Clover
Block 4	20ha	Wheat	Beans	Triticale	Clover	Clover
Block 5	20ha	Wheat	Beans	Triticale	Clover	Clover

Figure 12 A block cropping organic arable rotation

Table 22 Typical proportion of different crop types in arable and mixed rotations

Crop type	Min %	Max %
Legumes: forage/green manure/leys	20	100
Cereals	0	60
Grain legumes and oil seeds	0	25
Other forage crops	0	33
Roots and vegetables	0	50

The likely consequences of exceeding the limits suggested above are:

- insufficient fertility building versus cash-crop exploitation;
- problems with specific diseases and disease carryover between 'like' crops;
- a build-up of weeds;
- a reduction in soil organic matter due to over-cultivation.

Table 23 Recommended intervals or 'break periods' for common organic arable crops

Crop	Interval (years)	Self tolerant	Variable tolerance	Not self tolerant
Red clover	5–6			X
White clover	0–1	X		
Vetch	3–4		X	
Lucerne	5			X
Wheat	2		X	
Barley	2		X	
Oats	3–4		X	
Rye	1	X		
Peas & beans	4–5			X
Lupins	3–4		X	
Oil-seed rape	3–4		X	
Brassicas	4			X
Potatoes	5–6			X
Onions	6			X
Sugar beet	6			X
Maize	1–2	X		

In addition to considering the proportion of different crops in the rotation, good agronomic practice should form an integral part of the rotation, so that adequate intervals or 'break periods' between 'like' crops are built into the rotation to avoid the potential build-up of pest and disease problems, and which avoid creating carryover between crops. Recommended intervals or 'break periods' for common organic arable crops are shown in Table 23. These apply not only to cash crops but also to the legumes used to build fertility as green manures or leys – which should be rotated in order to avoid any potential pest or disease build-up in foliage, roots or soil.

In all cases, however, the aim of optimizing crop production is achieved through management of the resources within the farm ecosystem, rather than reliance on external inputs. This places additional constraints on the choice and combination of crops (and livestock enterprises) which can be produced, than would be the case in a non-organic system. This may have implications as to the overall performance of the farming system.

CROP CHOICE AND SEQUENCE

The factors acting on the actual crops grown in the rotation, and how these are sequences, are many and varied; they include soil type, fertility, pest and disease pressure, weed pressure, planting, harvesting and storage infrastructure, and market. Nick Lampkin in his book *Organic Farming* summarized schematically some of the characteristics of crops in the context of rotation design, and their suitability to different soils, climates and rotational combinations. These are such useful references that they have been reproduced here in the context of organic arable farming systems. Thus, Table 24 shows the characteristics of arable crops in the context of rotation design; Table 25 shows the suitability of crops for different soils and climates; and Table 26 illustrates the suitability of different crop combinations in the rotation.

There are many factors influencing rotation design, including the presence of livestock enterprises and forage requirements, soil type, weed incidence, climate and topography. These can broadly be categorized as resource, fertility or enterprise factors, and their influence will vary from site to site (*see* Table 27). During and immediately after conversion, the impact of historical land use and soil-borne pest or disease issues can limit the choices of crops in the rotation; but once an operational organic system has been developed, it is the interaction between these factors that determines an appropriate rotation and cropping sequence. This may vary at a farm level, or even at a field or part-field level. The other major influence will be the experience of the farmer in growing particular crops. Most farmers will know which crops grow well in their locality, in their climate and in their soils, and organic crops are no different in this respect.

Table 24 The characteristics of arable crops in context of rotation design

Crop	Rooting depth	Residual biomass (t DM/ha)	Soil structure	Organic matter contribution	Nitrogen balance	Weed control	Pest & diseases		Winter soil cover
							Self tolerant	Break (years)	
Wheat	O/+		-/+	-/O	-	-/O	-	2-4	-/+
Barley	O/+	0.9-1.7	-/+	-/O	-	-/O	-/O	2-4	-/+
Oats	O/+		-/+	-/O	-	-/O	-	5	-/+
Rye/spelt	O/+		-/+	-/O	-	O/+	+	1	-/+
Beans	O	0.5-2.3	O	O	+	-/O	-	4-5	-/+
Peas	O		O	O	+	-	-	6-7	-
Potatoes	-	0.6-1.0	-/O	-	-	-/+	-	5	-
Sugar beet	-		-/O	-	-	-/+	-	6	-
Green manures									
Non leguminous	-/+	0.9-3.0	O/+	O/+	-/O	+	+/-	-	+
Leguminous	-/+		O/+	O/+	+	+	+	-	+
Red clover	+	4.5-5.5	+++	++	+++	++	-	6	++
2-3 yr ley White clover	O	6.0-8.0	++	+++	++	++	O/+	-	++
3-5 yr ley Lucerne	++	6.0-8.0	+++	+++	+++	++	-	5	++

+++   Excellent
++   Very good
+   Good/deep/large
O   Neutral/average/medium
-     Bad/shallow/small
—   Very bad

Source: Lampkin N. Organic Farming

Within the operational arable rotation, the temptation is to try and squeeze in as much cash cropping as possible, as compared to fertility-building phases (often seen as non-productive). However, it should always be remembered that the fertility-building phase is the engine that drives the whole system. A shorter fertility-building phase, and poorer quality or less well managed leguminous green manures as part of this phase, will inevitably result in a shorter or less productive cash-cropping/exploitative phase of the rotation.

Table 25 The suitability of crops for different soils and climates

Soil/Climate	Suitable crops
Lightest sands	Rye, triticale, lupins, soya, carrots, vetch
Light soils	Barley, triticale, spelt, root crops, sugar beet, potatoes, peas, horticultural crops, clover, vetch
Light chalks	Barley, triticale, wheat, sanfoin, lucerne, clover, trefoil – not potatoes
Medium loams	Most crops
Heavy clays	Wheat, oats, triticale, spelt, beans, clover, trefoil, vetch, grass
Fens and silts	Wheat, potatoes, sugar beet, root crops, horticultural crops
Acid soils	Oats, rye, spelt, potatoes
Wet soils/areas	Oats, turnips, grass

Source: Lampkin N. Organic Farming

Within the cash-cropping years, yield, competitiveness and disease resistance are key components in the profitability of organic combinable crops. The current models used to produce organic grain crops rely heavily on a fairly cyclic rotation whereby fertility is built up over a period of typically one to three-plus years, and subsequently exploited with a sequence of cash grain crops. The crops placed first in the exploitative phase of the rotation are typically those where the highest yield and greatest profit is desired; crops later in the sequence typically result in reduced performance and profitability.

What is key to the success of any rotation is that it is tailored so as to be 'site specific'. For any given site, soil quality is one of the key determinants in crop and economic performance. On fertile silt soils at the ADAS Terrington research station in Norfolk, organic all-arable systems have shown the potential to out-perform similar non-organic all-arable rotations over a fifteen-year period. In these situations the ratio of cash crop to fertility building can be high. On more marginal soils the ratio may have to be much lower to cope with lower levels of inherent fertility, reduced crop competition, and greater pest and disease pressure.

Table 26 The suitability of different crop combinations in the rotation

Preceding crop → Following crop ↓		wh	wb	sb	r	o	m	pe	be	lr	ley	mc	ep	sbe
Winter wheat	(wh)	–	–	–	O	O	O	++	++	O	O	++	++	O
Spring wheat	(wh)	–	–	–	O	O	++	+	++	++	++	++	+	++
Winter barley	(wb)	O	–	–	O	O	–	++	–	O	O	–	++	–
Spring barley	(sb)	O	O	O	O	O	++	–	–	+	O	++	+	++
Winter rye	(r)	O	O	O	O	O	++	–	++	++	O	++	++	++
Spring rye	(r)	O	O	O	O	–	O	+	++	++	++	O	+	++
Oats	(o)	O	++	++	++	–	++	++	++	++	++	++	–	++
Maize	(m)	++	+	++	++	++	–	++	++	++	++	++	–	++
Peas	(pe)	++	+	++	++	++	++	–	–	–	++	++	–	++
Beans	(be)	++	+	++	++	++	++	–	–	–	–	++	–	++
Lucerne/Red Clov	(lr)	+	O	++	++	O	O	–	++	O	–	++	++	++
Ley (Grass/Clov)	(ley)	O	O	++	++	O	O	–	++	O	O	++	++	++
Potatoes Maincrop	(mc)	++	+	++	++	++	++	++	++	++	++	–	–	++
Potatoes early	(ep)	++	+	++	++	++	++	++	++	++	++	–	–	++
Sugar beet	(sbe)	++	++	++	++	++	++	++	++	++	++	++	–	–

++ Good

+ Good, but unnecessary. Other crops make better use of the preceding one. Could be used in combination with catch crop or green manure.

O Possible

– Limited applications – not advisable if preceding crop harvested late, in dry areas, if pest risk exists (mainly nematodes), or if danger of lodging (e.g barley after legumes).

| Inadvisable

(source: Lampkin N. Organic Farming)

Table 27 Factors influencing rotation design

Resource	Fertility	Enterprise
Soil type and quality	Soil type and quality	Soil type and quality
Climate		Livestock requirements
Farm geography & biodiversity	Fertility management	Balance of fertility & exploitation
Farm history/cropping	Legumes	Fertility-building choices
Weed levels from historic use may limit viability	Green manures	Pest/disease breaks
	Compost/FYM	Weed management
Farm management (farm/ contract managed)	Supplementary inputs	Autumn vs spring crops
Infrastructure and skills		Crop choices
Integration with other farm activities		
Other crops grown in locality		
Location to markets		

ORGANIC ARABLE ROTATIONS

The objective of a rotation within an arable farming context must be to enhance crop production by growing a sequence of crops that allows the farmer to manage soil fertility, soil condition, and pest, disease and weed competition. A suitable rotation should contain the following practical components:

- Alternating or combining shallow-rooting crops and deep-rooted crops. The use of deep-rooting crops allows air to get throughout the soil profile and will improve the rooting ability and hence nutrient scavenging ability of the following crop. It will also aid in the creation of fissures throughout the profile, which help the movement of water and earthworms and help improve soil structure.
- Alternating or combining high root-mass crops and low root-mass crops. A crop with a high root mass will increase the organic matter deep in the soil, and aid water retention and improve the structure of the soil.
- Alternating or combining competitive weed-suppressing crops and uncompetitive weed-susceptible crops. Weed-suppressing crops are crops that tend to form a dense canopy and reduce the impact of weeds by reducing light to the field floor, and which compete well for nutrients. For example, crops such as field beans are far more suppressive than cereals as they spread across the rows and grow far taller than the weeds.
   However, in the early stages of crop growth all arable crops are vulnerable to weed competition. The other main drawback of many crops is that as the crop canopy dies back in the summer, light penetration leads to increasing weed germination. This can be addressed to some extent by undersowing the crop with a suitable green manure (*see* Chapters 3 and 10).

- Alternating or combining nitrogen-demanding crops and nitrogen-fixing crops, or scheduling nitrogen-demanding crops in a rotational position close to incorporated leguminous fertility-building green manures and crops which are less nitrogen demanding, or nitrogen fixing later in the rotation from the incorporated leguminous fertility-building green manures.

Leguminous crops (clover, vetch, peas, beans, lupins, soya) fix nitrogen from the atmosphere. Clover leys can fix from 40–400kg of nitrogen per hectare (*see* Chapter 3). A percentage of this nitrogen is available for the following crop once the leguminous fertility-building green manure crop is incorporated. For example, once a red clover leguminous fertility-building green manure is incorporated, between 100–300kg N/ha is available to the following sequence of cash crops. In contrast, field beans may fix 50–200kg N/ha, but with only 20–30kg N/ha available for the following crop. This is due to the fact that the majority of nitrogen fixed by a field bean crop is removed as protein when the grain is harvested and removed for sale. As such, grain legumes that are harvested for their grain can be treated as net *neutral* in terms of nitrogen fixation, in that they fix what they need for their own production, and leave only a very small net surplus for following crops.

Legumes and Leys

Legumes form an essential component of fertility-building leys on organic farms. Their value is increasingly being recognized in non-organic agriculture, with their contribution to soil improvement, soil protection, and reduction in the need for high cost nitrogen fertilizers. The nitrogen-fixation ability of legumes is discussed in greater depth in Chapter 3. The greatest benefit of including legumes in the rotation is their ability to fix atmospheric nitrogen, which can be bound in the soil in organic complexes and then released via mineralization for use by subsequent cash crops. A wide range of leguminous covers and green manures is available to the organic farmer; these are discussed in detail in Chapter 3.

Clover

Red and white clovers are commonly used by organic farmers in the UK. **White clover** varieties with small, medium and large leaves are available, with different levels of persistence, resistance to grazing/cutting, and productivity. White clover spreads by means of a stolon, which spreads runners above ground. Care must be taken not to damage the stolons from over-grazing or mechanical operations. White clover is normally sown with a suitable grass mix, and is utilized for grazing and/or forage conservation.

Red clover varieties are typically much larger leaved, and produce a much greater and taller biomass for a period of two to four years before dying back, and are thus well suited to short-term fertility-building leys, and are ideal for an arable rotation. Red clovers have an upright growth habit from a single crown located at the base of the stem, and which acts

White clover showing stolons and spreading root pattern.

as a store of nutrients. Red clovers have a strong, single taproot from which finer roots arise. This is particularly adept at improving soil structure. Red clover can be sown with or without grass depending on use and management requirements. Red clover is more susceptible to clover rot disease (*Sclerotinia*) than white clover, and there have been limited occurrences of stem nematode infection. A careful approach must therefore be used, that ensures adequate break periods between successive red clover covers. This can be simply achieved by rotating red and white clover in the rotation. Red clover is also well known to contain high levels of oestrogen, which can affect fertility in some breeding livestock, so care needs to be taken in this respect.

Red clover showing upright growth and strong tap root.

Stockless versus Mixed Systems

The majority of farms that produce organic arable crops are likely to also operate livestock enterprises, principally beef cattle, sheep or dairy. Increasingly, pig and poultry systems are also being operated on organic arable farms as a result of increasing demand for pork, poultry and egg products. These farms therefore have a 'mix' of enterprises and resource requirements.

For the majority of farms that are 'mixed', the fertility-building phase is utilized by the livestock enterprises, and this is subsequently followed by a number of years of arable cropping. The fertility-building phase is commonly termed the 'ley' phase, and is often planted with a clover and perennial rye grass mixture and utilized for grazing, and/or forage production. This then provides a cropping rotation to minimize pest, disease and weed problems, and the rotation of fresh ground for livestock to maintain good health status, and avoid livestock pest and disease problems.

Some arable farms, especially the larger units, which have specialized in arable crops for many years or even decades, have no livestock enterprises, infrastructure (fencing, field water, winter housing and so on) or specific livestock management skills. Moreover, many arable farmers have no desire to operate livestock enterprises. These so-called 'stockless' arable farms are faced with special challenges when it comes to conversion to organic production, and in operating and maintaining suitable rotations.

Stockless Rotations

The main difference with all-arable or 'stockless' farms is that the period where fertility building is taking place is less likely to be utilized for livestock production. As such, the biomass produced by the fertility-building cover (typically clover, vetch and so on) has to be managed by cutting/topping and mulching.

Research has demonstrated that 'stockless' and predominantly arable organic systems are agronomically and economically viable, but that current conversion strategies are highly dependent on the use of annual green manures to provide sufficient soil nitrogen and income during the fertility-building phases of the rotation.

The old IACS regime in the UK allowed organic and converting farmers to grow clover and other legumes on set-aside, therefore utilizing it to build fertility whilst deriving some income during the non-cropping phase. On many stockless arable farms this could historically account for 30 to 50 per cent of the farm.

Under the Single Farm Payment Scheme (SPS) regime, farms have a lot more flexibility and can grow clover and other legumes as any part of the rotation to build fertility. However, there are restrictions on the use of legumes where land has to be set aside under the SPS rules.

Rotation Options

It is important that rotations are designed to suit farm conditions and marketing strategy. Below are some suggested rotations for stockless arable systems and mixed systems with livestock, for different soil types. The majority of grain comes from farms that also operate livestock enterprises, where the grazing land forms part of the rotation; however, a proportion of grain is produced from stockless holdings.

Light to medium soils incorporating root cropping

Year 1: Grass/clover or vetch (fertility building). The fertility-building crop may have to be in the ground longer where fertility has been depleted.

Year 2: Potatoes.

Year 3: Winter wheat, followed by winter cover of phacelia, vetch or mustard.

Year 4: Spring peas or beans. An early pea harvest may allow time for weed management or the sowing of a mustard green manure before sowing wheat. If pulses will be followed by a spring crop, then use winter cover green manure of grazing rye, mustard or phacelia.

Year 5: Winter wheat; follow with a winter cover green manure of rye mustard or phacelia.

Year 6: Spring barley, undersown with grass/clover.

Cultivation notes: Mulch green manures and cover crops, preferably with a flail topper for a better mulching effect, and incorporate principally with a plough and press-based system. This rotation requires a range of plough and cultivation depths, from deep for potatoes, to relatively shallow for pulses. Subject to soil, weather conditions and weed levels, it may be possible to establish cereals after potatoes and pulses without the use of the plough.

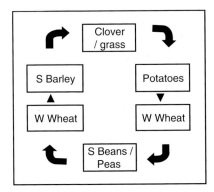

Figure 13 Rotation for light to medium soils incorporating root cropping

Light to medium soils with no root cropping

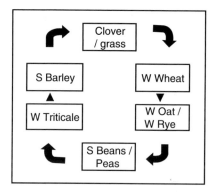

Figure 14 Rotation for light to medium soils with no root cropping

Year 1: Grass/clover or vetch (fertility building). The fertility-building crop may have to be in the ground longer where fertility has been depleted.

Year 2: Winter wheat.

Year 3: Winter oat or rye. These alternative winter cereals are less susceptible to take-all, followed by winter cover of phacelia, vetch or mustard.

Year 4: Spring peas or beans. An early pea harvest may allow time for a mustard green manure before sowing triticale.

Year 5: Winter triticale, long-strawed and competitive requiring less fertility, followed by winter cover green manure of vetch or phacelia.

Year 6: Spring barley – could be undersown with clover.

   Cultivation notes: Similar to the first example, use a plough and press-based system for incorporating green manures and cover crops and burying problem weeds. The use of deep cultivators other than a plough for crop establishment is less likely, unless crop residues or topsoil mixing is specifically required. Deeper subsoil management with a subsoiler may be required during the rotation. Non-inversion tillage techniques could be used to establish cover crops. Second cereal and first wheat after pulses will be subject to soil and weather conditions.

Rotation for medium to heavy soils

Year 1: Grass/clover or vetch (fertility building). The fertility-building crop may have to be in the ground longer where fertility has been depleted.

Year 2: Grass/clover ley – mulched or grazed.

Year 3: Spring or winter wheat (stubble cultivation according to weed species present).

Year 4: Winter beans.

Year 5: Winter wheat or triticale. Followed with mustard, phacelia or vetch over winter.

Year 6: Spring oats or barley.

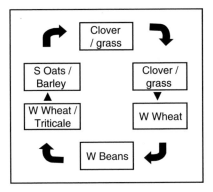

Figure 15 Rotation for medium to heavy soils

Cultivation notes: Plough after grass/clover ley and to establish beans; otherwise it may be possible to use non-inversion tillage methods to incorporate crop residues and create seedbeds, providing weed levels permit. If grass weeds increase, shallow plough where required to clean between cereal crops and to reduce brome and black-grass. Alternatively, consider a spring-sown break, such as beans, although this may not be ideal on heavier land; alternatively consider a summer fallow to help clean the field. Sowing a white clover or yellow trefoil undersown in first wheats will build fertility for second wheats. Including a second wheat before the beans could extend this rotation.

The incorporation of cereal crop residues will reduce nutrient removal and slowly increase organic matter return to the soil. However, it may also cause some short-term lock-up of mineralized nitrogen in the autumn and winter.

ROTATIONS WITH LIVESTOCK

Where the holding operates livestock enterprises in parallel to arable cropping, the rotation should contain a two- or three-year grass/clover ley. Instead of mulching the clover, which is what would take place on a stockless farm, the grass/clover ley can be used for grazing and/or forage conservation, with the manure applied improving the soil fertility further. Arable rotations are possible without the use of livestock, however the farmer would need to keep a careful eye on nutrient removal from the exploitative cropping part of the rotation.

Farms should avoid leaving grass/clover leys in place for too long, as not only will the total level of nitrogen fixed reduce over time, but pest problems such as leatherjacket and wireworm populations will increase with the age of the ley, and these can be a major problem for subsequent arable crops. Ideally grass/clover leys should be kept in place for no longer than two to three years to optimize their fertility-building capacity. However, this needs to be balanced against the cost of establishment and usefulness to the livestock enterprise.

Two rotations containing livestock are shown below.

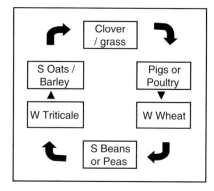

Figure 16 Livestock/arable rotation for light to medium soils

Livestock/arable rotation for light to medium soils
Year 1: Grass/clover. Cut for silage, graze aftermath.
Year 2: Grass/clover. Grazed by pigs or poultry.
Year 3: Winter wheat (stubble cultivation according to weed species present), followed by a cover of phacelia or mustard.
Year 4: Spring peas or beans.
Year 5: Winter triticale followed by a cover of vetch, phacelia or mustard.
Year 6: Spring barley undersown.

Cultivation notes: Remedial soil cultivations are likely after pigs or poultry to remove compaction and repair wallows, tracks and so on. Plough after grass/clover ley and to establish beans; otherwise it maybe possible to use non-inversion tillage methods to incorporate crop residues and create seedbeds, providing weed levels permit. If grass weeds increase, shallow plough where required to clean between cereal crops and to reduce brome and black-grass. Alternatively, consider a spring-sown break, such as beans, although this may not be ideal on heavier land; or consider a summer fallow to help clean the field. Sowing a white clover or yellow trefoil undersown in first wheats will build fertility for second wheats. Including a second wheat before the beans could extend this rotation.

Livestock/arable rotation for medium to heavy soils
Year 1: Grass/clover. Grazed by stock.
Year 2: Grass/clover. Cut for silage, graze aftermath.
Year 3: Winter wheat or spring wheat if winter grazing required (stubble cultivation according to weed species present).
Year 4: Winter beans.
Year 5: Winter wheat/ triticale. Overwinter cover of mustard, forage rape or phacelia.
Year 6: Spring oats or barley.

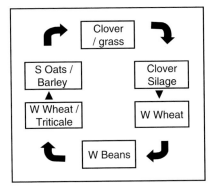

Figure 17 Livestock/arable rotation for medium to heavy soils

This rotation may be extended with, for example, oats in year six, then barley in year seven, though this would depend on the success of the previous crop and the fertility of the land. Some farmers leave the grass/clover in for a third year with the aim of increasing the fertility for the following crop even further.

It should be noted that grass/clover leys ploughed out after only one year will have a severely limited impact on fertility for the following crop due to the fact that the clover will still be relatively young and will have fixed only a small quantity of nitrogen for the following crop.

As well as providing fertility building for the following crop, clover/grass leys also provide forage for stock – and the symbiotic relationship doesn't stop there. Arable crops, as part of a stock rotation, can provide breaks and therefore reduce internal parasitic worm-burden build-up in livestock.

General Observations

Many organic arable rotations consist of a mixture of both spring- and winter-sown broad-leaved cereals plus fertility-building crops managed as green manures or grazing/forage leys. These crops act as pest and disease breaks for the cereals and provide different opportunities for weed control. It should also be noted that barley and oats are more effective at shading out broad-leaved weeds as compared to wheat and triticale, which allow more light penetration.

The main rotational concerns for cereal production are soil fertility and soil-borne pest and diseases such as take-all, leatherjacket, wireworm, frit fly and slugs. Cereal species vary in their susceptibility to some of these traits: for example, wheat is most prone to take-all, followed by barley, rye and triticale, with oats being least susceptible.

When taking site fertility into account, wheat shows the greatest response to high levels of soil nitrogen, and rye/triticale the least. Where a rotation imposes a range of potential drilling dates for cereals – that is, after root crops – then winter wheat and winter barley have the greatest flexibility in terms of drilling date, although for optimum crop performance, winter barley should be drilled before the middle of October.

Rotations that encourage mixed cropping can also reduce the potential risk of epidemics of foliar disease, which could develop where large areas of the same cropping occur.

ADAPTING ROTATIONS

Current rotation design is often sub-optimal with regard to soil nitrogen, phosphorus and potassium capture and utilization, with many commonly used rotations largely ignoring below-ground biomass production and diversity. The use of a greater diversity of fertility-building green manures within the rotation is a logical and simple next step, as explored in Chapters 2, 3 and 4.

Alternative strategies need to be considered which optimize soil nitrogen use, minimize disturbance to the soil microbial biomass, and break the reliance on subsidies to ensure the economic viability for stockless rotations that result in a highly cyclic production system. Where livestock enterprises are not an option, choices are more limited, especially when the system requires continuous organic crop production. The use of bi-crop or intercropping strategies, in which legumes provide the fertility for the simultaneous production of cash crops, may offer an appropriate alternative. But for these systems to operate successfully, innovative cultivation and planting techniques are required, careful management of green manures is critical (*see* Chapters 3 and 4), and selection of appropriate crop types and varieties is essential (*see* Chapters 6, 7 and 8). Currently only a limited range of crop varieties is available, and this is a limiting factor for the success of this option; at the time of writing, crops and varieties that are better suited to organic production conditions need to be bred. However, the expansion of the small but emerging European organic plant breeding initiative will play an important role in developing these options further.

From time to time it may be necessary to alter or modify a rotation, because for some reason it begins to under-perform. The reasons for this can be varied, from a change in the market, to poor soil conditions through inappropriate cultivations, or loss of fertility, or a build-up of a particular set of weeds. It cannot be stressed enough that the two key principles to bear in mind are, firstly, the need to adapt the general rules to your own farming situation (such as climate, soil type, machinery available, storage, markets, weeds, pest and diseases present, and so on); and secondly, the need to implement, and be flexible in your approach to, rotations and organic crop production.

The trend towards specialized all-arable farms means that many arable farmers are not in a position to consider converting to organic production if the reintroduction of livestock is involved. The use of this type of cut-and-mulch fertility-building green manure or ley therefore forms the backbone of the organic stockless rotation. Experience shows that when mulched, clover will grow back well. The implications of cutting and mulching on nitrogen fixation are discussed in more detail in Chapter 3. The rotations described, which operate a balance of grain legumes and leguminous green manures along with cereal cash crops, can provide sufficient nitrogen for a good productive and profitable rotation, with sufficient weed control. Straw should be incorporated wherever possible to ensure that excessive organic matter and potash is not lost from the system.

The main challenge is whether the financial viability of the rotation can be maintained with between 30 per cent and 50 per cent of a typical rotation down to a non cash-crop fertility-building cover at any one time. Only if the price achieved for the cash crops in the rest of the rotation is high enough, is viability assured. The rotation can be extended by the use of grain legumes, or by using combinations of legumes and cereals as bi-crops (*see* Chapter 9); however, the proportion of grain legumes in the rotation is limited by self-tolerance of crops and the desire to avoid a build-up of pests and diseases, as discussed earlier in this chapter. Alternatively, in the longer term the reintroduction of livestock, or renting the ley to a livestock producer, can help improve the long-term viability of the rotation.

It should also always be remembered that although legumes can potentially fix sufficient nitrogen to maintain a stockless arable system, great care has to be taken over the way the crop utilizes the nitrogen. A catch crop or overwinter green manure will not fix nearly as much nitrogen as a green manure grown over one or two years, as the nodules will not have had sufficient time to develop, and the cooler soil conditions in the winter will limit nitrogen fixation. Grain legumes will fix good amounts of nitrogen, but a large proportion of this is sold off the farm as protein in the grain of the cash crop. As such they should be treated as 'nitrogen neutral', as there is only marginal benefit to other crops in the rotation. Clovers and trefoils grown for herbage seed do not present the same problem, and it may be well worth investigating this option further on some farms. Relatively speaking there are few completely stockless organic farms in the UK; however, there are quite a number of farms where livestock inhabit only a small part of the farm and where manures are more or less in token supply. Furthermore some farms, such as Rushall Farms in Wiltshire, Luddesdown Farm in Kent and Chapel Farm in Evesham, have operated stockless organic systems for between twenty and thirty years, and whilst all have their own challenges, all are still operating functioning, profitable arable systems.

Chapter 6

Organic Cereal Production

Cereal grain as a food has its historical origins in the wild grass seeds that were harvested and stored by early man. Grass is thought to have been purposefully cultivated to produce wheat in the Middle East some 8,500 years ago, but the earliest archaeological evidence for wheat cultivation comes from the Levant and Turkey, from around 10,000 years ago. Many archaeological sites and ancient writings reveal the existence of early circular storage pits used for conserving the harvested grains, which would keep well from year to year. It is said that the Egyptian Pharaohs cut the first sheaf of wheat with a golden sickle as a sign that the harvest could commence.

While rice remains a staple food for about a third of the world's population, most Western diets use cereals as a staple in one form or another. In recent decades there has been a significant trend towards diets containing fewer whole grains and more refined grain products than in previous generations.

THE COMPOSITION OF A GRAIN AND ITS UTILIZATION

The grain has a number of components, and these are utilized in various ways. Wheat is the most common commodity grain produced, with the greatest demand from the human and livestock feed sector. While the components of all grains can be used in similar ways, for simplicity the components of wheat and how they are utilized is explained below.

The Endosperm

This is the name given to the interior of a wheat kernel, and it makes up about 83 per cent of the whole grain of wheat. Ground down to a powder, the endosperm becomes flour, and this has a multitude of uses including bread making and the production of semolina (which are, of course, graded flour particles).

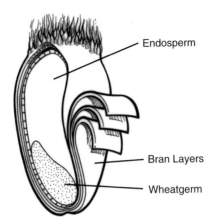

Figure 18 Components of the wheat grain

Bran

The outer layers of the wheat grain consist of bran, and are removed during the milling of white flour. About 14 per cent of the wheat kernel is bran. Some bran is rolled to make it flaky and is sold for human consumption; the rest is used in animal and poultry feed. Wholemeal flour contains all the naturally occurring bran.

The Germ

The germ is the part of the grain which would sprout if it were planted as a seed. It is packed with nutrients and protein with which to nourish a new plant. During processing the germ is usually separated from the rest of the wheat grain because its fat content limits the shelf life of the flour. It is occasionally used as a dietary supplement or sold as animal feed.

CEREAL CHOICE AND MANAGEMENT

Barley (*Hordeum vulgare*)

Cultivated barley (*H. vulgare*) is descended from wild barley (*Hordeum spontaneum*), which grows wild in the Middle East. Both forms are diploid. As wild barley is interfertile with domesticated barley, the two forms are often treated as one species, divided into *Hordeum vulgare* subsp. *spontaneum* (wild) and subsp. *vulgare* (domesticated). The main difference between the two forms is the brittle rachis of the former, which enables seed dispersal in the wild.

The earliest finds of wild barley come from Epi-Palaeolithic sites in the Levant, beginning in the Natufian. The earliest domesticated barley occurs at Aceramic Neolithic sites such as the layers of Tell Abu Hureyra in Syria. Barley was one of the first crops domesticated in the Near East, at the same time as einkorn and emmer wheat.

Alongside emmer wheat, barley was a staple cereal of ancient Egypt, where it was used to make bread and beer; together these were a complete diet in Upper Egypt. The ancient Greeks made flat, unleavened breads and a kind of barley porridge, which they called *Alphita*. On its own, barley produces close-textured but sweet-tasting bread.

The ritual significance of barley in ancient Greece possibly dates back to the earliest stages of the Eleusinian Mysteries. The preparatory *kykeon*, or mixed drink of the initiates, prepared from barley and herbs, was referred to in the Homeric hymn to Demeter, who was also called 'Barley Mother'.

Barley forms a staple food for humans and other animals the world over. It is more tolerant of soil salinity than wheat, which might explain the increase of barley cultivation on Mesopotamia from the second millennium BC onwards. Barley can still thrive in conditions that are too cold even for rye.

Today, considerable quantities of barley are produced by farmers, who sell their crop to the brewing industry or as animal feed. In 2005, barley ranked fourth in quantity produced and in area of cultivation of cereal crops in the world (560,000sq km), with the UK being the eighth largest producer of barley, as shown in Table 28.

Table 28 Top ten world barley producers – 2005 (million metric ton)

Russia	16.7
Canada	12.1
Germany	11.7
France	10.4
Ukraine	9.3
Turkey	9.0
Australia	6.6
United Kingdom	**5.5**
United States	4.6
Spain	4.4
World Total	**138**

Source: UN Food & Agriculture Organization (FAO)

Barley grows well in the British climate, and 2,000 years ago was the main cereal used for milling or boiling and eating as a grain or porridge. When barley is harvested it is encased in a fibrous outer hull which is inedible and must be removed to release the grain.

The organic market size in 2002/03 was 12,631 tonnes for feed and 5,408 tonnes for malting. By 2006 this had risen significantly for feed barley in line with the increase in organic livestock numbers to 21,346 tonnes; however, it remained fairly stable for malting at 5,336 tonnes.

Barley grain with its hull still on is called *covered* barley; once the grain has had the inedible hull removed, it is called *hulled* barley. At this stage, the grain still has its bran and germ, which are nutritious. Hulled barley is

Barley is an important domestic and international grain.

considered a whole grain, and is a popular health food. Pearl barley, or pearled barley, is hulled barley that has been processed further to remove the bran. It may be polished, a process known as 'pearling'.

Hulled or pearl barley may be processed into a variety of barley products, including flour, flakes similar to oatmeal, and grits. It may be malted and used in the production of alcoholic beverages. Malting barley is a key ingredient in beer and whisky production.

Agronomy

Barley can be divided by the number of kernel rows in the head. Three forms have been cultivated: two-row barley (traditionally known as *Hordeum distichum*), four-row (*Hordeum tetrastichum*) and six-row barley (*Hordeum vulgare*). In two-row barley only one spikelet is fertile; in the four-row and six-row forms, all three are fertile.

Two-row barley is the oldest form, wild barley having two rows as well. Two-row barley has a lower protein content than six-row barley, and thus a lower enzyme content. High protein barley is best suited for animal feed or malt that will be used to make beers with a large adjunct content. Two-row barley is traditionally used in English beers, six-row barley is traditional in German and American beers. Four-row is unsuitable for brewing.

In Britain barley is predominantly produced for animal feed, brewing and distilling. Historically over half the UK's non-organic spring barley was grown in Scotland. The popularity of spring barley increased in the mid- to late 1990s, and has remained a popular crop in organic arable rotations. Barley is often grown on thinner, shallower soils where wheat wouldn't perform to its maximum potential.

The market for organic barley has been weaker than for wheat and more volatile, with limited utilization by livestock feed compounders, however it has now become more robust, with a high demand from the organic livestock feed sector.

Demand for malting barley is strong, but from a very small market because only a few maltsters and breweries produce organic beer. Malting barley should be grown on contract, but if quality specifications are not met, then it has to be sold as feed at a significant discount. Spring barley may be more likely to achieve a good malting quality, and low soil nitrogen conditions may also be beneficial in this respect. Some barley for whole grain and/or flaking is required by the whole food market, and as an ingredient for muesli, but this demand is also very limited.

Organic Crop Management
Variety selection is discussed for individual crop types in Chapter 8.

Rotation   Low soil nitrogen availability in early spring, weed competition and disease susceptibility make winter barley a more difficult crop to grow under organic conditions. Spring barley's lower nutritional requirement and good productive potential on more marginal soils, or its position in the rotation, make it an ideal crop. It is a crop that is often grown at the end of the rotation and is frequently undersown with clover and grass (*see* variety notes in Chapter 8).

Barley is chiefly planted after turnips, sometimes after peas and beans, but less commonly after wheat or oats, unless under special circumstances. When sown after turnips it can be established directly after the turnips are consumed in the late winter or early spring. When sown after wheat, oats, beans and peas, ploughing is usually required so that good sowing conditions may be established. Rotation issues are discussed in more detail in Chapter 5.

Cultivation   The early harvest of barley opens up opportunities for repeated post-harvest mechanical operations in autumn for weed control. These may be limited prior to planting the barley crop by the harvest date of previous cereal.

Seeding and Establishment   Organic winter barley is normally sown from early October onwards, so as to reduce the early competition from weeds, and the threat from Barly Yellow Dwarf Virus and Barly Yellow Moziac Virus Spring barley is normally sown from February onwards up until March, but in exceptional conditions can be planted as late as April, especially in the far north where longer summer-day lengths allow longer growing periods. Its germination time is anywhere from one to three days.

The grain of barley is tender and easily damaged in any of the stages of its growth, particularly at drilling and during early establishment; a heavy shower of rain can almost ruin a crop on the best prepared land. A good seedbed is therefore essential for good barley establishment.

Seed rates in organic systems are typically higher to afford better competition. Winter barley should be sown at around 350–400 seeds/sq m = 160–200kg/ha (1.3–1.6cwt/ac); spring barley is typically sown at a higher rate of 375–425 seeds/sq m = 180–220kg/ha (1.4–1.8cwt/ac) to offer greater competition due to the reduced tillering capacity of spring varieties. Where there is a high weed competition, or where crops are drilled very early or very late, seed rates may need to be adjusted. Where there are localized problems such as slugs or poor seedbeds, higher seed rates should be used to compensate for losses.

Barley is normally established by drilling rows at 10–12cm spacings, with seed at 25–50cm depth depending on soil conditions, predation threat and so on. Barley can also be broadcast and incorporated, although this can lead to more uneven establishment. As with wheat, the production of barley on wide rows, typically on 20–25cm spacings (but up to 30cm spacing is workable), is gaining popularity with arable producers on heavier land and land with grass and tap-rooted weeds, so that the crop can be hoed between the rows with an inter-row hoe. (*See* Chapter 11 for more information on this system.)

Weeds Post-emergence mechanical weed control is possible using a harrow comb-type weeder or inter-row hoe (when planted on a wide row system). Aim to undertake a first weeding at the true three-leaf stage, when the plant is sufficiently anchored to withstand the weeding operation. Subsequent weeding operations can be carried out during tillering GS 25–30. (*See* more notes on mechanical weeding in Chapter 10.)

Harvesting Harvest of barley in the UK is normally between late July and late August, depending on location and the season. Prompt harvesting and drying (to <15 per cent) to ensure quality and to reduce storage problems is essential. Removal of contaminants (ergots, and so on) and non-crop debris is also essential for good marketing. When harvesting, careful attention must be taken when threshing, because the awn, which generally adheres to the grain, makes separation from the straw more troublesome. Grain can be damaged if over-threshed.

Oats (*Avena sativa*)

Oats are native to Eurasia and appear to have been domesticated relatively late. They are now grown throughout the temperate zones. They have a lower summer heat requirement and a greater tolerance of rain than other cereals such as wheat, rye or barley, so are particularly important in areas with cool, wet summers such as north-west Europe and the United Kingdom, even being grown successfully in Iceland. Oats are an annual plant, and can be planted either in the autumn or spring.

Historical attitudes towards oats vary. In England they were considered an inferior grain, firstly because they cannot be made into bread but only 'inferior' foods such as porridge or oatcakes, and secondly because they are associated with poorer areas where wheat cannot be grown, with less sun, more rain and less fertile soil, and where as a consequence the people were literally poorer. In Scotland they were, and still are, held in high esteem, as a mainstay of the national diet. A traditional saying in England was that 'oats are only fit to be fed to horses and Scotsmen', to which the Scottish riposte is 'and England has the finest horses, and Scotland the finest men'. Samuel Johnson notoriously defined oats in his dictionary as 'a grain, which in England is generally given to horses, but in Scotland supports the people'. While frequently seen as derogatory, this is no less than the literal truth. Oats are so central to traditional Scottish cuisine that the Scottish word 'corn' refers to oats (as opposed to meaning 'wheat' in England and 'maize' in North America and Australia). Oats grown in Scotland command a premium price throughout the United Kingdom as a result of these traditions.

Late nineteenth- and early twentieth-century harvesting of oats was performed using a binder. Oats were gathered into shocks, and then collected and run through a stationary threshing machine. Earlier harvest involved cutting with a scythe or sickle, and threshing under the feet of cattle.

A now obsolete Middle English name for the plant was *haver* (still used in most other germanic languages), although it survives in the name of the livestock feeding bag *haversack*. In contrast with the names of the other grains, 'oat' is usually used in the plural.

Since oats are unsuitable for making bread on their own, they are often served as a porridge made from crushed or rolled oats, oatmeal, and are also baked into cookies (oatcakes) which can have added wheat flour. As oat flour or oatmeal, they are also used in a variety of other baked goods – for example, bread made from a mixture of oatmeal and wheat flour – and cold cereals, and as an ingredient in muesli. Oats may also be consumed raw, and biscuits with raw oats are becoming popular. Oats are also occasionally used in Britain for brewing beer.

Oats are generally considered 'healthy', or a health food, being touted commercially as nutritious. Oat bran is the outer casing of the oat, and its consumption is believed to lower LDL ('bad') cholesterol, and possibly to reduce the risk of heart disease. After reports found that oats can help lower cholesterol, an 'oat bran craze' swept the US in the late 1980s, peaking in 1989, when potato crisps with added oat bran were marketed. The food fad was short-lived, however, and faded by the early 1990s.

Oats after maize has the highest lipid content of any cereal, for example >10 per cent for oats, and as high as 17 per cent for some maize cultivars, as compared to about 2–3 per cent for wheat and most other cereals. The polar lipid content of oats (about 8–17 per cent glycolipid and 10–20 per cent phospholipids, or a total of about 33 per cent) is greater than that of other cereals, since much of the lipid fraction is contained within the endosperm.

Oats are the only cereal containing a globulin or legume-like protein, avenalins, as the major (80 per cent) storage protein. Globulins are characterized by water solubility, and because of this property, oats may be turned into milk, but not into bread. The more typical cereal proteins, such as gluten, are prolamines; the minor protein of oat is a prolamine called avenin.

Oat protein is nearly equivalent in quality to soy protein, which has been shown by the World Health Organization to be the equal to meat, milk and egg protein. The protein content of the hull-less oat kernel (groat) ranges from 12–24 per cent, the highest among cereals.

Today, oats are still used for food for people and as fodder for animals, especially poultry and horses. Oat straw is used as animal bedding, and sometimes as animal feed.

The primary producers of oats are Russia, Canada, the USA, Poland, Finland, Australia and Germany. In 2005, according to the Food and Agriculture Organization (FAO), 24.6 million tons were harvested in over twenty-six countries across the world.

The UK does not feature in this list, producing only 91,000ha of oats (organic and non-organic) in 2005 (approximately 700,000t of grain).

Oats are predominantly grown for animal feed: approximately 18,144 tonnes were grown organically for this market in 2002/03, with a human consumption market of 4,536 tonnes, predominantly destined for the breakfast cereal market. This has remained more or less stable, and approximately 17,292 tonnes for animal feed and 4,323 tonnes for human consumption were produced in 2006, accounting for approximately 14 per cent of the total UK cereal production.

Table 29 Top ten world oat producers – 2005 (million metric ton)

Russia	5.1
Canada	3.3
United States	1.7
Poland	1.3
Finland	1.2
Australia	1.1
Germany	1.0
Belarus	0.8
China	0.8
Ukraine	0.8
World Total	**24.6**

Source: UN Food & Agriculture Organisation (FAO)

Oats are an important human consumption and feed grain.

Agronomy
Oats can be autumn or spring sown. An early start is crucial to good
yields, as oats can go dormant during very dry summers. Oats are cold tolerant and will be unaffected by late frosts or snow.

The market for oats for human consumption (flaking) is now fairly
robust, with a number of supply contracts available. There are some very
good direct production contracts to makers of breakfast cereals, muesli
bars and the specialized human consumption markets. Where human consumption specifications are not met, the crop can find a market as livestock feed. The quality specifications required for human consumption
flaking oats are a minimum of 47kg/hl, and up to 10 per cent screenings.
Appearance is important, as samples can be rejected for discoloration.
Variety choice is also important, as millers look for a good kernel content.

Oats are very good at weed suppression, and are easy to grow; they
make an ideal organic crop where it is to be fed back to stock on the farm.

Organic Crop Management
Variety selection is discussed for individual crop types in Chapter 8.

Rotation   High yields, disease resistance, weed competitiveness, and tolerance of lower fertility conditions than wheat or even triticale, make oats
a good second or third cereal crop. Oats should not be grown more than
once in succession because of the risk of cereal cyst nematode.

Oats remove substantial amounts of nitrogen from the soil. If the straw
is removed from the soil rather than being ploughed back, there will also
be removal of large quantities of potash. When the prior-year crop was a
legume, or where ample manure is applied, seed rates should be reduced
to avoid over-thick crops, which are prone to lodging.

Oats are generally very tall with a thick canopy, which can be of benefit
for weed competition; they are therefore well suited to weedier fields, and
for use later in the cereal rotation. Rotation issues are discussed in more
detail in Chapter 5.

Cultivation   As oats are early to harvest, there are opportunities for
repeated post-harvest mechanical operations in the autumn for weed
control, or for the early establishment of grass/clover leys and green
manures post harvest.

Seeding and Establishment   Organic winter oats are normally sown from
mid-October onwards, so as to reduce the early competition from weeds,
but they can be planted from early October until the new year. Spring oats
are normally sown from February onwards, up until late March.

Seed rates in organic systems are typically higher to afford better
competition. Winter oats should be sown at around 500–550 seeds/sq
m = 175–225kg/ha (1.4–1.8cwt/ac). Spring oats are typically sown at a higher
rate of 650–700 seeds/sq m = 220–270kg/ha (1.8–2.2cwt/ac), which offers
greater competition due to the reduced tillering capacity of spring varieties.
Where there is a high weed competition, or where crops are drilled very early

or very late, seed rates may need to be adjusted. Where slugs or poor seedbeds are a known local problem, higher seed rates should be used to compensate for losses; lower rates are used when under-seeding with a legume. Somewhat higher rates can be used on the best soils. Excessive sowing rates will lead to problems with lodging and may reduce yields.

Being very competitive, oats are normally established by drilling rows at 10–12cm spacings, with seed at 25–50cm depth depending on soil conditions, predation threat and so on. Oats can also be broadcast and incorporated, although this can lead to more uneven establishment. As with wheat, the production of oats on wide rows, typically on 20–25cm spacings (but up to 30cm spacing is workable), is gaining popularity with arable producers on heavier land and land with grass and tap-rooted weeds, so that the crop can be hoed between the rows with an inter-row hoe. (*See* Chapter 11 for more information on this system.)

Weeds The vigorous growth habit of oats will tend to choke out most weeds. A few tall broadleaf weeds can occasionally be a problem, as they can complicate harvest. If required, post-emergence mechanical weed control is possible using a harrow comb-type weeder or inter-row hoe (when planted on a wide row system).

Aim to undertake a first weeding at the true three-leaf stage when the plant is sufficiently anchored to withstand the weeding operation. Subsequent weeding operations can be carried out during tillering GS 25–30. (*See* more notes on mechanical weeding in Chapter 10.) Subsequent weeding is rarely required due to the competitive nature of oats. Weed management issues are discussed in more detail in Chapter 10.

Harvesting Harvest of oats in the UK is normally between late July and late August, depending on location and the season. Prompt harvesting and drying (to <15 per cent) is essential to ensure quality and reduce storage problems. The removal of contaminants and non-crop debris is also essential for good marketing.

Oats are normally combine-harvested direct. Large field losses can occur if the combine threshing mechanism is incorrectly set, or if the crop is over-ripe and very dry, as the grain can be threshed out by the header reel, falling from the heads and being lost from the combine table. Continental harvesting techniques include cutting, swathing into a windrow, and picking up with a draper header.

Naked Oats

In simple terms, naked oats are oats that thresh free from their husks when harvested. The end product is a grain that is very high in nutritional quality, especially in terms of oil content and metabolizable energy. They typically contain 10 per cent oil. Since oil contains 2.25 times more energy than an equivalent weight of carbohydrate, this confers a much higher energy content compared to other cereals. Naked oats have been demonstrated to decrease total milk fat and increase the proportion of monounsaturated fatty acids. Experimental varieties are being multiplied with up to 60 per cent more oil than existing oats. Poultry producers now appreciate that there is a high potential for the use of naked oats in poultry diets.

Many of the naked oat varieties have not been subject to the pressures of breeding for 'dwarf' characteristics, and as a result are very tall in structure, even more so than traditional oat varieties. However, they tend to have a much reduced yield potential as a result.

There is an embryonic organic naked oat market developing, mainly as a result of demand from the poultry sector which places great value on the high nutritional quality. There are a number of contracts now being offered for growing on a buy-back basis.

The production of naked oats is broadly similar to that of normal cultivated oats (*see* Oats *Avena sativa* above).

Rye (*secale cereale*)

Rye is a grass grown extensively both as a grain crop, and also as a forage crop for feeding livestock, especially over winter periods. It also has excellent attributes as a green manure (*see* Chapter 3). It is a member of the wheat tribe (*Triticeae*), and is closely related to barley and wheat.

Rye is a cereal, and should not be confused with *ryegrass* which is used for lawns, pasture, and hay for livestock.

The early history of rye is not clear. The wild ancestor of rye has not been identified with certainty, but is one of a number of species that grow wild in central and eastern Turkey, and adjacent areas. Domesticated rye occurs in small quantities at a number of Neolithic sites in Turkey, such as PPNB Can Hasan III, but is otherwise virtually absent from the archaeological record until the Bronze Age of central Europe, c. 1800–1500BC. It is possible that rye travelled west from Turkey as a minor admixture in wheat, and was only later cultivated in its own right. Since the Middle Ages, rye has been widely cultivated in central and eastern Europe, and is the main bread cereal in most areas east of the French-German border and north of Hungary.

During the seventeenth century, mixtures of wheat and rye were often planted in the same field so that, depending on the weather, one or the other would yield well: thus in a fine dry season the wheat would flourish, and in a cold damp season the rye would do better. The resulting crop would be taken to a local miller and ground into flour known as *maslin* flour. In Turkey, rye is often grown as an admixture in wheat crops. It is appreciated for the flavour it brings to bread, as well as for its ability to compensate for wheat's reduced yields in hard years.

Rye grain is used for flour, rye bread, rye beer, some whiskies, some vodkas, and animal fodder. It can also be eaten whole, either as boiled rye berries, or by being rolled, similar to rolled oats.

Rye grain produces dense, close-textured breads with distinctive characteristics. In Viking times, harvesting and storing grain in their inhospitable climate was difficult, so rye would be threshed and made into flat cakes with a hole in the middle, allowing the unleavened bread to be hung

Rye is a very competitive cereal, widely eaten in northern Europe.

up for storage. Rye bread, including pumpernickel, is a widely eaten food in northern Europe. Rye is also used to make the familiar crisp bread. Rye flour has a lower gluten content than wheat flour, and contains a higher proportion of soluble fibre.

Rye breads have remained a traditional fare in Germany and Scandinavian countries, and are now enjoying a revival in Britain.

Agronomy

Rye is a very hardy crop that will tolerate slightly acid soils and poor weather conditions, and so is widely grown in colder northern regions. It is more tolerant of dry and cool conditions than wheat, though not as tolerant of cold as barley. The crop is very tall (about 170–180cm) with very 'whippy' straw, and the grains are darker than wheat with a greenish tinge.

Rye is highly susceptible to the ergot fungus, and consumption of ergot-infected rye by both humans and animals results in a serious medical condition known as 'ergotism'. Ergotism can cause both physical and mental harm, including convulsions, miscarriage, necrosis of digits, and hallucinations. Historically, damp northern countries that have depended on rye as a staple crop were subject to periodic epidemics of this condition.

In the UK and northern Europe, rye grown for grain production is mainly destined for human consumption. Grain ryes grown in the UK are normally winter types, and cannot be sown after mid-February without risk of failure. Although hybrid rye varieties have higher yield potential they are more prone to brown rust, and the seed is more expensive. Rye varieties have the tallest structure of just about all cereal types, and this can be of benefit for weed competition. Rye is typically harvested in July, earlier than many other cereals, and this can be of benefit if early access to fields for weed control is desirable.

The main market for rye is into the human food chain as biscuits. Its value as livestock feed is limited. The non-organic market is small, with the organic market expanding but also still very small. Production contracts are essential, because once the milling market is supplied, the surplus has a limited value as a grain feed. Mills normally specify a minimum 180 Hagberg requirement for rye, however this can deteriorate rapidly with a late harvest after rainfall, and there is a risk that samples will fail milling specifications. Small mills often have a demand for limited quantities of rye. Rye may be sold in small, bagged lots for which an additional premium of £20/t may be available to cover the costs of the work done by the farmer.

Crop Management
Variety selection is discussed for individual crop types in Chapter 8.

Rotation   The ability of rye to compete against weeds and to thrive in lower fertility conditions than wheat makes this crop a good second or third cereal, although low yield can be a disadvantage (typically 2.5–3.5t/ha). Rye is far more self-tolerant than other cereals, and within reason can be grown in succession as a second or even third cereal. Rye has a very tall structure which can be of benefit for weed competition; it is therefore well suited to weedier fields, and to use later in the cereal rotation. Rotation issues are discussed in more detail in Chapter 5.

Cultivation   Early harvest in the south of the UK opens up opportunities for repeated post-harvest mechanical operations in autumn for weed control. These may be limited prior to the rye crop by the harvest date of previous cereal, and may not even be necessary as the crop is so competitive.

Seeding and Establishment   Organic rye is normally sown from mid-October onwards, so as to reduce the early competition from weeds, but can be planted from September to the end of the year. It is not suitable for grain production planting in the spring, as it requires vernalization (the process where the plant requires a period of cold in order to allow it to go from the vegetative phase to the reproductive phase). It can be planted virtually all year round for forage and grazing uses.

Seed rates in organic systems are typically higher to afford better competition. Seeding rates for rye can be lower than for other cereals due to its highly competitive nature. Rye should be sown at around 160–200kg/ha (1.3–1.6cwt/ac) assuming a thousand grain weight of 50kg with an aim of 400 seeds per square metre. Where there is a high weed competition, or crops are drilled very early or very late seed rates may need to be adjusted. Where there are known localized problems such as slugs or poor seedbeds, higher seed rates should be used to compensate for losses.

Being very competitive, rye is normally established by drilling rows at 10–12cm spacings, with seed at 25–50cm depth depending on soil conditions, predation threat and so on. Rye can also be broadcast and incorporated, although this can lead to more uneven establishment. As with wheat, the production of rye on wide rows – typically on 20–25cm spacings (but up to 30cm spacing is workable) – is gaining popularity with arable producers on heavier land and land with grass and tap-rooted weeds, so that the crop can be hoed between the rows with an inter-row hoe. (*See* Chapter 11 for more information on this system.)

Weeds Post-emergence mechanical weed control is possible using a harrow comb-type weeder or inter-row hoe (when planted on a wide row system). Aim to undertake a first weeding at the true three-leaf stage when the plant is sufficiently anchored to withstand the weeding operation. Subsequent weeding operations can be carried out during tillering GS 25–30. (*See* more notes on mechanical weeding in Chapter 10.) Subsequent weeding is rarely required due to the competitive nature of rye. The very early harvest of rye also helps with crop removal before the onset of late season weeds. Weed management issues are discussed in more detail in Chapter 10.

Harvesting The harvest of rye for grain in the UK is early, normally between mid-July and mid-August, depending on location and the season. Rye is very susceptible to sprouting, which will lead to rapid deterioration of Hagberg quality; it should therefore be harvested and dried, rather than delaying the harvest in the hope of dry conditions. Prompt harvesting and drying (to <15 per cent) to ensure quality and reduce storage problems is essential. Removal of contaminants (ergots and so on) and non-crop debris is also essential for good marketing. A very early harvest permits post-harvest cultivations and early ley/following crop establishment.

Triticale (*Triticosecale*)

Triticale is an artificial or man-made hybrid of rye and wheat first bred in laboratories during the late nineteenth century. The grain was originally bred in Scotland and Sweden. Commercially available triticale is almost always a second generation hybrid – that is, a cross between two kinds of triticale – and it has only recently been developed into a commercially viable crop. Depending on the cultivar, triticale can more or less resemble either of its parents: some have long-awned ears like rye, other varieties are more like tall wheat. It is grown mostly for forage or animal feed, although some triticale-based foods can be purchased at health food stores or are to be found in some breakfast cereals.

The triticale hybrids are all amphidiploid, which means the plant is diploid for two genomes derived from different species: in other words, triticale is an allotetraploid. Earlier work on octoploid and tetraploids showed little promise, but hexaploid triticale was successful enough to find commercial application.

The International triticale improvement programme has a remit to improve food production and nutrition in developing countries. Triticale is being used as it has potential in the production of bread and other food products such as pasta and breakfast cereals. The protein content is higher than that of wheat, although the gluten fraction is less. Assuming increased acceptance, the milling industry will have to adapt to triticale, as milling techniques used for wheat do not suit triticale.

Table 30 Top ten world triticale producers – 2005 (million metric ton)

Poland	3.7
Germany	2.7
France	1.8
China	1.3
Belgium	1.1
Australia	0.6
Hungary	0.6
Czech Republic	0.3
Sweden	0.3
Denmark	0.2
World Total	**13.5**

Source: UN Food & Agriculture Organisation (FAO)

The suitability of triticale as a ruminant livestock grain feed is better than other cereals due to its high starch digestibility. As a feed grain, triticale is already well established and of high economic importance. Triticale has also received attention as a potential energy crop, and research is currently being conducted on the use of the crop's biomass in bioethanol production in some countries.

The word 'triticale' is a fusion of the Latin words *triticum* (or wheat) and *secale* (rye). When crossing wheat and rye, wheat is used as the female parent and rye as the male parent (the pollen donor). The primary producers of triticale are Germany, France, Poland, Australia, China and Belarus. In 2005, according to the Food and Agriculture Organization (FAO), 13.5 million tons were harvested in twenty-eight countries across the world.

The top ten triticale producers in the world are shown in Table 30 The UK does not feature in this list, producing only 13,000 hectares of triticale (organic and non-organic) in 2005 (approximately 90,000t of grain). Triticale holds much promise as a commercial crop as it goes a long way towards addressing specific problems within the cereal industry.

Triticale is a wheat/rye cross, resulting in a high-yielding feed grain.

Its main use in the UK is as a feed grain, and in 2002/03 the organic volume stood at 14,303 tonnes, some 15 per cent of total UK production. This has remained more or less stable, and approximately 14,000 tonnes were grown in 2006 on approximately 3,000ha, with organic triticale representing 23 per cent of all the triticale grown in the UK in 2006.

In the UK the main challenge is market acceptance by the feed-compounding industry, which still prefers wheat, mainly on the basis of ease of purchase of domestic and imported crop.

Agronomy

Triticale is a cross between wheat and rye. As a rule, triticale combines the high yield potential and good grain quality of wheat with the disease and environmental tolerance (including soil conditions) of rye. It can be grown as an alternative to wheat and barley, particularly on the lighter soils, or on more marginal land where its performance can equal or even better that of wheat.

Although there may be some use for more specialized human consumption markets, experience has shown that market opportunities can be limited for this cereal, and care should be taken to secure a customer before growing it if it is to be sold off the farm. Triticale is more popular in organic farming than in non-organic farming. It is very good at weed suppression, it is easy to grow, and is an ideal organic crop where it is to be fed back to stock on the farm.

Crop Management

Variety selection is discussed for individual crop types in Chapter 8.

Rotation High yields, disease resistance, weed competitiveness, and more tolerance of lower fertility conditions than wheat, all make triticale a good second or third cereal crop. As triticale is half rye in make-up, it is more self-tolerant than wheat, and within reason can be grown in succession as a second or even third cereal. Triticale has a very tall structure which can be of benefit for weed competition. It is therefore well suited to weedier fields and later in the cereal rotation. Rotation issues are discussed in more detail in Chapter 5.

Cultivation Early harvest in the south of the UK opens up opportunities for repeated post-harvest mechanical operations in autumn for weed control. These may be limited prior to the triticale crop by the harvest date of previous cereal, and may not be necessary. Some winter varieties can be sown up until the end of February, thus allowing late autumn cultivations for weed control within the rotation.

Seeding and Establishment Organic winter triticale is normally sown from mid-October onwards so as to reduce the early competition from weeds, but can be planted from September to the end of February without any significant yield loss. Spring triticale is normally sown from January/February onwards, up until March.

Seed rates in organic systems are typically higher to afford better competition. Winter triticale should be sown at around 160–220kg/ha (1.3–1.8cwt/ac) assuming a thousand grain weight of fifty with an aim of 400 seeds per square metre. Spring triticale is typically sown at a higher rate of 225–230kg/ha (1.8–2.0cwt/ac) to offer greater competition due to the reduced tillering capacity of spring varieties. Where there is a high weed competition or crops are drilled very early or very late, seed rates may need to be adjusted. Where localized problems such as slugs or poor seedbeds are known, higher seed rates should be used to compensate for losses.

Being very competitive, triticale is normally established by drilling rows at 10–12cm spacings, with seed at 25–50cm depth depending on soil conditions, and predation threat. Triticale can also be broadcast and incorporated, although this can lead to more uneven establishment. As with wheat, the production of triticale on wide rows, typically on 20–25cm spacings (but up to 30cm spacing is workable) is gaining popularity with arable producers on heavier land, and land with grass and taprooted weeds, so that the crop can be hoed between the rows with an inter-row hoe. (*See* Chapter 11 for more information on this system.)

Weeds Post-emergence mechanical weed control is possible using a harrow-comb type weeder or inter-row hoe (when planted on a wide row system).

Aim to undertake a first weeding at the true three-leaf stage, when the plant is sufficiently anchored to withstand the weeding operation. Subsequent weeding operations can be carried out during tillering GS 25–30. See more notes on mechanical weeding in Chapter 10. Subsequent weeding is rarely required due to the competitive nature of triticale. Weed management issues are discussed in more detail in Chapter 10.

Harvesting Harvest of triticale in the UK is normally between early August and late September, depending on location and the season. Prompt harvesting and drying (to <15per cent) is essential to ensure quality and reduce storage problems. The removal of contaminants (ergots and so on) and non-crop debris is also essential for good marketing.

Spelt (*triticum speltum*)

Spelt is an ancient relative of modern common wheat. There is documentary evidence that spelt was cultivated by several ancient civilizations both in Europe and Asia, following its early origins in the fertile Middle East. This means that the unique characteristics and special qualities of spelt were highly valued, and traders carried the sought-after spelt grain to communities far and wide. Preserved grains have been found throughout Europe, including Britain, in many Stone Age excavations. Spelt is mentioned in the *Book of Ezekiel* in the Old Testament. It was a major cereal crop for the Roman Empire, and several recipes referring to spelt were written by the Roman epicurian, Apicius. In the twelfth century the abbess Hildegard of Bingen wrote about the restorative qualities of spelt in her *Causa Medica*, and *Gerard's Herbal*, written in 1597, also refers to spelt.

The exact history and origins of the grain itself are complex and confused. As there are no records of spelt being found naturally in the wild, its origins lie in early agricultural practices involving the intentional cross-pollination of grasses by the earliest farmers. The parentage of spelt is attributed to a cross-pollination between emmer wheat (*triticum dicoccoides*) and goat grass (*aegilips squarrosa*).

Spelt grain (*triticum spelta*) comes from the same genus as common wheat (*triticum aestivum*). It is a hexaploid wheat with forty-two chromosomes and several distinctive physical and nutritional characteristics that differentiate it from the common wheat grown today. These unique characteristics were key to the popularity of spelt, which was widespread until the industrial revolution. Developments in agricultural methods, engineering, and animal husbandry led to the widespread preference for common wheat, which was easy to thrash, and the fall from popularity of spelt.

Spelt has a tough hull, or husk, that makes it more difficult to process than modern wheat varieties. However, the husk, separated just before milling, not only protects the kernel, but helps retain nutrients and maintain freshness.

Modern wheat has changed dramatically over the decades as it has been bred to be easier to grow and harvest, to increase yield, and to have a high gluten content for the production of high-volume commercial baked goods. Unlike wheat, spelt has retained many of its original traits and remains highly nutritious and full of flavour. Also, unlike other grains, spelt's husk protects it from pollutants and insects, and usually allows growers to avoid using pesticides.

Some 800 years ago Hildegard von Bingen (St Hildegard) wrote about spelt:

> The spelt is the best of grains. It is rich and nourishing, and milder than other grain. It produces a strong body and healthy blood to those who eat it, and it makes the spirit of man light and cheerful. If someone is ill, boil some spelt, mix it with egg, and this will heal him like a fine ointment.

These twelfth-century texts refer to the vitality and digestive desirability of spelt, and interest in it continues today in its use as an alternative to wheat, which has a nutty, wheaty flavour. Spelt's 'nutty' flavour has long been popular in Europe, where it is also known as '*farro*' (Italy) and '*dinkle*' (Germany). In Roman times it was '*farrum*', and origins can be traced back to early Mesopotamia.

But it's not just good taste that has caught the attention of consumers: there is much interest in the suitability of spelt in various allergy diets. From the ancient medicinal healing texts of St Hildegard to modern studies, the special nutritional properties of spelt have been considered beneficial. The grain is naturally high in fibre, and contains significantly

Spelt is an ancient husked wheat grain suitable for some allergy diets.

more protein than wheat. Spelt is also higher in B-complex vitamins, and both simple and complex carbohydrates. It also contains special carbohydrates called 'mucopolysaccharides', which play a decisive role in stimulating the body's immune system, helping to increase its resistance to infection. Due to spelt's high water solubility, the grain's vital substances can, like liquid nutrients, be absorbed quickly by the body, which aids its digestibility. Another important benefit is that some gluten-sensitive people have been able to include spelt-based foods in their diets.

The area of spelt produced in the UK (organic and non-organic) is very small indeed, with probably less than twenty-five growers currently. It should only be grown under contract for specific markets. A small number of specialist organic mills, bakers and human breakfast cereal producers specialize in spelt for the dietary and health food markets.

Agronomy
Spelt can tolerate very cold winters, lying dormant if necessary (depending upon climate and soil conditions); it normally requires between 270 and 290 days between planting and harvest. Only winter-sown varieties are available.

Spelt requires dehulling prior to use, resulting in losses of about 60 percent of the harvested weight and costs of about £60 per tonne. The quality specifications are for 12 per cent protein, 240 hagberg and 77kg/hl bushel weight. Spelt is apparently an excellent calf feed, as the fibre content of the grain is good at stimulating rumen development.

Organic Crop Management
Variety selection is discussed for individual crop types in Chapter 8.

Rotation Spelt is only an autumn-sown crop with no spring varieties. It can be very susceptible to lodging if grown under medium or high fertility conditions, and prefers low fertility conditions; it is therefore suited to later in the organic rotation.

Spelt can be grown directly after wheat with a negligible risk of take-all. Due to its very competitive nature, rapid emergence and profuse tillering, it is well suited to weedier situations. Rotation issues are discussed in more detail in Chapter 5.

Cultivation The crop is suited to a range of soils, particularly those not suitable for winter wheat. It will perform correspondingly better on good soils. It is a very tall (circa 1.7m), aggressive crop, suppressing weeds and with no significant pest or disease problems known so far.

Seeding and Establishment Organic spelt wheat is normally sown in late October or November, to reduce any early competition from weeds. Seed rates can be lower than wheat due to its competitive nature and vigorous tillering. Spelt should be sown at around seed rate 180kg/ha (1.4cwt/ac). Thousand grain weights are difficult to calculate as the seed is sown with the husk attached; this can cause problems with some drills, creating blockages and requiring a slower forward speed and forced air systems to be used. Where slugs or poor seedbeds are a known localized problem, higher seed rates should be used to compensate for losses.

As with wheat, the crop is normally established by drilling rows at 10–12cm spacings, with seed at 25–50cm depth depending on soil conditions, predation threat and so on. Spelt can be broadcast and incorporated, although this can lead to more uneven establishment, or sown on wide rows. (*See* notes on wheat sown on wide rows, and Chapter 11 for more information on this system.)

Weeds Post-emergence mechanical weed control is possible using a harrow comb-type weeder or inter-row hoe (when planted on a wide row system).

Aim to undertake a first weeding at the true three-leaf stage when the plant is sufficiently anchored to withstand the weeding operation. Subsequent weeding operations can be carried out during tillering GS 25–30, although weeding is rarely carried out later as the crop is very competitive once growing through stem elongation stages to maturity. (*See* more notes on mechanical weeding in Chapter 10.) The height of spelt helps with weed suppression. Hand-rogueing of wild oats and docks can sometimes be a significant cost. Weed management issues are discussed in more detail in Chapter 10.

Harvesting The harvest of spelt in the UK is normally early, between mid-July and mid-August, depending on location and the season. Prompt harvesting and drying (to <15 per cent) is essential to ensure quality and reduce storage problems. The removal of contaminants (ergots and so on) and non-crop debris is also essential for good marketing.

Crop storage requirements are the same for wheat, although the area required to store the same volume of grain is likely to be double or even treble that of wheat, as spelt is harvested with the husk attached, which significantly increases the size of the grain. As a general rule, the harvested grain will be 33 per cent grain and 66 per cent husk. Prior to milling the husk will need to be removed via a dehulling machine. There are a small number of these machines in the UK operating on a commercial basis; however, transport costs and logistics should be assessed before growing the crop.

Wheat (*triticum aestivum*)

Wheat is a grass that is cultivated worldwide. Globally it is the most important human food grain, and ranks second in total production as a cereal crop behind maize; the third is rice. Wheat grain is a staple food used to make flour, and for its fermentation properties to make beer, alcohol, and more recently biofuel. Wheat is planted to a limited extent as a forage crop for livestock, and the straw can be used as fodder for livestock (*see* later in this chapter) or as a construction material for roofing thatch.

Wheat is a very adaptable crop, and is grown from the borders of the Arctic to the equator. Hard wheat (used for bread making) is best produced in regions of extreme climatic variance, with cold winters and hot, dry summers. Soft wheats are best produced in areas with milder climates such as the UK, which is well known for its high quality soft and biscuit wheat varieties.

In the UK, farmers select specific wheat varieties to grow depending on the local climate, the crop's location and its intended end use. Winter wheat varieties are planted in the autumn and spring wheat varieties are sown in the spring, and both are harvested during the late summer. The crop is ready to harvest when the ripe ears of wheat can be cut from their stalks and thrashed to loosen the individual grains of wheat.

The top ten wheat producers in the world are shown in the following table. The UK does not feature in this list, having only 4,593,000 hectares of agricultural land, of which 1,868,000 hectares is under wheat cultivation.

As in many other parts of the world, wheat is the most widely grown, and is the most widely adaptable cereal produced in the UK. It has many domestic markets including bread and biscuit flour, animal feed, distilling, seed and export.

Table 31 Top ten world wheat producers – 2005 (million metric ton)

China	96
India	72
United States	57
Russia	46
France	37
Canada	26
Australia	24
Germany	24
Pakistan	22
Turkey	21
World Total	**626**

Source: UN Food & Agriculture Organisation (FAO)

Wheat is one of the most important cereals in the world.

Organic feed wheat volumes in 2002/03 were at approximately 42,605 tonnes, while organic milling wheat was at approximately 10,651 tonnes. By 2006/07 feed wheat volumes had risen to 55,386 tonnes, while organic milling wheat had only risen slightly, to approximately 13,846 tonnes. Organic wheat accounts for approximately 46 per cent of the total organic combinable crop area in the UK.

There are winter and spring wheat types, although the majority of wheat grown in the UK is winter sown. Spring wheats tend to have a lower yield potential but slightly better grain quality. This is partly due to genetic parentage and breeding strategies. Spring wheat varieties typically produce grain which is 1 to 2 per cent higher in protein than winter wheat varieties under organic production conditions, though some new spring wheats can be up to 3 per cent higher than traditional spring wheats. This can be at the expense of yield.

Spring wheats typically attain a 1kg/hl higher specific weight over winter types. The hectolitre measurement gives the weight of the grain in a given volume, that is, a hectolitre (the apparatus used is usually a 1ltr container). For example, a wheat with a specific weight of 74 means that in a litre container the grains weigh 740g – the higher the reading, the larger the grain. Buyers like grain with a high hectolitre as they will be purchasing more flour than husk, thus buying more energy. It is also more efficient to transport heavier grain than grain with a low specific weight.

Agronomy

While winter wheat can lie dormant during the very coldest winters, winter wheat – depending upon climate, seed type and soil conditions – normally requires between 300 and 330 days between planting and harvest.

Crop management decisions require a knowledge of the stage of development of the crop, for example, the timing of weeding and/or application of approved trace elements to rectify a specific deficiency (namely manganese).

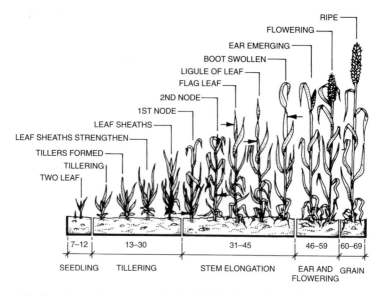

Figure 19 Cereal growth stages – using the Zadoks decimal code for cereals

Farmers also benefit from knowing when the flag leaf (last leaf) appears, as this leaf represents about 75 per cent of photosynthesis reactions during the grain-filling period, and as such should be preserved from disease, physical damage or insect attacks to ensure a good yield.

Several systems exist to identify crop stages, with the 'growth stage' scales being the most widely used. Each scale is a standard system describing the successive stages reached by the crop during the agricultural season. The Zadoks decimal code for cereals is shown in Figure 19.

Milling grain values are directly related to the quality of the sample. Bread-making specifications are >12 per cent protein at 14 per cent moisture, >250 Hagberg, >76kg/hl specific weight. For each 0.1 per cent fall in protein, deduct £1/t down to a minimum of 11 per cent. High protein levels are difficult to achieve, requiring yield and quality considerations to be balanced. Spring wheat varieties are more consistent at achieving milling quality on many sites, but are less suited to flaking. There is a small biscuit wheat and malt wheat market.

A strong feed-wheat market exists, which has been expanded by recent changes to livestock feeding standards; higher-yielding varieties are often grown especially for this market on land less suited to milling wheat. Grain should be cleaned before sale; screenings (5–10 per cent) may be paid at conventional prices or fed to livestock.

Organic Crop Management
Variety selection is discussed for individual crop types in Chapter 8.

Rotation First cereal after grass/legume break or residual fertility from crops with manure applications (for example, potatoes/maize) and where good weed control can be obtained. Wheat should not be grown more than twice in succession due to a decline in soil mineral nitrogen levels (which reduces tillering and protein level), to the increased risk of take-all, and weed competition. Rotation issues are discussed in more detail in Chapter 5.

Seeding and Establishment   Organic winter wheat is normally sown from early October onwards, so as to reduce the early competition from weeds and reduce pressure from BYDV and so on. Spring wheat is normally sown from January/February onwards up until March, but many spring varieties can be sown as early as November and produce good yields.

Seed rates in organic systems are typically higher to afford better competition. Winter wheat should be sown at around 400–450 seeds/sq m = 180–220kg/ha (1.5–1.8cwt/ac), assuming a thousand grain weight of 50 with an aim of 400 seeds per square metre. Spring wheat is typically sown at a higher rate of 500–550 seeds/sq m = 225–275kg/ha (1.8–2.2cwt/ac) to offer greater competition due to the reduced tillering capacity of spring varieties. Higher seed rates may be required due to later drilling (for weed and disease control) and reduced establishment. Where there is a high weed competition or crops are drilled very early or very late, seed rates may need to be adjusted. Where there are known localized problems such as slugs or poor seedbeds, higher seed rates should be used to compensate for losses. However, too high a seed rate may reduce specific grain weight.

Remember that in organic systems where fertilizers cannot be used to increase tillering, it is always preferential to have too many plants (which can be thinned by weeding vigorously) rather than insufficient plant numbers.

Wheat is normally established by drilling rows at 10–12cm spacings, with seed at 25–50cm depth depending on soil conditions, predation threat and so on. Wheat can also be broadcast and incorporated, although this can lead to more uneven establishment. The production of wheat on wide rows, typically on 20–25cm spacings (but up to 30cm spacing is workable) is gaining popularity with arable producers on heavier land and land with grass and tap-rooted weeds, so that the crop can be hoed between the rows with an inter-row hoe. (*See* Chapter 11 for more information on this system.)

Attention must be paid to variety choice, especially relating to protein levels for the milling market.

Weeds   Post-emergence mechanical weed control is possible using a harrow comb-type weeder or inter-row hoe (when planted on a wide-row system).

Aim to undertake a first weeding at the true three-leaf stage when the plant is sufficiently anchored to withstand the weeding operation. Subsequent weeding operations can be carried out during tillering GS 25–30. (*See* more notes on mechanical weeding in Chapter 10.) The number of passes required depends on weed competitiveness and crop density. Spring wheat is a poor competitor with fewer tillers, more weed problems, and higher cultivation costs. Long-strawed varieties can help with weed suppression. Hand-rogueing of wild oats and docks can sometimes be a significant cost. Weed management issues are discussed in more detail in Chapter 10.

The principal demand for an organic cereal crop is for wheat.

Harvesting The harvest of wheat in the UK is normally between mid-August and mid-September, depending on location and the season. Prompt harvesting and drying (to <15 per cent) to ensure quality and reduce storage problems is essential. Removal of contaminants (ergots and so on) and non-crop debris is also essential for good marketing.

THE HISTORICAL IMPORTANCE OF WHEAT

Grass is thought to have been purposefully cultivated to produce wheat in the Middle East some 8,500 years ago, but the earliest archaeological evidence for wheat cultivation comes from the Levant and Turkey, from around 10,000 years ago. Cultivation and repeated harvesting and sowing of the grains of wild grasses led to the selection of mutant forms with tough ears which remained attached to the ear during the harvest process, and larger grains. Selection for these traits is an important part of crop domestication. Because seed dispersal mechanisms have been lost, domesticated wheats cannot survive in the wild.

The cultivation of wheat began to spread beyond the Fertile Crescent during the Neolithic period. By 5,000 years ago, wheat had reached Ethiopia, India, Ireland and Spain. A millennium later it reached China.

Agricultural cultivation using horse collar-leveraged ploughs (3,000 years ago) increased cereal grain productivity yields, as did the use of seed drills, which replaced the broadcasting sowing of seed in the eighteenth century. Yields of wheat continued to increase, as new land came under cultivation and with improved agricultural husbandry involving the use of fertilizers, threshing machines and reaping machines (the 'combine harvester'), tractor-draw cultivators and planters, and better varieties. With population growth rates falling, while yields continue to rise, the acreage devoted to wheat may now begin to decline for the first time in modern human history.

Genetics and Breeding

Wheat genetics is more complicated than that of most other domesticated species. Some wheat species occur as stable polyploids, having more than two sets of diploid chromosomes. Einkorn wheat (*T. monococcum*), however, is a diploid, having two chromosomes.

Most tetraploid wheats (for example, durum wheat) are derived from wild emmer, *T. dicoccoides*. Heterosis or hybrid vigor (as in the familiar F1 hybrids) occurs in common (hexaploid) wheat, but it is difficult to produce seed of hybrid cultivars on a commercial scale as is done with maize, because wheat flowers are complete and normally self-pollinate. Commercial hybrid wheat seed has been produced using chemical hybridizing agents, plant growth regulators that selectively interfere with pollen development, or naturally occurring cytoplasmic male sterility systems. Hybrid wheat has been a limited commercial success, in Europe (particularly France), the USA and South Africa. F1 hybrid wheat cultivars should not be confused with the standard method of breeding inbred wheat cultivars by crossing two lines using hand emasculation, then selfing or inbreeding the progeny many (ten or more) generations before release selections are identified to release as a variety or cultivar.

There are many taxonomic classification systems used for wheat species. Within a species, wheat cultivars are further classified by growing season, such as winter wheat versus spring wheat; by gluten content, such as hard wheat (high protein content) versus soft wheat (high starch content); or by grain colour (red or white).

There is an expansion of the small but emerging European organic plant-breeding initiative, which, in addition to placing primary importance on crop yield and disease resistance, will focus on characteristics such as the utilization of slow-release soil nitrogen, early crop development, crop architecture; localized suitability and adaptability will allow the further development of such systems. If more widely adopted, these have the potential to optimize soil nitrogen use, minimize disturbance to the soil microbial biomass (by reducing the need for cultivations), and provide a mechanism for continuous organic grain crop production in all-arable based systems.

Major Cultivated Species of Wheat

Common wheat or bread wheat (*T. aestivum*) A hexaploid species that is the most widely cultivated in the world.

Durum (*T. durum*) The only tetraploid form of wheat widely used today, and the second most widely cultivated wheat today.

Einkorn (*T. monococcum*) A diploid species with wild and cultivated variants. One of the earliest cultivated, but rarely planted today.

Emmer (*T. dicoccon*) A tetraploid species, cultivated in ancient times but no longer in widespread use.

Spelt (*T. spelta*) Another hexaploid species cultivated in limited quantities.

Kamut ® or QK-77 (*T. polonicum or T. durum*) A trademarked tetraploid cultivar grown in small quantities that is extensively marketed. Originally from the Middle East.

Kaploid (*T. Kapioto*) This is a type of wheat that is grown only in the tropical regions of Australia.

GROWING FOR THE MARKET

The market demand for organic cereals in the UK is roughly one half feed wheat and one half all other crops, as shown in Figure 20. Although wheat represents the largest potential, market opportunities with the other cereals should not be overlooked, especially if there is local market potential. Once an appropriate market is identified, then the cereal type can be matched to site and rotational requirements.

ORGANIC STRAW

Straw has many uses and values on an organic farm. On mixed farms with livestock, straw is valuable for livestock bedding, and subsequently as farmyard manure with which to supplement the fertility of the rotation. Straw, especially barley straw, can also be a useful feed for ruminants, and can form part of the organic feed ration. In a mixed system the nutrients are retained on the farm and recycled via manure applications.

On stockless farms, straw should normally be incorporated to retain potassium and other minerals on the farm. Where it is exported, the minerals should be monitored and replaced on a routine basis. (*See* Chapter 3 for details on soil analysis and supplementary nutrients.)

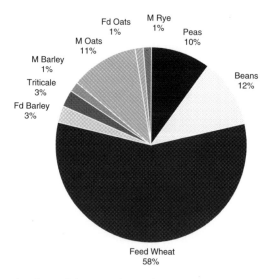

Figure 20 UK market demand for organic cereals

The straw of organic crops is also a commodity product. The two main outlets for straw crops are for organic mushroom production, and for thatching.

Organic Mushroom Production

Organic mushrooms are produced from a compost made from organic poultry manure and organic straw. Wheat and triticale straw is the preferred straw substrate, but spelt and occasionally rye are used also. Organic mushroom producers need a year-round supply of straw in large quantities; this is normally supplied in large square bales (to minimize haulage costs and for ease of handling). Straw sold for organic mushroom production can be worth £60–£70 per tonne.

Thatching Straw

Thatchers like organic straw. Straw produced in non-organic conditions with agrochemical inputs can suffer from early degradation in the thatch; this is often seen as a thatch with lots of mould and decaying straw. Organic straw is preferred, because without agrochemical inputs its longevity is far greater. Thatchers are one of the few remaining professions with an obsessive attention to detail and quality, who get very upset if their work does not last half a century! The main challenge for organic producers is that the straw needs to be cut, bound, stacked and then put through a threshing machine to remove the grain (a bi-product as far as the thatcher is concerned). There is potential to use a stripper header harvester cutting very high, or a modified combine harvester, but many thatchers have access to cutting and binding equipment. But don't underestimate the work and time involved in thatching straw. If these hurdles can be overcome, good quality, long straw can command £500–£600/t threshed. Long-strawed wheat, triticale, rye or spelt is preferred.

USE OF CONTRACTORS AND MACHINERY ON ORGANIC AND NON-ORGANIC LAND

Many farms use contractors for cultivations, drilling and harvesting operations. There are no problems in operating these systems within the constraints of the organic standards, however there are specific requirements that need to be met where machinery is involved with organic and non-organic crop production systems. This also applies to farms which operate organic and non-organic parcels of land. Organic standards require that any machinery used on non-organic land should be thoroughly cleaned prior to being used on organic land or on organic crops, to remove any prohibited residues (such as of seed treatments, artificial fertilizers or agrochemicals). Evidence of cleaning dates, procedures and any cleaning products used must be documented and made available as part of the organic inspection process; this applies to all mobile and fixed machinery, and also to grain handling and storage equipment. Examples include the following:

- The use of a contractor to drill or combine harvest: the machines should be cleaned and vacuumed out to remove any residues of non-organic crop and seed treatments prior to use on the organic unit.
- The use of a non-organic neighbour's fertilizer spreader or crop sprayer to spread a trace element amendment permitted under organic production: the spreader or crop sprayer should be cleaned and washed out to remove any residues of synthetic products prior to use on the organic unit.
- Using a neighbour's batch drier to dry some harvested grain in a wet season: the drier should be thoroughly cleaned and emptied prior to putting the organic crop through the machine.

Contamination of the organic crop with non-organic crop may lead to the loss of organic status of the crop, with a heavy loss in crop value.

CROP MIXTURES

Organic farmers often plant mixtures of crops, working on the principles that greater diversity brings benefits in terms of reduced risk and improved pest and disease control. Mixtures of cereals, cereals and grain legumes, or cereals and legumes such as clover, are all realistic and workable options.

Cereal Mixtures

Cereal mixtures are receiving increasing attention in organic systems. On-going research by Professor Martin Wolfe is demonstrating that mixtures of different wheat varieties grown together can result in a more robust population. Whilst the mixture may not be as good as the best yield in the best season, the mixtures result in a greater degree of reliability in good, moderate and poor seasons. They achieve this by allowing the different components to achieve their best potential in different conditions and seasons.

When grown for home feeding or for the feed market, mixtures of varieties from the same species are acceptable to the market. For example, growing two or three wheat varieties together, or two or three oat, triticale or barley varieties together, can work very well in organic systems. This is currently not an option if producing wheat or oats for the human consumption market or barley for the malting market, with millers and maltsters being very specific in their grain variety requirements.

Cereal and Grain Legume Mixtures

Mixtures of cereals and grain legumes such as spring barley and peas, oats and peas, wheat and field beans, are grown by some organic farmers. These mixtures are true bi-crops, being two crops grown simultaneously.

These provide greater pest and disease control for both the cereal and the legume. The cereal also provides a physical support to the pea, acting like a 'pea stick', helping to keep the pea elevated from the soil surface, which aids harvesting.

When beans and wheat are sown together as a bi-crop, there is a notable but small benefit in terms of nitrogen fixed by the legume, and this can lead to higher protein contents in cereals. Provided planting dates are adjusted, and varieties with similar maturity dates used, the crops, which are less dense in their component parts, but thicker than a single crop stand in combination, come to maturity for harvest at the same time. Experience suggests that there is a level of maturity compensation by the two crops, with the beans maturing slightly earlier (being a thinner crop than normal) and the wheat maturing slightly later (being held back by the thicker bean crop in late summer), resulting in a harvest with both components ripening together. The bi-crop is best established by deep drilling the beans at 10cm deep, and then drilling the wheat shallower and over the top of the beans at 3–4cm depth. The crop can be harvested together with appropriate settings and adjustments made to the combine harvester, and either fed as a mixture to livestock, or passed through a sieve cleaner/dresser to separate the smaller cereal grains from the larger bean grain.

Dr Hugh Bulson completed a Phd on wheat/bean bi-cropping, and demonstrated that when a winter wheat and bean were sown at 75 per cent of their normal sowing densities as individual crops, the resulting crop provided a greater output in terms of yield and financial return, whilst at the same time a far improved weed suppression.

Barley (or oats) and peas can also be bi-cropped in a similar way. The barley will mature for harvest just before, or at the same time as, the peas. The barley helps stimulate podding, and helps keep the mix drier than normal by pulling moisture from the peas into the barley. Being a shallower mix, the best results are achieved with the peas drilled and then the barley broadcast over the top and harrowed in, followed by a firm rolling and consolidation.

SEED PRODUCTION

Organic crops must be established by sowing seed produced under organic conditions. There is therefore an opportunity for some farmers to grow organic crops for seed production. This should always be under contract to a recognized seed merchant, and is subject to seed multiplication legislation and entry on the UK seed production listings with DEFRA under the UK seed certification scheme. All seed production crops are subject to a programme of field inspections and testing. In the UK, seed is graded as pre-basic (PB), basic (B), certified seed of the first generation (C1), and certified seed of the second generation (C2). Basic and C1 seed is generally grown for the production of further seed crops, and C2 seed is used for commercial crop production. Pre-basic seed is generally not available for sale, basic seed is the most expensive seed, and C2 the least expensive. Certified seed has a minimum germination of 85 per cent and is required to meet standards for purity.

Seed merchants usually offer a premium of £20–£30 per tonne over and above normal sale values for crops produced for seed. These crops also often have the benefit of early dispatch from the farm after harvest as they are required for drilling in the season that follows.

CEREALS FOR ARABLE/ WHOLE-CROP SILAGE

For those farms that operate mixed organic systems with livestock, cereals can also be grown for whole crop/arable silage. Wheat, triticale, oats and barley are best suited to this option, are normally harvested at 25–30 per cent grain moisture content, when the grain is 'cheesy' ripe. This is normally four or more weeks earlier than normal grain harvest, depending on location, local climate and season. Cereals can be grown and taken for arable silage as single stands, as mixtures of varieties, and as mixtures of cereals and grain legumes for improved feeding value (*see* earlier sections in this chapter on cereal mixtures and cereal and grain legume mixtures).

There are a number of benefits of arable silage, which can be useful to the organic farmer: namely an early harvest

- spreads the workload;
- is normally before many weeds set viable seeds, thus reducing the potential seed return to the weed seed bank;
- provides an opportunity for post harvest cultivations and weed management, or to undertake a cultivated summer fallow;
- allows an early entry for the establishment of green manures or clover leys.

Stockless farms, although not operating livestock enterprises, can consider arable silage for the benefits detailed above, but need to establish links with local livestock units. The establishment of a 'linked farm' can be particularly useful when the arable farm supplies forage from arable silage or from clover leys, and in return receives manure or compost from the livestock producer.

Chapter 7

Organic Pulse Production

As well as being protein crops, pulses also act as break crops, which should form a part of any rotation. They provide a disease break for cereals, have different rooting architecture, and provide an opportunity to improve soil structure and control weeds through the use of spring- and winter-sown crops that have different sowing and harvesting dates.

Organic standards recommend that where a crop rotation permits, it should include:

- a balanced use of fertility-building and depleting crops;
- crops with various rooting systems;
- a legume crop (for example, clover or beans); and
- leave enough time between crops with similar pest and disease risk.

Pulses form an important part of any organic rotation. For most arable farms their inclusion enables the proportion of cash crop to fertility building to be expanded, effectively lengthening the rotation. Their inclusion also provides a valuable break between cereals and provides a larger break before going back into clover, therefore reducing the risk of a build-up of pests and diseases.

The main challenge for the production of most pulses in organic systems is weed management. Most pulses are not very competitive, and careful consideration needs to be paid to crop type, rotation position, site election, establishment type and timing, and weeding methods used. Approaches for weed management for pulses are discussed in more detail in Chapter 10.

The most important pulses grown in the UK are peas and beans. Approximately 240,000 hectares of peas and beans were grown in the UK in 2005, representing approximately 5.5 per cent of the available cropping area. The area of organic peas and beans grown in 2005 was 6,172 hectares, representing 14 per cent of the available organic cropping area. The far higher percentage area of pulses in organic production clearly demonstrates their importance in the organic arable system.

The Processors and Growers Research Organisation (PGRO) estimates that the market for all UK pulses (organic and non-organic) is for about 2.2 million tonnes per annum. In 2005, production was only about 0.88 million tonnes, leaving huge scope or expansion to replace imports and develop exports for value-added products. An estimate of production for the 2000 to 2005 period is given below.

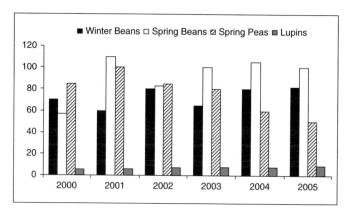

Figure 21 UK pulse crop

Source PGRO

This chapter will cover the main pulse crops of peas and beans, along with crops such as lupins and soya that, although not a regular feature of some farms, may prove useful production alternatives to those organic farmers with suitable production conditions and soils.

PULSE CROP CHOICE AND MANAGEMENT

Field Beans (*Vicia faba*)

Beans have a long tradition of cultivation in Old World agriculture, being among the most ancient plants in cultivation and also among the easiest to grow. It is believed that along with lentils, peas and chickpeas, they became part of the eastern Mediterranean diet in around 6000BC or earlier. They are still often grown as a cover crop to prevent erosion because they can overwinter, and because as a legume, they fix nitrogen in the soil.

In much of the world, the name 'broad' bean is used for the large-seeded cultivars grown for human food, while 'horse' bean and field bean refer to cultivars with smaller, harder seeds (more like the wild species) used for animal feed, though their stronger flavour is preferred in some human food recipes, such as falafel. The term 'fava' bean (from the Italian name *fava*) is commonly used in the United States (especially for beans grown for human consumption), but is also seen elsewhere, especially in Mediterranean recipes.

Approximately 240,000 hectares of peas and beans were grown in the UK in 2005, representing approximately 5.5 per cent of the available cropping area. The area of organic beans grown in 2005 was 5,863 hectares, representing 13.6 per cent of the available organic cropping area.

Agronomy

The field bean (*Vicia faba*), also known as the broad bean, fava bean, faba bean, horse bean, field bean or tic bean, is a species of bean (*Fabaceae*) native to north Africa and south-west Asia, and extensively cultivated elsewhere. Although usually classified in the same genus *Vicia* as the vetches, some botanists treat it in a separate monotypic genus as *Faba sativa Moench.*

The bean is a rigid, erect plant, with stout stems with a square cross-section. Depending on variety and location, plants can be between 0.5–1.7m tall. The leaves are 10–25cm long, with two to seven leaflets, which are a distinct grey-green colour. Unlike most other vetches, the leaves do not have tendrils for climbing over other vegetation, and therefore plants have stand-alone stems. The flowers are 1–2.5cm long, with five petals. The fruit is a broad leathery pod which is green until it matures to a blackish-brown, with a densely downy surface.

In their wild state, bean pods are 5–10cm long and 1cm in diameter. In cultivars developed for agricultural food use, depending on variety and location, pods can be 15–25cm long and 2–3cm thick. Each pod contains three to eight seeds that are round to oval, usually flattened, and up to 20–25mm long, and 5–10mm thick in food cultivars.

Winter and spring bean types are grown in the UK. Winter beans are the classic low-input pulse crop for heavy land that is difficult to work in the spring. Sowing can be either by drilling or by ploughing in, and winter beans usually show great vigour during germination and early crop development. Winter beans do not have a vernalization requirement, although they are more winter hardy than spring types.

Winter beans are generally large seeded, with a thousand-seed weight normally above 530g. Spring varieties are generally smaller seeded than this.

Spring bean production has fluctuated wildly, the success of the crop being largely linked to early summer rainfall. In dry years, yields can be disappointing, but in wet years much better results can be expected. Vulnerability to drought can be reduced by growing on more moisture-retentive soils, and by sowing early. Although good yields can be achieved in higher rainfall areas, the late maturity of spring beans needs to be considered. However, they are now being successfully produced in arable areas of northern England and southern Scotland. There is a wide range of spring bean varieties, including early maturing types.

The main market for field beans in the UK is for animal feed, where it makes up the protein part of many rations, although there are varieties grown for pigeon feed, and small export markets for human consumption. There are limits as to the inclusion level of field beans in ruminant diets due to their protein profile and tannin content, which limit the amount that can be fed.

There is a very small human consumption market for broad beans. These, however, are picked green as 'picking beans' and as such are treated as a horticultural crop, rather than as an arable field crop.

Organic Crop Management
Variety selection is discussed for individual crop types in Chapter 8.

Field beans are traditionally successful on heavy soils or soils where ample moisture is available, though this varies, depending on where they are grown in the country. They will grow on lighter soils, though not as

Field beans are an important grain legume crop for organic farmers.

well as peas due to their sensitivity to drought and dry years, which can cause unstable yields.

In wet years beans can become very tall and dense. This can reduce pollination and increase disease incidence and harvesting difficulties, and there is no benefit for having taller bean plants as they do not produce more pods. The balance which must be achieved is sixteen to seventeen plants per square metre in the spring. This is high enough to allow good competition against weeds, and low enough to allow good pollination and pod set.

Winter beans are to be preferred where spring cropping is not possible, if the land is slightly drought prone, or an early harvest is a priority.

If you are considering spring beans, remember that they can be a risky crop, relying on a dry period in February/March for sowing, followed by wetter conditions to ensure good establishment, and a good summer to ripen the crop in reasonable time. A crop planted in April followed by a very dry May and June will be prone to drought stress and this will result in poor yields. A tall crop is beneficial as this characteristic should make it more competitive against weeds; selecting varieties with a long straw is therefore advantageous.

The value of organic beans has improved considerably over recent years because of the importance of these crops as feed in organic animal production. There is also a strong market for beans produced from land undergoing organic conversion (the twenty-four-month period between conversion start and land achieving full organic status), and growing beans as an in-conversion crop may prove a useful strategy for some farms when seeking to establish a rotation. Rotation design and conversion strategies are discussed in Chapter 5 and Chapter 13.

Significant increases in market values can be achieved for low tannin or white/coloured varieties in the organic market, for their higher inclusion in animal feeds.

Pollination Field beans usually benefit from the activity of pollinating insects, so encouraging the location of beehives at the farm can aid pollination. Plants are typically about 60 per cent self-fertile, and a high level of flower/pod abortion is inevitable with both winter and spring crops.

Rotation Beans are an important break cash crop in any organic arable rotation. For many stockless arable farms their inclusion in the cropping programme enables the proportion of cash crop to fertility building to be expanded, effectively lengthening the rotation.

Beans are an ideal break crop after a first or second cereal crop, and can be followed with a subsequent cereal crop. Where spring beans are programmed after harvesting a cereal, a green manure can be sown following the preceding cereal, and prior to planting spring beans.

To reduce the risk of a build-up of persistent soil-borne diseases such as foot rots caused by *Fusarium solani* and *Phoma medicaginis var. pinodella*, field beans, broad beans, peas and green beans should be considered as forming a single crop group, and, from the point of view of rotation, no more than one of these crops should be grown on any field every five years. An additional risk to spring beans is stem rot caused by *Sclerotinia sclerotiorum*, and therefore other host crops such as linseed, oilseed rape and lupins should not be grown closely in the same rotation.

Where phosphorus and potassium levels in soils are adequate, there is little or no response to applications of appropriate phosphate P205 or potash K20 fertilizers permitted under organic standards. It should, however, be borne in mind that beans remove large amounts of potash and phosphate in the harvested crop, and this will need to be replaced by soil reserves, or from the application of approved fertilizers.

Seeding and Establishment Winter beans should not be sown too early – that is, not before the second week of October. Crops which are too forward are more prone to disease and to the effects of severe winter weather and frost damage to early emerging plants.

Sowing from mid-October to early November is usually the optimum time, but acceptable crops have been produced from early December establishment.

Harvesting tends to be from early September onwards. Winter beans tend to be very hardy under organic production conditions. Competitiveness against weeds is generally good, although when the crop foliage dies back in midsummer, late weed development can be problematic.

Spring beans are normally sown from March onwards so that the last frosts do not damage emerging plants. Spring beans tend to be a thinner crop and require a dry early spring for planting and a good supply of moisture in late spring. Spring beans are prone to drought damage if a late dry spring prevails.

In wet years where moisture arrives in time for pod fill, beans will yield higher than peas; they will also suffer less pigeon damage, because the pigeons find it easier to feed on the peas. Field beans are also easy to harvest, with pod-set well off the ground, allowing easy access for the combine harvester.

Thousand-grain weight for beans is very variable, so seed rate should be estimated directly each time prior to sowing, and seed rates adjusted accordingly.

Seed rates in organic systems are typically higher to afford better competition. Winter beans should be sown at around 200kg/ha (1.7cwt/ac) with an aim of twenty-five seeds per square metre. Spring beans are typically sown at a higher rate of 250kg/ha (2cwt/ac) with an aim of forty-five seeds per square metre, so as to offer greater competition over weeds. Where there is a high weed competition or crops are drilled very early or very late, seed rates may need to be adjusted. Where there are localized problems such as high rook, pigeon or pheasant numbers, or poor seedbeds, higher seed rates should be used to compensate for losses.

Dense crops of winter beans are more likely to suffer from disease and early lodging. A final target of eighteen plants per square metre appears to be the optimum for winter beans, which produce several tillers.

Where beans are ploughed in, a 15 per cent field loss is assumed. For spring-sown beans a final population of forty plants per square metre is required because spring beans produce few tillers and therefore greater plant numbers help to out-compete weeds. About 5 per cent seedbed loss, depending on soil type and conditions, is assumed for drilled spring crops.

The seed rate can be calculated from the following formula:

$$\text{Seed rate kg/ha} = \frac{\text{thousand seed weight} \times \text{target population plants/m}^2}{\% \text{ germination}}$$
$$\times \frac{100}{100 - \text{field loss}}$$

Establishment techniques used vary with conditions (weeds) and the seedbed required. Weedy fields may require stubble cultivation (discing) before ploughing. Winter beans are often broadcast or drilled on to the soil surface and then covered by shallow ploughing to 15cm (6in). On heavy, wet soils it is seldom possible to obtain a suitable seedbed for drilling, and broadcasting is the most convenient way of handling large seeds which do not flow easily through cereal drills. Ploughing also places seed at depth and may prevent seedlings being pulled out by pheasants, pigeons or rooks.

Trials have shown that the ploughing-in method is less successful for spring beans, and better yields are achieved where they are drilled conventionally. Sowing depth is important, and the seed should be covered by a minimum of 3cm of soil, but up to 7.5cm is acceptable.

Beans do not require a particularly fine seedbed and will tolerate cloddy conditions (although weed control may be poor), and over-cultivation should be avoided, which can impair emergence on soils that slump. Beans are sensitive to soil compaction, but are more tolerant of consolidation and waterlogging than peas. When beans are either ploughed in or drilled, in organic situations it is recommended that the soil after drilling is levelled (with a power harrow or shallow tines) to allow mechanical weeding to be undertaken in the juvenile crop. Leaving the soil surface rough after drilling can prohibit mechanical weeding or lead to unacceptable levels of crop damage from the mechanical weeder 'jumping' about and damaging the young crop.

Being reasonably competitive, and producing a dense mid-season crop cover, winter beans are normally established by ploughing or drilling rows at 10–12cm spacings, with seed at 15–18cm depth depending on soil conditions, predation threat and so on. As with cereals, the production of beans on wide rows, typically on 20–25cm spacings (but up to 30cm spacing is workable) is gaining popularity with arable producers on heavier land and land with grass and tap-rooted weeds, so that the crop can be hoed between the rows with an inter-row hoe. (*See* Chapter 11 for more information on this system.) This system works better with spring-planted beans that are typically drilled and where weed competition can be more of a problem. Drilling on wide-row spacings provides an opportunity for later inter-row cultivations and possibly undersowing, while narrow-row spacings or broadcasting can help establish a dense crop to smother weeds. Undersowing is possible and can assist with the control of late-emerging weeds.

Weeds A useful technique in beans is to plant the seed deeper than normal and carefully monitor pre-emergence germination. Where soil conditions permit, a 'blind harrowing' can be undertaken by using a spring-tined weeder to undertake a shallow weeding prior to plant emergence, thus creating a reduced weed population for the emerging plants.

Post-emergence mechanical weed control is possible using a harrow comb-type weeder or inter-row hoe (when planted on a wide-row system). Aim to undertake a first weeding at the six-to-seven true leaf stage when the plant is sufficiently anchored to withstand the weeding operation, and using a light harrow pressure to minimize crop damage. Subsequent weeding operations can be carried out during plant tillering up until the crop is 35–40cm tall. The number of passes will depend on weed competitiveness, crop density, and how easy it is to penetrate the crop with minimal damage. Winter beans can be harrowed hard in the spring and will tiller, providing a good means of weed control. (There are more notes on mechanical weeding in Chapter 10.)

Subsequent weeding is rarely required (or possible) due to the rapid development and competitive nature of beans. Weed management issues are discussed in more detail in Chapter 10.

Harvesting Bean harvesting dates will depend upon weather, variety and crop location, but are typically in September or early October. Differences in harvest dates are greatly influenced by growing conditions and are usually extended in the north and west compared to the south and east.

Spring-sown crops are often earlier to mature than winter ones; harvesting usually follows that of winter wheat. The bean crop is much less affected by wet weather at harvest than cereals or other pulses such as peas.

Beans grown for animal feed should be dry and free from moulds. However, beans grown for seed, human consumption or specialist markets must be harvested and handled with care to avoid damage.

Bean leaves usually fall during ripening and therefore do not impede harvest. Bean pods blacken, and seed becomes dry and hard first, but stems usually remain green for longer. The pods will be easily threshed and the seed is fit for combining at 18 per cent Moisture content(MC), but to avoid combine blockages it is best to wait until only a small percentage of green stem remains. If the seed is very dry, however, it may be damaged and seed crop quality may be reduced. If the crop is very dry and there is a risk of it 'shelling out' when harvested, losses can be reduced if the beans are combined when slightly damp in the early morning or evening.

Prompt harvesting and drying (to 14 per cent MC) is essential to ensure quality and reduce storage problems. The removal of contaminants and non-crop debris is also essential for good marketing.

Peas (*Pisum Sativum*)

Historically peas were grown mostly for their dry seeds. Along with broad beans and lentils, these formed an important part of the diet of most people in Europe during the Middle Ages. By the 1600s and 1700s it became popular to eat peas 'green', that is, while they are immature and right after they are picked. This was especially true in France and England. New cultivars of peas were developed during this time which became known as 'garden peas' and 'English peas'. With the invention of canning and freezing of foods, green peas became available year-round, not just in the spring.

In continental Europe, fresh peas are often eaten boiled and flavoured with butter and/or spearmint as a side dish vegetable. Fresh peas are also used in pies, salads and casseroles. Pod peas (particularly sweet cultivars called mangetout and sugar peas) are used in stir-fried dishes. Pea pods do not keep well once picked, however, and if not used quickly are best preserved by drying, canning or freezing within a few hours of harvest.

Dried peas are often made into a soup or simply eaten on their own. In Japan and other East Asian countries including Thailand, Taiwan and Malaysia, the peas are roasted and salted, and eaten as snacks. In the UK, marrowfat peas are used to make pease pudding (or 'pease porridge'), a traditional dish. In North America a similarly traditional dish is split pea soup. In Chinese cuisine, pea sprouts (*dou miao*) are commonly used in stir-fries, and their price is relatively high due to their agreeable taste.

In the United Kingdom, dried, rehydrated and mashed marrowfat peas, known as *mushy peas*, are popular, originally in the north of England but now ubiquitously, and especially as an accompaniment to fish and chips. In 2005, a poll of 2,000 people revealed the pea to be Britain's seventh favourite culinary vegetable.

Processed peas are mature peas that have been dried, soaked and then heat treated (processed) to prevent spoilage – in the same manner as pasteurizing.

Today, peas are not only widely used for human consumption as fresh, processed, and dried/split peas, they also have a valuable place as a high protein component of animal feeds. The UK area expanded during the

1980s, when the combination of new varieties and agronomic development-ments made production easier and more reliable.

The first step in planning a pea crop is to decide which market to aim for. Many types of high quality peas are suitable for a range of niche markets, but all types are suitable for animal feeds. Current marrowfat human consumption varieties are relatively low yielding and they are often more expensive to produce, but in recent years they have commanded a high premium price. The suitability of varieties for the different markets is included in the variety descriptions. The production of seed is another option.

Approximately 240,000 hectares of peas and beans were grown in the UK in 2005, representing about 5.5 per cent of the available cropping area (it is not possible to separate peas from beans in national DEFRA statistics). The area of organic peas grown in 2005 was only 309 hectares, representing less than 1 per cent of the available organic cropping area. Despite a large demand for organic peas, this represents the difficulties and challenges in the production of peas in an organic arable system.

Agronomy
Spring peas are a very versatile crop, and most varieties are now semi-leafless with high yields and enhanced standing ability. However, heavy rainfall in June and July can lead to the production of tall crops that are then prone to lodging. Growing peas on lighter soils can reduce the lodging risk, and the crop's tolerance to drought stress allows good yields in low rainfall areas. Spring peas mature early enough to allow production as far north as central Scotland, and a number of premium markets can be serviced through the use of special varieties.

The production of winter peas has all but disappeared from the non-organic sector and it is unlikely to be popular in an organic production system, where overwinter weed management would be very difficult.

The pea plant is an annual plant, with the pea growing in a pod on a leguminous vine. The average pea weighs between 0.1 and 0.36g. The seeds may be planted as soon as the soil temperature reaches a minimum of 6°C, with the plants growing best at temperatures of 10°C to 18°C. They do not thrive in the summer heat of warmer temperate and lowland tropical climates, but do grow well in cooler high altitude tropical areas. Peas grow best in slightly acid, well drained soils.

Spring peas can be categorized into white peas, large blue peas, small blue peas, marrowfat and maple. A description of white-flowered pea types is given in Table 32. White peas are grown for compounding for animal feed, though there is a small market for human consumption. Large blue peas may also enter the animal feed market, though in addition to this they may be micronized, producing a high protein feed which goes into the pet food market. Small blue peas may be grown for the processed canning market, though this market is very small. Marrowfat peas are the most important pea for human consumption.

All peas for human consumption should be free from staining and mould, and should be a clean green/blue colour free from blemishes. In addition to this they will need to pass a soaking and cooking test. Nearly all peas that are grown and harvested dry are semi-leafless, meaning that some leafiness has been bred out of the plant and replaced with tendrils, allowing the plant to bind together more in the field and reduce lodging. Semi-leafless types will generally be less competitive against weeds than normal leaved types.

Table 32 A description of white-flowered combining pea types and quality criteria

Type	Description	Quality criteria
White peas	Seed coat white/yellow, smooth and round. Primarily of use in animal feeds, but small quantities of white peas are used for the human consumption canning market as '*pease pudding*' and as split peas in ingredients for soups and prepared meals. Suitable for a wide range of soil types.	Samples for the human consumption markets should have smooth skin of a bright, even colour.
Large blues	Seed coat blue/green, smooth, large and round. In addition to the animal compounding market, large blue varieties can be sold for micronizing and for human consumption for export or UK packet sales. The micronizing process produces a high protein feed for use in certain dried animal rations and some pet foods (limited). Breeding programmes are now producing a number of high-yielding large blue varieties with different agronomic characteristics suited to a range of soil types.	Sample colour is one of the more important quality criteria for micronizing, with the higher premiums being offered for samples of green, large, even-sized seed.
Small blues	Seed coat blue/green, smooth, round and small. Varieties are available for use on a limited scale for canning as small processed peas. Varieties are low yielding and require adequate premiums to make them competitive.	Canning samples must be free from waste and stain, and pass cooking tests. A good even, green colour is necessary for acceptance.
Marrowfats	Seed coat blue/green, large, dimpled seeds. Varieties in this group are the most	A good blue/green colour, free from blemishes, is also required for packet sales.

Table 32 (continued)

Type	Description	Quality criteria
	important for human consumption, being used for both dry packet sale and canning as large processed peas. They are suited to a wide range of soil types and some are relatively late maturing.	Samples for canning must be free from waste and stain, and must pass soaking and cooking tests.
Maple	Coloured flowered. Seed coat is brown, often with flecked markings. Principally used in the pigeon trade. Limited market in the human health food sector.	Samples should be blemish free, usually small, round, smooth, and brown seeded.

Source: Adapted from UK Processors and Growers Research Organisation (PGRO)

Growers should remember that the desired market should be sought prior to drilling, and peas for human consumption are lower yielding than those for animal feed, though they can attract a premium where specifications are met.

Cultivation and Site Selection

Peas are traditionally grown in relatively dry areas of the country, the main reason for this being that they will not withstand wet conditions in the later stages of crop development. They tend to have a low, dense canopy, and naturally lodge as they get towards harvest; this makes them vulnerable to disease and infection, and they may become blemished. They are also sensitive to waterlogging, and if this occurs at flowering, up to 75 per cent losses can occur. Growing peas on thin, light, drought-prone soils should also be avoided. Peas will perform best on free-draining soils with a pH of 5.9 to 6.5; soils with a higher pH than 6.5 can cause manganese deficiency.

Techniques used vary with conditions, but normally post-Christmas ploughing is followed by power harrowing and other cultivations to create a fine seedbed. Good soil structure is particularly important. A false seedbed and weed strike greatly assist weed control.

Often land is ploughed in the autumn, using the winter to break down the soil and reduce wheelings in the spring. Whilst this allows natural weathering to aid in the production of an adequate tilth in the spring with minimal cultivations, it will also result in prolonged exposure of bare soil, and increase the potential of overwinter nutrient leaching and weed development. Cultivations that combine using an overwinter green manure such as phacelia or grazing rye should be considered, which can be incorporated prior to planting in the late spring. On lighter soils, spring ploughing is an option where overwintered stubbles are required for environmental stewardship or conservation objectives.

Peas are very sensitive to compaction and waterlogged soils, and often this is seen where there are excessive tractor wheelings, resulting in poor establishment and growth. As long as farmers do not travel excessively over the land in the spring, and reduce compaction by using floatation tyres or dual wheels to spread machine weight, compaction problems experienced by the peas should be avoided. The crop prefers a reasonably fine tilth with an adequate phosphorus and potassium supply.

There may be small responses to applications of appropriate phosphate P205 or potash K20 fertilizers permitted under organic standards where phosphorus and potassium levels are low. It should, however, be borne in mind that peas remove large amounts of potash and phosphate in the harvested crop, and this will need to be replaced by soil reserves or from the application of approved fertilizers.

Organic Crop Management

Variety selection is discussed for individual crop types in Chapter 8.

While there is a burgeoning organic fresh pea or 'vining pea' market in the UK, it is severely constrained by the limited number of processing facilities in the UK which handle organic crops. Because of the higher feed value and low tannin content of peas compared to field beans, there is a large and expanding market for dried organic peas, which are likely to achieve a considerably better sale value than beans.

Pollination   Peas usually benefit from the activity of pollinating insects, so encouraging the location of bee hives at the farm can aid pollination.

Rotation   Peas prefer free-draining soils. Their position in the rotation should be chosen depending on nutrient supply and weeds. Peas can be used as a break crop after cereals, but fields with high weed pressure

With human markets, and also their higher inclusion rates in livestock feeds, there is a strong demand for organic peas.

should be avoided, as peas are not competitive during early and late stages of development. As peas clear the field early they can be followed by a winter cereal or overwintering green manure that would further improve their rotational value.

It is recommended that the rotation carries no more than a single crop of the following group every five years: peas, field beans, green beans, vetches, oilseed rape, linseed and lupins. This four-year break is the minimum that can be allowed without increasing the risk of building up persistent soil-borne pests and diseases. Where one or more of these crops are grown on a routine basis, it is recommended that a predictive test for the presence of soil-borne, root-infecting diseases is undertaken.

Seeding and Establishment Seed size is very dependent on growing conditions and variety, and in seed for sowing, will be influenced by the cleaning process. As such, the published 1,000 seed weight data should be taken only as a guide to the relative seed size of varieties, and it is recommended that a thousand-grain weight test be carried out prior to establishing the seeding rate. The target populations should be set according to the plant type of the pea variety being sown.

As a general rule, peas need a minimum of 5 degrees soil temperature for rapid germination and emergence, and this normally dictates drilling dates. Organic spring peas are generally sown from early April onwards to ensure rapid emergence and good weed competition. Where cold or wet springs prevail it may be better to delay drilling until late April or even early May to ensure good, rapid establishment. The Processors and Grower Research Organisation (PGRO) conducted a great deal of research into drilling dates for non-organic peas, and concluded that for each delayed week after the first of March, yield fell by 125kg per ha. Peas sown early mature more quickly in poor seasons and suffer less from pest and disease during high pressure incidence. This research was, however, under conditions where weeds could be managed with herbicide inputs, and whilst yield may be higher from early drilling, experience of organic systems suggests that later drilling will result in similar yields and far better weed control.

High seed rates should be used to compensate for damage from birds and for rapid canopy cover to assist with weed control. Typical seed rates are 150–300g per 1,000 seeds, with the aim to establish sixty to eighty plants per square metre (or approximately 300–350kg/ha) for medium soils. On light soils, a seed rate of 265–285kg/ha can be used.

Bird damage can be a significant problem and precautions should be taken.

If mechanical weeding is to take place, growers increase seed rates by 10 to 15 per cent to compensate for mechanical damage.

Most cereal drills are suitable for peas. The drill should be accurately calibrated for each seed lot before sowing. Peas should be drilled at approximately 4–6cm (2in) deep. Seeds should be sown so that they are covered by at least 3cm of settled soil after rolling.

The seed rate can be calculated from the following formula:

$$\text{Seed rate kg/ha} = \frac{\text{thousand seed weight} \times \text{target population plants/m}^2}{\% \text{ germination}}$$

$$\times \frac{100}{100 - \text{field loss}}$$

On most soil types it is necessary to roll the field to depress stones so as to permit weed control in the developing crop and to allow harvesting close to the ground without causing damage to the combine. Rolling should be done soon after sowing and well before plant emergence. Leaving the soil surface rough after drilling can prohibit mechanical weeding or lead to unacceptable levels of crop damage from the mechanical weeder 'jumping' about and damaging the young crop.

Where maximum competition is desirable, rows should be no more than 20cm apart; any wider than this and weed competition can occur further into the season. Narrower rows result in higher yields and tend to give more even crops, easier combining and better competition with weeds. A good plant population is essential, since low populations are more difficult to harvest, are later maturing and more prone to bird damage.

As with cereals, the production of peas on wide rows, typically on 20–25cm spacings (but up to 30cm spacing is workable) is gaining popularity with arable producers with grass and tap-rooted weeds, so that the crop can be hoed between the rows with an inter-row hoe. (*See* Chapter 11 for more information on this system.) Drilling on wide-row spacings provides an opportunity for post-emergence inter-row cultivations until the pea canopy closes when the tendrils meet and mesh in between the rows.

Experience has shown that sowing a small volume of cereal (barley or oats) with organic peas can help keep the peas standing well to harvest, with the cereal component acting as 'pea stick' supports. The cereal can be cleaned out after harvest and helps the pea to stand tall and be competitive, and aids harvesting.

Weeds   Dried peas are very susceptible to competition from weeds, so they should only be grown on relatively clean fields.

Weed control during the initial stages of plant growth using spring-tined harrows or inter-row hoeing (depending on row width) is important. Avoid passes from shortly before germination until the three-leaf stage to avoid crop damage. Once the plants are 4–5cm long they are sufficiently anchored to withstand the weeding operation, and weeding can be undertaken using a light harrow pressure to minimize crop damage. Subsequent weeding operations should be carried out on a seven to ten day basis until the plant tendrils meet and mesh in between the rows and prevent further weeding.

The number of passes will depend on weed competitiveness, crop density, and whether it is possible to penetrate the crop with minimal damage. Efficient control will ease combining and facilitate rapid drying in addition to increasing yield. (*See* more notes on mechanical weeding in Chapter 10.)

Harvesting Most soils should be rolled after seeding to produce a flat, even surface, with any protruding stones buried. As peas grow very close to the ground this is important to aid harvesting close to the ground whilst avoiding damage to the combine. If the crop is free from weeds and is dying back evenly it can be left until it is dry enough to direct combine. Efficient lifters are helpful with badly lodged crops, and it may be necessary to combine in one direction only. If the crop is very weedy or uneven in maturity, it may have to be combined when it is not evenly ripened to avoid a total crop loss, and subsequent drying and cleaning may be required.

Pea harvesting dates will depend upon weather, variety and crop location, but are typically in early to midsummer. Differences in harvest dates are greatly influenced by growing conditions, and are usually extended in the north and west compared to the south and east.

It is possible for peas to pass through most combines without damage when the seed moisture content (MC) is about 20 per cent, and early harvesting at 18 per cent avoids bleaching, shelling out losses and splitting, or the deterioration in quality of human consumption or seed crops during wet weather. For animal feed peas, harvesting later, when they are about 16 per cent MC, will reduce drying costs. If the crop is very dry and there is a risk of it 'shelling out' when it is harvested, losses can be reduced if the beans are combined in the early morning or evening when they are slightly damp. Prompt harvesting and drying (to 15 per cent MC) is essential to ensure quality and to reduce storage problems. The removal of contaminants and non-crop debris is also essential for good marketing.

Once harvested, peas for animal feed should be dry (about 15 per cent MC) and free from moulds. Split or stained peas do not adversely affect the crop value. In contrast, the price that a sample of combining peas intended for human consumption can command is largely dependent upon its colour and its freedom from defects. Since both aspects are greatly influenced by the method of harvesting, every care must be taken to ensure success. Peas for micronizing, or those sold loose in packets for human consumption, should be relatively dark green, whereas marrowfats for canning as processed peas should be of lighter but even colour. Peas for canning must not be left too long in the field so that they become too dry, because this may inhibit rehydration. They may not be acceptable if the moisture content is less than 15 per cent.

Lupins (*Lupinus angustifolius, Lupinus albus* and *Lupinus luteus*)

Lupin, often spelled *lupine* in some parts of the world, is the common name for members of the genus *Lupinus* in the family *Fabaceae*. The genus comprises between 150–200 species. Most species are herbaceous perennial plants 0.3–1.5m tall, but some are annual plants, and a few are shrubs up to 3m tall. The species used for agricultural production are annual plants.

Lupins have a characteristic and easily recognized leaf shape, with soft green to grey-green or silvery leaves divided into five to seventeen finger-like leaflets that diverge from a central point. In many species the leaves are hairy with dense silvery hairs. The flowers are produced in dense or open whorls on an erect spike, each flower 1–2cm long, with a typical pea-flower shape with an upper 'standard', two lateral 'wings' and two lower petals fused as a 'keel'. The fruit is a pod containing several seeds.

Lupins have a wide distribution in the Mediterranean region and the Americas. As wild plants they are found throughout Greece including on the islands, in Albania, on the islands of Sicily, Corsica and Sardinia, as well as in Israel, Palestine and western Turkey (European and the western part of Asia Minor). Lupins occur naturally in meadows, pastures and grassy slopes, predominantly on sandy and acid soils. They are cultivated over the whole Mediterranean region and also in Egypt, Sudan, Ethiopia, Syria, central and western Europe, the USA and South America, tropical and southern Africa, Russia and the Ukraine.

Lupins are also cultivated as forage and grain legumes. Three Mediterranean species of lupin, *Lupinus angustifolius* (blue lupin), *Lupinus albus* (white lupin) and *Lupinus luteus* (yellow lupin) are cultivated for livestock and poultry feed, and also for human consumption.

The Andean lupin *Lupinus mutabilis* and the Mediterranean *L. albus, L. angustifolius* and *L. hirsutus* are edible after soaking the seeds for some days in salted water. These lupins are referred to as 'sweet' lupins because they contain smaller amounts of toxic alkaloids than the bitter varieties.

Both sweet and bitter lupins in feed can cause livestock poisoning if they are not processed prior to feeding. Lupin poisoning is a nervous syndrome caused by the alkaloids in bitter lupins, similar to neuro-lathyrism. Mycotoxic lupinosis is a disease caused by sweet lupin material that is infected with the fungus Phomopsis leptostromiformis: the fungus produces mycotoxins called phomopsins, which cause liver damage.

Lupins are a relatively new arable crop to the UK, though interest in them has increased steadily in recent years, with the total UK area standing at approximately 6,400ha in 2005. This increase has come about in part due to the EU ban on animal protein and fears over the GM status of some bought-in protein sources. This has coincided with new breeding programmes that have produced new varieties of lupin that can be fed at increased levels to livestock. The actual area of organic lupins grown in the UK is not precise, but is very small. In 2005 approximately 250ha of oil-seed crops and 1,094ha of unspecified crops were grown, and a small percentage of this area will be lupin crop.

Agronomy

There are three species of lupin being developed in the UK: white lupin (*Lupinus albus*), blue lupin (*Lupinus angustifolius*) and yellow lupin (*Lupinus luteus*). Flower colour should not be used to identify the type of lupin, however, since most whites have a blue flower, and many blues have a white flower. These all have very different characteristics, and should be regarded as separate from each other.

Like most members of this family, lupins fix nitrogen from the atmosphere improving the fertility of the soil for other plants. The genus *Lupinus* is nodulated by the soil microorganism *Bradyrhizobium sp.* (*Lupinus*).

Lupins are an accepted pulse crop in many countries, and livestock feed compounders are aware of their value. There is a choice of winter and spring varieties, but winter crops are very susceptible to pest damage and diseases, and may mature later than the spring-sown varieties; as a result the vast majority of the current UK crop is spring sown. Spring varieties are available of all three types, but the plant architecture varies between types and between varieties. The newer lupin varieties offer growers a pulse crop with a significantly higher protein content than peas or beans; however, growers need to match the variety to their particular situation, since certain varieties can be later maturing, or sensitive to alkaline soils.

Lupins generally prefer acid, free-draining soils below pH 7.0 down to pH 5.0. Blue and yellow lupins will not tolerate alkaline soil types: a pH of 6.5 or less is ideal. White lupins are more tolerant of alkaline conditions, growing well at pH 7.5. While it would seem that each lupin has a different ideal pH range for production, it is actually their tolerance to soil *calcium* levels that is the determining factor. High levels of soil calcium can severely restrict the development and performance of lupins, and lupins should *not* be considered for soils with high levels of free calcium. It is therefore important to check the level of free calcium in the soil (which is not the same at the pH).

Lupins also prefer light, free-draining soils with good traficability. With a relatively late harvest in the early autumn good soil access with machinery is important. There are both 'branched' and 'spikelet' varieties of the different lupin types. These refer to their foliage structure, with the branched varieties tending to produce lateral limbs and leaf at a lower level than the spikelet type, which can be useful for weed competition.

Lupin Growth Habit

There are three types of growth habit found in lupins: determinate, where there is no branching; semi-determinate, where the plant produces a certain number of stems; and indeterminate, where the plant requires sun or drought in order to reach maturity.

The question is, which type of lupin to grow, and this will depend on whether the grower wishes to grow for silage, crimping or grain. Growers who wish to grow silage would grow a semi-determinate lupin due to the larger biomass, whereas grain growers may prefer less biomass if drilling in wide rows, and would use an inter-row hoe for weed control.

White Lupins

White lupins can be distinguished by their large flat-shaped seed. All white lupins currently grown in the British Isles are semi-determinate, forming a tall, thick canopy, and are competitive against weeds. White lupins tend to yield higher than blue and yellow lupins, though they do have a longer growing season. White lupins can be drilled earlier than blue or yellow lupins, though this will depend on location. White lupins are normally harvested around the middle of September, though where drilling has been delayed, harvest will come later.

Organic Pulse Production

White lupins have a greater tolerance of compaction, which may be of use to growers on heavy land; however, they should remember that if drilling is late, then harvest will be delayed.

Blue Lupins

Blue lupins can be either semi-determinate or determinate. Yield is not as high as white lupins, though harvest may be up to ten to fourteen days earlier. In some parts of the country blue lupins would be preferred where white would give an unacceptably late harvest.

Yellow Lupins

Yellow lupins have a yellow flower, with the seed having a distinctly speckled appearance. In the south of the country harvest will be earlier than white lupins, though in the north they may be harvested at the same time. New breeding programmes are creating earlier maturing varieties all the time. Yellow lupins can perform well on light acidic soils, and even those with a pH of below five. Most yellow lupins are semi-determinate and can form a relatively dense stand compared to determinate types.

In all cases, farmers should ensure that they specify to their seed suppliers whether they wish to grow for grain, crimping or silage.

Organic Crop Management

Variety selection is discussed for individual crop types in Chapter 8.

The major attraction in growing lupins in an organic system is their very strong market for livestock feed due to their high protein content. This will vary between 30 and 45 per cent depending upon species, variety and growing conditions, and they often also provide a useful level of oil. The protein, oil, energy and typical yield of beans and lupins (on a dry basis) are compared in Table 33.

Table 33 Protein, oil, energy and typical yield of beans and lupins (dry grain)

	Field Peas	Spring Beans	White Lupins	Blue Lupins	Yellow Lupins
Protein	22.5%	25%	36–40%	31–35%	34–42%
Oil Content	1.9	1.8	10	6	4
EnergyME (MJ/kgDM)	13.5	12–13.5	15.5	13.5	13
Yield t/ha	3.5	2.8	3.5	2.5–3t/ha	2.5t–3t/ha
t/acre	1.4	1.2	1.4	1	1
pH tolerance	5.9–6.5	6.5–7.5	5.0–7.9	5.0–7	4.8–7

The crop therefore provides opportunities as a home-grown pulse in animal feeding rations, but care should be exercised by growers because lupins are still a new crop, and a number of production factors have yet to be fully established. Variety trials are currently being

Lupins can provide high grade protein, for which there is a high demand for organic livestock feed.

undertaken with organic spring lupins in a number of locations throughout the UK.

Growing lupins north of the Midlands or in the west of the country may not be advisable due to the late harvest associated with the wetter conditions in these areas, and a reduced chance of the crop ripening in time for grain harvesting. In these situations growing lupins for forage as a whole-crop silage, or as part of an arable silage mixture containing cereals and lupins, is more successful.

Rotation Lupins provide a useful break crop between cereals, but should not be grown in close rotation with other host crops including peas, beans, oilseed rape and linseed, due to the potential of disease carryover. Infection by *Sclerotinia sclerotiorum*, which causes a stem rot, can occur particularly in warm, wet periods.

The other major rotation determinant is the weed level in a particular location. Grown in rotation after a first cereal that was preceded by a clover ley offers the best chance of success and the lowest potential weed problems. Growing at the end of a rotation where weed levels have built up is likely to result in a crop dominated with weeds, and one which may be unharvestable.

Seeding and Establishment Early seedbed establishment in March/April is important to allow weed control prior to drilling. Spring varieties should be sown as early as conditions allow, with an optimum time of mid-March to mid-April. Later sowing necessitates a finer seedbed using spring-tined cultivations, but in many locations, later drilling into warmer soils achieves more rapid and even emergence. This is of benefit because of weed competition in a relatively uncompetitive crop.

Lupins require a similar seedbed to peas, and over-cultivation and compaction should be avoided. The seedbed can be rolled to depress stones and clods to aid mechanical weeding and harvesting, also to ensure good soil:seed contact, or where the seedbed is dry to conserve moisture and aid germination. The requirement for phosphorus and potassium is minimal, and normal levels of fertility are sufficient for the crop.

High seed rates are needed to compensate for pest damage (birds and rabbits) and to ensure rapid canopy cover for weed suppression. Seed rates of between 185–225kg/ha are commonly used.

Lupin seed is similar in size to pea seed, though the seed size of the different varieties does vary. The seeds of white lupins are flat in appearance. Aim for a drilling depth of 3–5cm. A conventional cereal drill can be used.

Target plant populations are thirty-five plants per square metre for white lupins, fifty per square metre for yellow lupins, fifty-five per square metre for semi-determinate blue, and for determinate blues seventy per square metre for the branching varieties and up to one hundred plants per square metre for the non-branching varieties. To enable the sowing rate to be calculated, information on thousand seed weight and percentage germination should be obtained from the seed supplier.

Although lupins 'fix' their own nitrogen, in order for them to do this they must first be inoculated with a suitable rhizobium bacteria strain (*Bradyrhizobium lupini*); this is because UK soils do not contain the strain of rhizobium required.

The inoculant is added to the seed in the drill, and the crop can typically fix around 180 to 200kg of nitrogen per hectare per year. Once inoculated, soils are unlikely to require subsequent inoculation unless the crop is not grown for some considerable time. Most commercial seed suppliers supply the seed together with inoculant and recommended sowing rate for the particular location.

The seed rate can be calculated from the following formula:

$$\text{Seed rate kg/ha} = \frac{\text{thousand seed weight} \times \text{target population plants/m}^2}{\text{\% germination}}$$
$$\times \frac{100}{100 - \text{field loss}}$$

Weeds   Lupins are very susceptible to early weed competition, and should only be grown on clean fields and early in the rotation. Early seedbed establishment in March/April is important, as is weed control prior to drilling. Stale seedbeds and weed strikes should be used prior to drilling. Weed control during the initial stages of plant growth using spring-tined harrows and/or inter-row hoeing will be required, with passes from shortly before germination until the four-leaf stage when the plant is sufficiently anchored to withstand the weeding operation; weeding can be undertaken using a light harrow pressure to minimize crop damage. Subsequent weeding operations should be carried out on a seven to ten day basis until canopy closure prevents further weeding. The number of passes will depend on weed competitiveness, crop density, and whether it is possible to penetrate the crop with minimal damage. Efficient control of weeds will ease combining and facilitate rapid drying, in addition to increasing yield. (*See* more notes on mechanical weeding in Chapter 10.)

Fields with a history of deadly nightshade should be avoided; these are often where maize has been grown as a forage crop. Experience has shown that the deadly nightshade ripens at the same time as the lupin seed and is of a very similar size and shape, so it is nearly impossible to clean it out of a harvested crop; it can therefore render the crop useless for feeding to livestock (and unmarketable) due to the dangers of the deadly nightshade.

Harvesting Most soils should be rolled after seeding to produce a flat, even surface with any protruding stones buried. As with peas, lupins grow relatively close to the ground, so rolling is important in order to aid harvesting close to the ground whilst at the same time avoiding damage to the combine.

The maturity of lupins will depend on their location and the season, heat being the main determinant for ripening. For most southern UK crops maturity is reached at approximately 150 days, with harvest in early September. If the crop is free from weeds and is dying back evenly it can be left until it is dry enough to direct combine. If, however, it is very weedy or uneven in maturity, combining may have to be undertaken when the crop is not evenly ripened to avoid a total loss, and subsequent drying and cleaning may be required. If the crop is harvested damp, it should be dried to 14 per cent MC, and stored at a low temperature to ensure quality and reduce storage problems. The removal of contaminants and non-crop debris is also essential for good marketing.

Whole Crop Options Lupins can be harvested as whole crop/arable silage or at 30 per cent moisture content for crimping. Lupins can be cut early with a forage harvester for silage, or with a combine harvester as a moist grain for ensiling, or for passing through a crimping machine (indented roller crimper) and ensiling. These options offer particular benefits to mixed organic farms seeking to home-feed the lupin as a high grade protein.

There are many benefits associated with whole-crop silage and crimping, which can be useful to the organic farmer; namely, an early harvest:

- spreads the workload;
- is normally prior to many weeds setting viable seeds, thus reducing the potential seed return to the weed seed bank;
- provides an opportunity for post harvest cultivations and weed management, or to undertake a cultivated summer fallow;
- allows an early entry for the establishment of green manures or clover leys.

Lupins can also be grown in combination with cereals to improve the quality and feeding value of the arable silage (*see* the sections in Chapter 6 on cereal mixtures, and cereal and grain legume mixtures).

Stockless farms, although not operating livestock enterprises, can consider arable silage for the benefits detailed above, but need to establish links with local livestock units. The establishment of a 'linked farm' can be particularly useful when the arable farm supplies forage from arable silage (or from clover leys) and in return receives manure or compost from the livestock producer.

Soya (*Glycine max*)

Like wheat and some other crops of long domestication, the relationship of the modern soya bean to wild-growing species can no longer be traced with any degree of certainty. It is a cultural variety with a very large number of cultivars. However, it is known that the progenitor of the modern soya bean was a vine-like plant that grew prostrate on the ground.

Beans are classed as pulses, whereas soya beans are classed as oilseeds. The word 'soy' is derived from the Japanese word *shoyu* (as used in soy soya sauce).

Soya beans have been a crucial crop in eastern Asia since long before written records, and they are still a major crop in China, Korea and Japan today. Soy was not actually used as a food item until fermentation techniques were discovered around 2,000 years ago. Prior to fermented products such as soy sauce, tempeh, natto and miso, soy was considered sacred for its use in crop rotation as a method of fixing nitrogen. The plants would be ploughed under to clear the field for food crops.

Soy was introduced to Europe and the United States in the 1700s, where it was first grown for hay. Soya beans did not become an important crop outside Asia until about 1910. In America, soy was considered an industrial product only, and not utilized as a food before the 1920s.

Soya beans are one of the 'biotech food' crops that are being genetically modified, and GMO soya beans are being used in an increasing number of products. In 1995, Monsanto introduced the 'Roundup Ready' (RR) soya bean: this bean has had a complete copy of a gene plasmid from the bacteria Agrobacterium sp. strain CP4 inserted into its genome, which allows the transgenic plant to survive being sprayed by the Monsanto non-selective, glyphosate-based herbicide Roundup. Roundup kills conventional soya beans.

In 1997, 81 per cent of all soya beans cultivated for the commercial market were genetically modified. However, as with other 'Roundup Ready' crops, concern is expressed over damage to biodiversity and the threat to the integrity of organic production.

Soya is an important global crop grown for oil and protein. The bulk of the crop is extracted for vegetable oil, and then defatted soy meal is used for animal feed. A very small proportion of the crop is consumed directly for food by humans. Soya bean products, however, appear in a large variety of processed foods.

The total non-organic market size is 170 million tonnes annually. The EU imports 32 million tonnes and is the largest buyer. The UK imports 1 million tonnes of soya beans as well as 1 million tonnes of soya bean products per year.

At present the UK-grown soya market is very small, although the area of soya increased in the late 1990s with the introduction of new varieties.

The UK organically grown soya market is even smaller, but demand for the crop is very high. This increase has come about in part due to the EU ban on animal protein, and fears over the GM status of some bought-in protein sources. The actual area of organic soya bean grown in the UK is imprecise, but is very small. In 2005 approximately 250ha of oilseed crops and 1,094ha of unspecified crops were grown. A small percentage of this area will be soya bean crop.

Agronomy
Soya can be grown in the UK, though at present the growing of organic soya is extremely limited. The seed cannot be directly fed as it contains trypsin inhibitors and can upset the digestive system of cattle. In the medium term, the area of soya may increase as demand increases for organic protein.

The soya bean is a species of legume native to eastern Asia. It is an annual plant that may vary in growth habit and height. It may grow prostrate, no higher than 20cm, or stiffly erect up to 1m in height. The pods, stems and leaves are covered with fine brown or grey pubescence. The leaves are trifoliate (sometimes with five leaflets), and the leaflets are 6–15cm long and 2–7cm broad, and fall from the plant before the seeds are mature. The small, inconspicuous, self-fertile flowers are borne in the axial of the leaf and are either white or purple. The fruit is a hairy pod that grows in clusters of three to five, with each pod 3–8cm long and usually containing two to four (rarely more) seeds 5–11mm in diameter.

Soya beans occur in various sizes, and in several hull or seed coat colours, including black, brown, blue, yellow and mottled. The hull of the mature bean is hard and water-resistant, and protects the cotyledon and hypocotyl from damage. If the seed coat is cracked, the seed will not germinate.

Soya beans are an accepted pulse crop in many countries, and livestock feed compounders are aware of their value, especially for monogastric animals such as pigs and poultry. All varieties grown in the UK are spring planted.

Organic Crop Management
Variety selection is discussed for individual crop types in Chapter 8.

The major attraction in growing soya in an organic system is its very strong market for livestock feed due to its high protein content. Variety trials are currently undertaken with organic soya in a number of locations throughout the UK.

Cultivation is successful in climates with hot summers, with optimum growing conditions in mean temperatures of 20°C to 30°C (68°F to 86°F); temperatures of below 20°C and over 40°C (68°F, 104°F) retard growth significantly. This is the case for much of the UK. The soya plant requires large amounts of heat to ripen, and when combined with a relatively late harvest in the early autumn and good soil access with harvesting machinery, its production area is rather limited to the extreme south and east of the UK. Growing even as far north as the Midlands or in the west of the country may not be advisable due to the late harvest associated with wetter conditions in these areas, and a reduced chance of the crop ripening in time for grain harvesting.

Soya can grow in a wide range of soils, with optimum growth in moist alluvial soils with a good organic content. Natural and free-draining soils with a pH of 5.8+ are ideal. Free-draining soils with good traficability are important. With a relatively late harvest in the early autumn, good soil access with machinery is important.

Soya beans, like most legumes, perform nitrogen fixation by establishing a symbiotic relationship with the bacterium *Bradyrhizobium japonicum*. However, for best results an inoculum of the correct strain of bacteria should be mixed with the soya bean seed before planting. Modern crop cultivars generally reach a height of around 1m (3ft), and take between eighty and 120 days from sowing to harvesting

Rotation Soya beans provide a useful break crop between cereals, but should not be grown in close rotation with other host crops including peas, beans, oilseed rape and linseed due to the potential of disease carry-over. Infection by *Sclerotinia sclerotiorum*, which causes a stem rot, can occur, particularly in warm, wet periods.

The other major rotation determinant is the weed level in a particular location: grown in rotation after a first cereal that was preceded by a clover ley offers the best chance of success and the lowest potential weed problems. Growing at the end of a rotation where weed levels have built up is likely to result in a crop dominated with weeds, which may be unharvestable.

Seeding and Establishment Early seedbed establishment in March/April is important to allow weed control prior to drilling. Soya beans require a similar seedbed to peas and lupins, and over-cultivation and compaction should be avoided. The seedbed can be rolled to depress stones and clods to aid mechanical weeding and harvesting. The requirement for phosphorus

Organic soya can be produced in the warmer, drier areas of the UK.

and potassium is minimal, and normal levels of fertility are sufficient for the crop. The crop has a similar level of susceptibility to manganese and magnesium shortages, as with cereals.

Soya should be sown into a fine seedbed 2–4cm deep in late April or early May, and into a warm soil (at a minimum of 5°C) to achieve rapid and even emergence. This is of benefit for weed competition in a very uncompetitive crop. An even drilling depth is essential. Never exceed 5cm of planting or this will impair emergence and crop development. A conventional cereal drill can be used.

Careful seed handling is required prior to planting, because if the seed coat is cracked the seed will not germinate. As a guideline, dropping the seed from a height of 3m can reduce germination by 5 per cent.

The soil conditions for drilling are far more important than the sowing date, so if conditions are sub-optimal, too cold or too wet, it is best to wait until conditions improve. Seedbeds should be rolled to ensure good soil:seed contact, or where the seedbed is dry to conserve moisture and aid germination.

An inoculant containing the strain of rhizobium bacterium (*Bradyrhizobium japonicum*) should be added to the soya at seeding to allow them to 'fix' their own nitrogen. The inoculant is added to the seed in the drill, and the crop can typically fix around 200kg of nitrogen per hectare per year. Once inoculated, soils are unlikely to require subsequent inoculation unless the crop is not grown for some considerable time. Most commercial seed suppliers supply the seed together with inoculant and recommended sowing rate for the particular location.

Soya seed is similar in size to a large pea, though the seed size of varieties does vary. Seed rates of between 120–150kg/ha are commonly used to ensure rapid canopy cover for weed suppression. Higher seed rates to compensate for pest damage (pigeons, pheasants, rabbits) should be used where these are a problem. Soya is slow to develop and prone to pest attack by pigeons and rabbits, and requires vigilant predator management for good establishment.

After developing from emergence, the crop should develop rapidly in warm conditions. The crop puts on 75 per cent of its growth and total bulk between the start of flowering and the start of leaf fall, over a period of six to eight weeks.

Weeds   The soya plant is relatively short and open, and not at all competitive against weeds; good weed management is therefore critical to the success of the crop. Soya beans are particularly susceptible to early weed competition, and should only be grown on clean fields and early in the rotation. Early seedbed establishment in April is important, as is weed control prior to drilling. Stale seedbeds and weed strikes should be used prior to drilling. Weed control during the initial stages of plant growth using spring-tined harrows and/or inter-row hoeing will be required. Passes should be delayed until the plant is sufficiently anchored to withstand the weeding operation, using a light harrow pressure to minimize crop damage. Weeding should then be carried out on a seven to ten day basis until canopy closure prevents further weeding.

The number of passes will depend on weed competitiveness, crop density, and whether it is possible to penetrate the crop with minimal damage. Efficient control will ease combining and facilitate rapid drying, in addition to increasing yield. (*See* more notes on mechanical weeding in Chapter 10.)

Harvesting Most soils should be rolled after seeding to produce a flat, even surface with any protruding stones buried. As with peas and lupins, soya beans grow relatively close to the ground, so rolling is important to aid harvesting close to the ground whilst at the same time avoiding damage to the combine.

When the pods are 3–5cm in length and the seeds have set inside the pods, the crop will start to die back, normally about three weeks after the end of flowering. Once the crop starts to die back the colour changes increasingly from green to bronze/yellow, and the leaves progressively fall from the stems. After two weeks from the start of die-back the plants will be 75–90 per cent defoliated and the crop stems and pods yellow. After 90 per cent of the leaf fall, the pods will turn a deep brown colour, with the seeds in the top pods turning a chestnut brown. The crop then needs seven to ten days (depending on the climate) to dry and for the seeds to loosen in the pods (they can then be rattled in the pods), which are then ready for combine harvesting. The one benefit of soya beans is that the pods do not shatter when harvested with a combine. The combine drum should be set slow, the fan speed on a medium to high setting, and the concave set very open. All rasping plates and threshing extras should be removed to avoid over-threshing and damaging the crop.

The maturity of soya depends on location and season, heat being the main determinant for ripening. For most southern UK crops, maturity is reached in mid-September. If the crop is free from weeds and is dying back evenly, it can be left until it is dry enough to direct combine. If the crop is very weedy or uneven in maturity, it may have to be combined when it is not evenly ripened to avoid a total loss, and subsequent drying and cleaning may be required. If the crop is harvested damp, it should be dried to 14 per cent MC and stored at a low temperature to ensure quality and to reduce storage problems. The removal of contaminants and non-crop debris is also essential for good marketing.

As soya cannot be fed direct to livestock, and because it requires treating and processing before being fed, it is unlikely that it will be used on farm for feeding livestock. It therefore tends to be purchased by specific end users, and careful marketing is a prerequisite.

SEED PRODUCTION

Organic crops must be established by sowing seed produced under organic conditions. There is therefore an opportunity for some farmers to grow organic crops for seed production. This should always be under contract to a recognized seed merchant, and is subject to seed multiplication legislation and entry on the UK seed production listings with DEFRA under the UK seed certification scheme. All seed production crops are subject to a programme of field inspections and testing. In the UK, seed is graded as 'Pre-Basic' (PB), 'Basic' (B), 'Certified Seed of 1st Generation' (C1) and 'Certified Seed of 2nd Generation' (C2). Basic and C1 seed is generally grown for the production of further seed crops, and C2 seed is used for commercial crop production. Pre-Basic seed is generally not available for sale, Basic seed is the most expensive seed, and C2 the least expensive. Certified seed has a minimum germination of 85 per cent, and is required to meet standards for purity.

For field beans, seed-borne leaf and pod spot (*Ascochyta fabae*) and stem nematode (*Ditylenchus dipsaci*) can be very damaging, and any seed planted or harvested for seed should be tested. The UK Processors and Growers Research Organisation (PGRO) recommend that basic seed should not contain more than 0.2 per cent infection, C1 seed should not contain more than 0.4 per cent, and C2 seed should not contain more than 1 per cent infection.

For peas, certified seed is required to meet a minimum germination of 80 per cent, and to achieve a standard of purity. Leaf and pod spot (*Ascochyta*) caused by *Mycosphaerella pinodes* and *Ascochyta pisi* is a potentially serious seed-borne disease, which can affect both quality and yield. However, growers should note that there are no minimum infection standards specified by the statutory certification scheme for this disease. It is always recommended that seed testing is undertaken prior to planting and after harvesting.

The potential to produce lupins and soya for organic seed is fairly limited, as the total volume of crop produced commercially is still very small, and the challenges of crop production demanding. There will be seed production opportunities for those farms whose location, resources and infrastructure are well suited to these crops.

Seed merchants usually offer a premium of £20–£30 per tonne over and above normal sale values for crops produced for seed. These crops also often have the benefit of early dispatch from the farm after harvest, as they are required for drilling in the season that follows.

Chapter 8

Variety Choice

VARIETY OPTIONS FOR ORGANIC PRODUCTION

The primary consideration for choosing which type of cereal or pulse to grow is the requirements of the customer or market for which the crop is being grown. Second to this is the resource availability on the farm – soil type, fertility levels, weed challenge – and what productive potential is desired, and what rotational position the crop will be grown in. The choice in relation to rotation is discussed in Chapter 5.

Once a crop has been chosen, selecting the correct variety for the farm needs careful consideration and should be undertaken taking into account a wide range of criteria. Variety choice is discussed later in this chapter for different types of cereals and pulses.

Production Systems

Most organic farming systems grow cereals in rotation with leguminous green manures (used for forage or mulched as green manures) and grain legume crops. (*See* Chapter 5 on rotations, Chapter 6 on cereals and Chapter 7 on pulses.) Cereals and pulses are in the main established after primary cultivations (ploughing) and secondary cultivations (spring tines), and drilled in rows at 10–12cm spacings or similar. Most cereals can also be broadcast and incorporated, although this can lead to more uneven establishment.

The production of cereals and pulses on wide rows, typically on 20–25cm spacings (but up to 30cm spacing is workable) is gaining popularity with arable producers on both heavier land and land where grass and tap-rooted weeds are a problem, so that the crop can be hoed between the rows with an inter-row hoe. (*See* Chapter 11 for more information on this system.) Experience suggests there is no negative yield impact by growing cereal and pulse crops on a wide row system, and that where specific weed problems exist, this type of approach results in viable crop production whilst simultaneously allowing weed populations to be managed in the crop and the subsequent return of weed seed to be reduced, thus controlling the weed seed bank.

There is also emerging evidence that when late hoeing (at GS 50–60) cereals destined for the bread-making market, the action of soil movement by the inter-row hoe blades stimulates soil nitrogen mineralization, which can be transferred through the plant and into the developing grain, resulting in higher grain protein levels.

Organic Seed

Organic standards require that seed produced under organic conditions to produce organic certified seed should be used for planting on organic land. Many varieties of organic seed with a wide range of varietal characteristics are available to the UK organic farmer, with good levels of seed availability. Only rarely should there be a need for an organic farmer to seek a derogation to plant a non-organic seed. In these situations only a recleaned, untreated (and undressed) seed can be used, and a derogation must be sought from the appropriate organic certification body prior to planting. Not obtaining a derogation for planting could result in the loss of organic status of the crop, and therefore a significant loss in crop value.

VARIETY CHARACTERISTICS AND SELECTION

When choosing cereal and pulse varieties for non-organic production, there are many sources of information, including the recommended lists for cereals and pulses drawn up by the UK National Institute of Agricultural Botany (NIAB), the Home Grown Cereals Authority (HGCA), and the Processors and Growers Research Organisation (PGRO). The criteria used for selecting non-organic varieties are mainly based upon end market, yield, pest and disease resistance, and response to artificial inputs.

When selecting varieties for organic production systems, the criteria used need to take into account a much wider array of factors, which amongst others includes market, yield, rotational position of the crop and level of fertility present, disease resistance, localized pest problems, crop height (for competition and potential straw value for livestock enterprises), early development characteristics and establishment rate (for competition), canopy cover, and leaf shape size and attitude to stem (for shading effect and ground cover in late season).

The challenge for organic crop producers is that not only are the criteria used for selecting varieties somewhat different to the criteria used by non-organic producers, but currently there is no official recommended list of varieties suitable for organic production; and moreover, there are no UK breeding programmes where varieties suited to organic production systems are being developed, with the aforementioned criteria in mind. So where does this leave the organic farmer?

As a first step the NIAB, HGCA and PGRO recommended lists for cereals, pulses and oilseeds can be used to narrow the choice from varieties produced for non-organic production systems using a 'best fit' approach. Thereafter it is largely down to individual farmers to trial a range of varieties on their farms to see which perform best.

Some organic trials have been performed by levy bodies such as HGCA and PGRO, and independent organizations such as NIAB, ADAS, Elm Farm Research Centre and Abacus Organic Associates, but there is currently no single source for organic farmers to access data. The amount of trial results available under organic production for current varieties is very limited compared to that being grown non-organically.

A number of trials have shown that extrapolating data and transferring it from non-organic to organic production should be done with a great deal of caution, as varieties of cereals and pulses tend to perform very differently under conventional and organic production systems. However, as is often the case, the more extreme characteristics of any given variety under either production system are still relevant. The main considerations to be taken into account when selecting a variety for organic production are noted below, along with some of the following favourable characteristics in a cereal which may be desirable:

- Strong market demand
- Yield
- Good rooting
- Good all-round disease resistance
- Resistance to lodging
- Speed of development from germination (vigour)
- Early vegetative characteristic and tillering capacity, with a prostrate growth habit (for early competition)
- Ripening days
- Crop architecture, height, leaf characteristic and attitude – large, wide leaves are beneficial for improved photosynthesis and for mid- and late season competition: a cereal planophile leaf angle (not erect) is ideal for the latter. Pulses which do not lie flat on the ground aid drying and harvesting
- Resistance to sprouting or grain degradation in the field (especially from late harvesting situations)
- Straw or stem length (important on a mixed farm where straw is a valuable commodity), ideally tall (for late season competition)
- 'Robustness' and reliability of yield in an organic system
- For milling wheats, Nabim (National Association of British and Irish Millers) group varieties according to their milling potential. For barley, consideration should be given as to whether the variety is suitable for malting
- Grain quality characteristics

As well as the HGCA, PGRO and NIAB recommended lists, growers should also speak to neighbours and local organic farmers who may have experience of a certain variety grown in similar conditions to their own, and should visit organic farms and trial sites or programmes. These are often arranged by seed companies for growers to visit, or there are a number of independent demonstration and training events focused on varietal selection.

As discussed above, crops with vigorous early development characteristics and tall stature (such as oats, triticale, rye) are more appropriate to this type of system, whereas less competitive crops (wheat, peas, lupins, soya) may provide greater challenges for integration into such systems. This is largely as a result of the growth characteristics of the commercially available crop varieties, which have been bred for artificial fertilizer and chemical-based, non-organic systems.

The aim should be to establish crop ground cover as early as possible and provide crop competition with weeds throughout the growing season. The relative importance of these characteristics will vary according to the situation in which they are being grown: soil type, climate, position in the rotation, soil fertility and nutrient levels, rooting characteristics, crop architecture, weed competition, pest and disease resistance and the growing system used.

Within the organic research fraternity there has been a long-running and robust debate highlighting the benefits of older varieties, especially older cereal varieties with more weed-suppressive and higher quality attributes. However, research has shown that as a general rule, modern cereal varieties can be superior to old varieties, particularly in terms of disease resistance and yield.

The variety selection of a particular crop in the rotation should be based upon a range of criteria as outlined above (including soil type, climate, crop fertility requirements, weed burden and spectrum, disease resistance requirements, the growth characteristics of the variety and its 'robustness' in organic systems, and of course, end market demand and requirements). These are discussed in more detail below.

Yield Variety yield potential is always important, with many modern varieties performing well under organic and conventional management. However, high yields alone should not be the basis for choice.

Quality In the absence of comprehensive variety trials on organic farms, the NIAB, PGRO and HGCA guides for grain quality should be used. Alternatively, you should consider some research with millers and merchants to obtain up-to-date specifications and requirements.

Competitiveness The major pests to most organic crops are weeds. Therefore, it is important to select varieties that are competitive against weeds. Varieties that develop quickly, provide maximum ground cover, and are tall are often better suited to organic systems. Selecting varieties with a low score for shortness of straw or crop height on the NIAB, PGRO or HGCA lists is a useful start, but knowing the physiology during the early growth stage is also important.

Disease resistance The level of disease in organic crops is not normally a problem. Organic crops typically have a lower nitrogen content and are often less susceptible to disease than highly fertilized non-organic crops. However, it is a sound policy to select for good disease resistance in all crops in order to minimize risk.

Single or multi varieties? There is potential for minimizing disease risk, for greater biological diversity (above and below ground), and the improved exploitation of soil resources by planting two varieties together. For home use, a logical step may be two cereal varieties or a combination of cereals and pulses. For marketable crops, many millers/merchants are prepared to accept mixed variety loads when 'like' classes are grown together (for instance, two class II or two class I cereal varieties, or two bean varieties grown together).

Winter milling wheat Millers prefer a sample for milling with a protein content of 10.5 per cent or greater, and high gluten quality (gluten being the elastic protein in wheat which allows bread to rise). Samples with lower proteins are likely to receive a lower premium (–£15/t at 10 per cent, –£25/t at 9.5 per cent in 1999). This is difficult to achieve on many light soils, and will be beyond all but first wheat after a fertility-building phase for many producers. Other criteria include variety selection, autumn or spring cropping and Hagberg falling number, which is normally required to be in excess of 250 and a minimum specific weight of 76kg/hl. Higher proteins are often achieved in spring wheats, however their yield potential is lower and establishment can be problematic on heavier soils.

PLANT POPULATION AND CROP ARCHITECTURE

Plant population is a major factor in determining yield. Where plant populations are low, strong tillering is particularly important for competition; but where plant populations are satisfactory, shading is more important than tillering. This is demonstrated by the reduced light penetration in a long-strawed planophile variety (Naturastar) which has relatively few tillers (*see* photo below), compared to a shorter-strawed erect variety with a lower tiller number (Hereward) as in the photo on page 179.

High seed-sowing rates are typically used on organic farms as a method of providing additional early competition against weeds. Very high plant populations and subsequent tiller numbers are typically recorded during the early spring on organic farms. In most cases the available nitrogen is too low to support the high level of tillers, and subsequent loss is inevitable. The benefit of high early tiller numbers is much improved by

Long-strawed planophile leaf variety (Naturastar) and reduced light penetration, with a reduced potential for weed growth.

A shorter-strawed erect leaf variety (Hereward) with more light penetration to the ground and therefore more potential weed growth.

early weed suppression through better crop competition, especially in the varieties that exhibit a more erect growth habit, or in less competitive crops such as peas, soya or even wheat.

Experience suggests that in most organic situations plant populations in late spring are typically only 55 to 60 per cent of seeds drilled, the large loss in plant numbers being attributed to some predation loss by slugs, beetles and birds, but more commonly as a result of dieback in plant and tiller numbers in response to available nutrients.

Unlike non-organic production, where additional nitrogen can be applied to support or even encourage plant growth and tillering, in organic situations it is always better to start with too many plants or too many tillers for better competition (these can be thinned out by weeding if necessary), with subsequent numbers falling back in line with nutrient availability, than to start with a poor plant population and few tillers, as this will allow weeds to develop more easily and can result in very poor crops. It is always worth remembering that it is more or less impossible to stimulate additional plant numbers and tillering in organic situations, so better to start with as many plants as possible (within reason) and adjust numbers down to the nutrients available.

Beans

There are relatively few field bean varieties, and the differences in crop architecture and leaf size and area are not hugely significant compared to their differences in pest and disease resistance and ripening ability. Most bean varieties are tall and competitive. The main challenge in organic systems is maintaining green leaf for as long as possible for competition and photosynthesis. Early loss in green leaf leads to reduced yields and more weed growth.

Where selecting bean varieties, desirable characteristics are a larger leaf, a tall growing habit, and those that mature early.

A tall and competitive crop of field beans.

A crop of field beans with a lot of leaf loss due to disease, with reduced photosynthesis and more weed competition.

Peas

Pea varieties are grouped into different end-market uses: vining (for use as fresh or frozen peas), white peas, large blue peas, small blue peas, marrowfat and maple. A description of white-flowered pea types is given in Table 32, Chapter 7 (page 156). White peas are grown for compounding for animal feed, though there is a small market for human consumption. Large blue peas are grown mainly for the animal feed market, though some are grown for micronizing, which produces a high protein feed that goes into the pet food market. Small blue peas are grown for the processed canning market, though this outlet is very small. Marrowfat peas are the most important pea for human consumption.

Marrowfat pea with full leaf area and better competitive advantage.

When selecting pea varieties farmers should consider the desirable characteristics discussed above, alongside issues such as rotational constraints (*See* Chapter 5), resistance to pea wilt or downy mildew, standing ability and ease of combining (some varieties lie very close to the ground and are very difficult to combine when weeds are present), earliness of ripening (more weeds will grow in later-maturing varieties) and leaf type. Most modern pea varieties have been bred to be 'semi-leafless': this results in a better harvest index (more pod, less leaf/stem), but also in a pea that is very uncompetitive against weeds. Growers should consider this when considering pea production organically, as weed management is one of the single biggest challenges in obtaining a successful pea crop. Some varieties, mainly marrowfat types, have normal leaves (they are not semi-leafless), and as such tend to be more competitive.

Semi-leafless pea variety with reduced competition against weeds.

Lupins

There are three species of lupin being used in the UK: white lupin (*Lupinus albus*), blue lupin (*Lupinus angustifolius*) and yellow lupin (*Lupinus luteus*). Flower colour should not be used to identify the type of lupin, since most whites have a blue flower and many blues have a white flower. These all have very different characteristics, and should be regarded as separate from each other.

When selecting varieties for use in organic systems the main characteristics to consider are suitability to soil type (discussed later in this chapter), earliness of ripening, and branching versus spiked types. The branching types tend to provide better competition against weeds, but there is a great deal of variation between varieties, and careful selection to local conditions is required.

Spike-type lupin with reduced competitive advantage.

Branched-type lupin with better competitive advantage.

Soya

Trials of soya have only been undertaken with a small number of varieties in the UK and on a relatively low number of sites. Consequently data is limited, and even more so for crops grown under organic conditions. Some of the newer varieties have had some success in some parts of the country in trial years with warm summers. As a general rule, varieties that are taller and that mature earlier are better suited to organic situations. Soya variety selection is discussed in more detail later in this chapter.

Cereals

There is a great deal of variation between the crop architectural characteristics of different cereal types, from the tall, slender spelt with a husked grain, to tall bushy oats, and the short drop-headed barley with awned ears. All have undergone a process of breeding and development at one level or another, yet most organic farmers seek to achieve a balance between the productive traits of highly developed varieties and the competitive characteristics of some of the older or native varieties.

Whatever the cereal type, when selecting varieties, the farmer should consider the desirable characteristics discussed above alongside issues such as rotational considerations and positioning (*See* Chapter 5), market and competitive ability. As discussed earlier, due to their tall stature, spelt, rye and triticale tend to be inherently more competitive than wheat and barley. Oats, although with a very different crop architecture, can equally be very competitive.

Where the cereal type is inherently more competitive, variety selection for architecture is less important; but where the crop is by nature less competitive, as with wheat, these characteristics become more important.

Wheat varieties with more erect early leaf development and which tiller less, such as Hereward, Naturastar, Magister and Exsept, tend to be less competitive early on (GS 25–31), whereas varieties that are more prostrate in early season and tiller more, such as Claire, Alchemy, Gladiator and Istabraq, produce more early ground cover and are more competitive against early weeds.

Despite the absence of this useful characteristic, varieties such as Hereward can still yield well; conversely, some varieties with good early ground cover can deliver poor yields. So poor early growth habits do not always reflect production potential – although it can help guide variety management. For instance, the poor early ground cover and poor tillering of Hereward and Exsept suggest that these varieties should not be sown at lower seed rates, or in poor seedbeds or weedy situations. On the other hand it would be safer to plant at lower seed rates with varieties that tiller more and are competitive in early season, such as Claire, Alchemy and Istabraq.

The spring varieties function very differently with respect to tillering and growth habit, and tend to produce much fewer erect tillers. With less time in the ground, early competition is sacrificed for rapid development and growth.

The prostrate early growth habit and strong tillering of Claire aids early competitiveness.

The erect early growth habit and limited tillering of Hereward reduce early competitive advantage.

CEREAL VARIETY CHOICE

The current cereal varieties available for organic producers are almost completely determined by breeding programmes developed for non-organic systems. Short-strawed wheats can struggle to give their best when environmental variables cannot be modified by pesticides and fertilizers. And while older varieties may offer agronomically desirable characteristics, they can still be out-yielded by newer varieties. Until such time as breeding for organic production is further developed and proven, the choices will largely remain with 'non-organic' varieties that can best cope with organic conditions.

Abacus Organic Associates undertook a cereals evaluation programme over three years (2003–2006): it was called the Organic Crops Demonstration Project (OCDP), and was funded by DEFRA (*see* photo below). This has been helping farmers evaluate different arable crop management strategies and techniques to optimize productivity and produce profitable crops that meet market requirements.

With their demand foremost, the focus was on wheat, with pea and lupin varieties also evaluated. Field-scale plots of wheat, pea and lupin

Organic Crops Demonstration trials 2003–2006.

varieties (1 acre of each variety) were demonstrated on five organic farms in Wiltshire, Cambridgeshire, Yorkshire, Lincolnshire and Staffordshire, and monitored throughout the growing season to harvest. Although the project was not designed for statistical analysis it has yielded valuable information on organic arable crop production in an area of limited data.

It evaluated a number of European wheat varieties bred for low input conditions (Naturastar, Socrates, Amaretto) alongside current commercial UK winter and spring wheat varieties such as Claire, Alchemy, Exsept, Hereward, Istabraq, Welford, Brompton, Paragon and Tybalt. The performance was monitored of fifteen winter-sown versus five spring-sown wheat varieties in different seasons and rotational positions, weeded and unweeded, and with and without undersowing. For most of the sites the rotational position was as a first crop after a two-year red clover green manure, and at one site as a second wheat crop after a first wheat, which was preceded by a two-year red clover green manure. Performance results

There are fewer late season, in-crop weeds when crops are undersown.

Barton
Not undersown
post harvest
22 September 2004
oblique view

A typical post-harvest stubble with weeds (not undersown).

over the three-year programme demonstrated that second wheats are possible in organic systems, provided the fertility levels are sufficient and weed competition is not severe.

Undersowing is used by many farmers to establish a following ley or green manure. In the OCDP plots, undersowing spring or winter wheat with trefoil did not have a positive or negative impact upon crop yields over the three-year programme. However, late season in-crop weeds were noticeably fewer when undersowing was practised, as compared to a normal post-harvest stubble (*see* picture above and below) and this may help reduce a long-term weed build-up and provide greater long-term soil fertility. Where fertility levels are low and the annual weed challenge is significant, farmers can consider undersowing after a mid-season weeding in May to improve weed management.

CULTIVATIONS AND ESTABLISHMENT METHODS

Ploughing followed by secondary cultivations and drilling is the common approach employed for crop establishment on most organic farms. Drilling after non-inversion cultivations is undertaken by some farmers, and was evaluated as part of the Organic Crops Demonstration Project programme. Limited experience suggested that, at least in a dry autumn, non-inversion methods can produce good crops with comparable yields. However, whilst such practice reduces soil disturbance and helps to increase soil-dwelling pest predators, it should be used cautiously, as it can increase slug damage risk, allow aphids on grass, or grass weeds, or volunteers to survive, and may allow some diseases (namely bunt spores) to survive to reinfect subsequent crops.

Late Sowing Implications

As part of the OCDP programme, winter variety drilling was severely delayed in 2004/5 at one of the project sites in Cambridge, with all winter and spring varieties drilled in February. This led to some interesting results: varieties such as Hereward, Claire, Njinski and Naturastar failed to fully vernalize and remained in vegetative growth far longer, reducing ultimate yields by between 50–70 per cent, whereas varieties such as Istabraq and Welford performed with near-normal yields comparable to, or even better than the spring varieties sown at the same time. Whilst not normal practice, the late drilling did identify varieties that are very flexible with regard to drilling date, and provided information that could be useful in adverse seasons.

WINTER OR SPRING CROPPING?

Winter Cereals

Winter wheat, barley, triticale, rye and spelt cereals need a cold period to become vernalized, the process where the plant requires a period of cold in order to allow it to go from the vegetative phase to the reproductive phase. Many varieties can be sown during early spring with reasonable success, provided field conditions are satisfactory; however, the yield potential of such crops is likely to be lower than the potential of spring varieties sown at the same time. Winter oats, some winter wheat varieties and most of the winter triticale varieties can be sown in the spring, but when sown after the first week of March they tend to be lower yielding and later maturing than spring varieties. Spelt and rye should only be sown in the autumn.

Winter cereals can be sown from August to the end of February and will be harvested the following July/August. This typically represents a period of up to nine months in the ground. However, if winter type varieties of wheat and barley are sown after the latest safe sowing date in the spring there is a risk of crop failure due to lack of vernalization. It should also be noted that some varieties of winter crops could be sown as late as early spring, though some will require more vernalization than others. As a rule of thumb, farmers should follow the Home Grown Cereal Authorities (www.HGCA.com) latest safe sowing date for particular varieties.

The advantages of winter sowing over spring sowing are that:

- Winter-sown crops have the potential to be higher yielding than spring-sown crops.
- Some soils may not be workable in the spring.
- A mixed cropping pattern of winter and spring varieties helps spread risk and workload.

Winter cereals, especially those sown in the early autumn in late September or early October, spend a greater proportion of their time at a growth stage where they are more susceptible to weed competition, often leaving a large proportion of soil bare and open to weed competition during the late autumn and winter period. Once they fully tiller in the spring their competitive ability reduces this effect. However, if weeds have become established, mechanical intervention may be required. Winter cereals are also potentially exposed to pests and diseases for longer periods.

Generally, winter-sown cereals have the greatest yield potential, although this is not always the case on lighter soils or on less fertile sites, where spring varieties can yield equal to, or in some cases outyield, the winter varieties. This is partly due to the longer period in the ground and the greater genetic potential derived from the breeding effort in improving traits such as the harvest index (the ratio of biomass required in order to produce the grain, or the plant:grain ratio). A plant with a good harvest index requires less biomass for the production of the same given amounts of grain.

Winter-sown cereals generally produce a larger plant with greater root volume and above-ground structure. The rate of maturity of winter types is slower compared to spring types, which makes them potentially more sensitive to moisture stress during the summer and can make harvesting dates more uneven between varieties.

Spring Cereals

Spring cereals are normally sown from January through to April, though any crop drilled after late March will have a severely decreased yield. Many of the modern spring varieties of wheat have a lower vernalization requirement than older varieties, and many of these can be planted in the late autumn (November to January), which can benefit establishment where access to soils is achievable.

Spring crops are usually in the ground for six to eight months, and are harvested slightly later than their winter counterparts. To maximize yields from a spring-sown crop, drilling should take place as early as possible.

Many of the spring varieties tend to have far fewer tillers and a more rapid development through the growth stages, literally making up for less time in the ground, but with much bigger flag leaves, to capture as much photosynthesis as possible in the shorter lifespan. Farmers should be mindful of this, and when it comes to seed rate, not skimp on seed, as the lack of early tillers can lead to poorer early competition until the plant starts to develop fully.

The advantages of spring sowing over winter sowing are that:

- The workload may be spread over the year with a mixture of spring- and winter-sown crops.
- Yield may exceed that of winter-sown crops in poor years, on light soils or in lower fertility conditions.
- Protein levels for milling quality cereals are more reliable and achievable.
- It provides the opportunity to flush weeds after harvest until drilling in the spring.
- It provides an opportunity for autumn-germinating weeds to be controlled.
- It provides additional overwinter habitat for beneficial insects, invertebrates and ground-nesting birds in the form of overwinter stubble.

Varieties for Milling Quality

Historically the majority of wheat varieties grown for the milling market were winter varieties. However, experience and industry information suggests that protein levels are typically lower in winter milling varieties and higher in spring wheat varieties, which are more reliable at achieving milling specification. Comments from the trade are that 'spring wheat varieties typically produce 1–2 per cent higher protein levels with acceptable Hagberg and specific weights.' There are many reasons for this, but notably spring variety breeding and potentially better matching of soil-nitrogen release to the demand of a later developing spring variety.

Experience from the Defra-funded Organic Crops Demonstration Project during 2003–2006 has resulted in a significant swing into spring varieties, which with the current varieties available seem better suited to producing a milling quality sample under organic management conditions. Results from this project suggested that the protein levels in the winter varieties Naturasar and Magister are generally better than Hereward, but in most years most winter samples are not of an acceptable quality for milling. Conversely, spring varieties such as Paragon, Amaretto and Chablis have proved better able to produce a milling specification sample. Where crops are grown as second wheats in the rotation, not surprisingly protein levels are always lower and less able to produce a milling specification sample.

Yield Performance

A competitive growth habit and good disease resistance is all very well, but what matters to a farmer is crop yield, quality and the reliability that these can be reproduced. Previous trials by NIAB and Elm Farm Research Centre show that varieties produced for non-organic production systems such as Claire, Exsept, Deben and Hereward perform with reasonable consistency in organic production.

Research results from three years of wheat variety trials undertaken by NIAB show relatively little difference in mean yields between spring (3.56t/ha mean) and winter (3.83t/ha mean) varieties (Figure 22 and Figure 23). These trials showed that spring varieties are generally better able to achieve bread-making quality (with a protein count of 12 per cent +). With the potential for better retention of soil nitrogen from using a combination of overwinter green manure covers and spring planting, for farms with soils that permit spring cropping the choice would seem clear. However, competitive pressure from weeds also needs consideration.

Relative yield results from the Organic Crop Demonstration Project (Table 34) between 2003 and 2006 show similar results, and confirm the variety Claire as a consistent yielding variety; they have also identified some new varieties (such as Alchemy, SW Tatorus, Glasgow, Brompton) which also perform with reasonable consistency, and would seem to have a better yield potential, especially in more fertile situations. Actual yield at the different sites can be seen in Tables 35 (winter varieties) and 36 (spring varieties).

The yields achieved are always a function of soil type and fertility, rotation position and weed competition. Winter varieties tend to have consistently higher yields in good and high fertility situations. However, in rotational positions with more moderate fertility or on less fertile soils, spring wheats can perform at the same level as winter varieties, and in some cases higher than winter varieties, often having the benefit of lower weed populations that are associated with spring drilling.

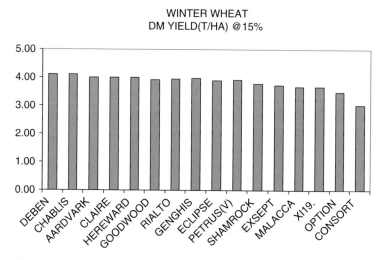

Figure 22 Three years of NIAB winter wheat variety trials

Source NIAB

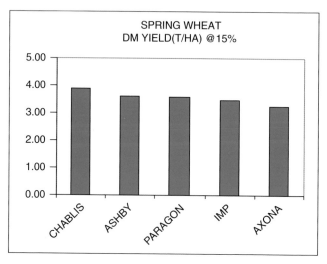

Figure 23 Three years of NIAB spring wheat variety trials

Source NIAB

Table 34 Organic Crop Demonstration Project 2003–2006 relative yields, three-year average performance

	All sites	High fertility sites	Low fertility sites
(Average) Winter Wheat	(100)	(100)	(100)
Alchemy	121	122	120
SW Tatorus	112	122	102
Glasgow	111	117	105
Brompton	112	112	113
Claire	106	106	107
Welford	104	97	111
Istabraq	101	90	113
Nijinsky	101	98	104
Sokrates	100	95	78
Magister	99	101	97
Exsept	96	92	101
Hereward	95	98	92
Deben	92	92	93
Gladiator	87	84	76
Naturastar	84	78	89
(Average) Spring Wheat	(100)	(100)	(100)
Chablis	113	108	118
Tybalt	112	110	114
Paragon	105	108	102
Amaretto	103	106	99
ADS 75	66	64	68

Table 35 Organic Crop Demonstration Project 2003–2006 yield results – winter varieties

	2003/04 Yield t/ha			2004/05 Yield t/ha			2005/06 Yield t/ha		
Site	low	Av.	high	low	Av.	high	low	Av.	high
Yorks	n.a	n.a	n.a	5.8	7.0	8.2	7.0	7.7	9.2
Cambs	4.4	5.1	7.5	2.7	4.7	6.9	5.5	6.1	7.2
Wilts	2.4	3.2	3.8	5.9	6.1	6.3	3.5	4.1	5.1
Staffs	3.0	4.5	5.3	n.a	n.a	n.a	n.a	n.a	n.a
Lincs	6.5	9.0	10.3	n.a	n.a	n.a	n.a	n.a	n.a

Table 36 Organic Crop Demonstration Project 2003–2006 yield results – spring varieties

	2003/04 Yield t/ha			2004/05 Yield t/ha			2005/06 Yield t/ha		
Site	low	Av.	high	low	Av.	high	low	Av.	high
Yorks	n.a	n.a	n.a	7.2	7.3	7.4	*3.4	6.2	6.4
Cambs	5.8	6.5	7.2	6.1	6.2	6.2	*3.7	5.5	6.5

Table 36 continued

	2003/04 Yield t/ha			2004/05 Yield t/ha			2005/06 Yield t/ha		
Wilts	3.5	4.1	4.6	3.5	3.7	3.8	*2.8	3.5	4.8
Staffs	3.1	3.5	4.1	n.a	n.a	n.a	n.a	n.a	n.a
Lincs	6.3	7.1	8.2	n.a	n.a	n.a	n.a	n.a	n.a

*ADS75 yields in 2005/6 (3.65t/ha Cambs, 2.76t/ha Wilts, 3.38t/ha Yorks) lowered overall averages

CEREAL VARIETY SELECTION

The following sections provide information on choosing suitable varieties of different crop types for organic production. Rather than a definitive and static list of varieties, the notes provide a guide as to what characteristics to look for in a variety, and what to avoid, so as to be better able to select a crop type and variety for the organic production situation in question.

Wheat

Careful selection of varieties is necessary for optimal wheat production, taking into account criteria such as crop architecture, height, tillering, disease resistance, emergence characteristics and so on. Newer varieties can be as productive as some of the older varieties. In very fertile situations winter-drilled wheats generally outyield spring-sown crops. In less fertile and weedier situations, spring-sown crops can perform just as well. Yields in organic systems are wide-ranging, depending on rotational position and site fertility. Typical yields are between 3.75–5.5t/ha depending on variety, rotational position, soil type.

Bread-making Wheat

As discussed above, spring groups 1 and 2 varieties have repeatedly proved more reliable at delivering bread-making quality specifications than winter varieties. The spring varieties also have the desirable characteristics of rapid development, tall structure and big, broad, planofile leaves for good shading and weed competition. Some of the spring bread-making varieties suitable for organic production in 2006 were Paragon, Chablis and Amaretto. The variety Tybalt (Group 2) has also performed well under organic conditions. Spring varieties tend to tiller far less than winter varieties, but with less time in the ground and more rapid development, tillering is less important in the spring varieties.

Where winter varieties are used, they should be grown close in rotation to a good source of fertility, such as after a clover ley. They tend to tiller less than feed-type varieties, so seed rates should be kept high to compensate. Some of the winter bread-making varieties better suited to organic production in 2006 were Hereward, Xi19, Einstein, Magister and Naturastar.

Biscuit Wheat

These tend to be soft endosperm types (Group 3) used for biscuit, cake and other flours. Some grain is also used in the organic brewing and distilling industries. Characteristics such as good disease resistance, rapid development, strong tillering and tall stature are important in these wheats. Some of the winter varieties that are better suited to organic production in 2006 were Claire, Deben, Nijinski and Robigus.

Feed Wheat

These are principally group 4 varieties, and offer the highest yields for wheat under organic conditions; they are therefore the best for maximizing output per unit of area, especially where the ability to achieve bread-making specifications is limited or impossible. The specifications for the livestock feed sector are generally much easier to achieve in organic situations. Characteristics such as good disease resistance, rapid development, strong tillering and tall stature are important in these wheats. Some of the winter varieties that are better suited to organic production in 2006 were Alchemy, Istabraq, Brompton, Welford, Exsept and SW Tatorus.

Orange Blossom Midge

Where this is a problem, for example in the south and east of the UK, some of the varieties that exhibit orange blossom midge (OBM) resistance, such as Brompton and Welford, have performed reasonably well under organic conditions. However, other varieties often yield higher in organic conditions, despite having been attacked by OBM.

Market

There are plenty of opportunities for all types of organic wheat in most parts of the UK. It is important to bear in mind that most of the bread-making market outlets are based in the south of England or in Yorkshire, whereas the feed market outlets are predominantly in Wales, Scotland, and the south-west, west and north-west of England, closer to the main livestock areas.

Barley

There are both winter and spring types available for planting in the UK. In recent years winter-sown barley has decreased in relation to spring-sown barley. Very little organic winter barley is grown, with the majority of organic barley produced being spring sown.

Spring varieties can be broadly divided into two groups, malting and feed. The feed varieties tend to be taller and higher yielding, and are suitable for undersowing with a following ley.

The malting varieties tend to be very short (65–75cm), and considerable difficulties in harvesting can occur if undersowing with a grass/clover ley mixture, because the clover can become taller than the barley, especially in wetter seasons.

There is a considerable difference in the growth habit of some barley varieties. The variety Doyen is significantly more prostrate with more tillers than the varieties Riviera and Dandy early in the season. This evens out somewhat later on, but Doyen is very short and tends to remain in the sheath and upright, whereas Riviera and Dandy are much taller. This is an important consideration when undersowing and harvesting.

Many of the modern varieties have a good disease-resistance profile. Select varieties carefully where barley yellow dwarf virus (BYDV), barley yellow mosaic virus (BYMV) and rhynchosporium are known problems, especially for the western and south-western areas of the UK, where disease pressure will be higher for barley.

Malting varieties selected should ideally be recommended by the brewing market prior to use, to ensure acceptable quality.

Some of the winter varieties that are better suited to organic production in 2006 were Pearl and Fanfare for malting, and Sequel and Cannock for feed.

Some of the spring varieties that were better suited to organic production in 2006 were, for malting, Westminster, Cellar and Optic, and for feed, Waggon, Tocada, Doyen, Riviera, Dandy and Hart.

Market

There are plenty of opportunities for all types of organic barley in most parts of the UK. Most of the malting market outlets are based in the north and east, with one or two specialist outlets in the South West. The feed market outlets are predominantly in Wales, Scotland, the south-west, west and north-west of England, closer to the main livestock areas. In these areas the taller varieties are also preferred due to their better straw yield, which is important for organic livestock.

Oats

There are winter and spring varieties, though winter types have a larger market share. There are strong markets for human consumption and breakfast markets, as well as the livestock feed sector.

Oats perform well on thinner, more marginal soils and under lower fertility conditions than other cereals. Many of the oat varieties are very tall in structure and have not been subject to the pressures of breeding for 'dwarf' characteristics and better harvest index. The tall structure of oats makes them ideal for sites with a higher weed pressure. In the north of the UK, winter oats can be prone to winter kill and so varieties having a higher level of winter hardiness should be selected. Some varieties suffer from crown rust, and these are not suited to the wetter south and west of the UK.

There is increasing evidence to support the allopathic effect of oats on weed populations, which can be useful to organic producers.

Some of the winter varieties better suited to organic production in 2006 were Millennium, Gerald, Jalna and Kingfisher.

Some of the spring varieties better suited to organic production in 2006 were Firth, Banquo, SW Argyle, Atego and Dula.

Naked Oats

Many of the naked oat varieties have not been subject to the pressures of breeding for 'dwarf' characteristics and as a result are very tall in structure, even more so than traditional oat varieties. However, they tend to have a much reduced yield potential as a result.

The demand for naked oats is because of their improved feed efficiency, with high oil and protein, for use in the organic poultry sector.

Current varieties available for use in the organic sector include Grafton, a naked oat with similar length straw to Gerald, Saul and Expression, which has higher potential yields than other naked varieties.

Rye

Only winter varieties are available, and there is often confusion as to whether these are 'forage' or 'grain' rye types. All rye plants are grain types, as they all produce seed if left to grow. In reality some varieties produce more seed, and some more foliage and less seed. For cereal grain production, the varieties that produce more grain are the preferred choice. For grain production, they should not be sown after mid-February lest they risk failure from the lack of vernalization.

All varieties are very tall (circa 170–180cm), and thrive in low fertility conditions. Rye tillers very profusely and covers the ground well, and is very competitive against weeds early on and when fully developed.

All varieties are prone to ergot. Rye is very unpalatable to rabbits, and organic farmers often plant rabbit-prone fields or headlands with rye as a deterrent. Triticale, being half rye, also has similar rabbit-resistant traits.

Hybrid types are the main types used in the UK non-organic sector, but these have not found favour in organic systems where the conventional rye types are preferred due to their better competitive advantage and lower fertility requirement. Although hybrids can offer higher yield potential they tend to be more disease prone, and the seed cost is very high and not available as organic seed currently.

Some of the winter varieties better suited to organic production in 2006 were Hacada, Matado, Eho and Admiral.

Triticale

Triticale is a species cross between wheat and rye in an attempt to combine desirable characteristics of both species. Winter and spring types are available, although there is a wider range of winter than spring varieties currently available. The small number of spring varieties is proving very popular and successful with organic farmers.

Some of the older winter varieties are very tall (*circa* 170cm) and yield less in more fertile situations than some of the newer varieties which are shorter and with a better harvest index. In low fertility conditions the older varieties can out-perform the newer, more developed varieties.

Triticale tillers very strongly and covers the ground well, and is very competitive against weeds early on and when fully developed.

Some varieties are more open flowering than others, and can be prone to ergot. Some varieties are more prone than others to sprouting in the ear if harvest is late in the season, and these should be avoided in the wetter west and north of the country where harvest is later. Triticale is less palatable to rabbits than some other cereals, and organic farmers often plant rabbit-prone fields or headlands with triticale as a deterrent.

Winter triticale does have a vernalization requirement and should not be sown after the end of February. In these situations it would be better to switch to a spring variety.

Some of the winter varieties better suited to organic production in 2006 were Bellac, Ego, Purdy, Modus, Partout, Taurus, Tremplin, Tricolor and Trinidad.

Some of the spring varieties better suited to organic production in 2006 were Logo, Legalo, Nilex and Trimour.

Spelt

There are no UK varieties of spelt. Most of the varieties have been developed in Germany and Austria, the main producers of the grain crop and where the market is largest.

Spelt does have a vernalization requirement, and should not be sown after the end of February. All varieties are very tall (*circa* 1,780cm +) and thrive in low fertility conditions. Spelt tillers very strongly and covers the ground well, and is very competitive against weeds both early on and when fully developed.

It is important to grow pure spelt varieties for human consumption. Some varieties have been crossed with wheat to improve yield and ease of hulling, but there are concerns that this introduces some of the genes responsible for wheat intolerance, and so such varieties are not acceptable for the specialist market.

Some of the winter varieties better suited to organic production in 2006 were Schwabenkorn, Ebners Rotkorn, Austrozerb, Obercomel, Frankencorn , Alcar and Fielding.

Hubble is a spelt variety with some wheat genes, and is therefore not a pure spelt.

PULSE VARIETY SELECTION

Beans

Winter and spring bean varieties are available, with some spring beans maturing earlier than winter beans. As well as being earlier maturing, spring varieties are less prone to indeterminate growth. Some varieties have small, rounded seeds, which may be suitable for the pigeon trade.

Winter-bean seed is generally larger than spring-bean seed, and the flowers may be either white or coloured. The majority on the National Institute of Agricultural Botany list (NIAB) and PGRO are coloured types.

The coloured-flowered varieties are less attractive to some users than the white-flowered tannin-free varieties. In white-flowered tannin-free types, the seed coat tannins are absent or at very low levels, leading to the possibility of higher inclusion rates of bean meal in certain non-ruminant feed compounds. Low tannin (white flowered) spring beans, although widely available and suitable for inclusion at higher rates in livestock rations, have proved to be difficult to grow organically. Since tannin acts as a natural fungicide, the low tannin beans can suffer from poor establishment, increased disease risk and lower yield potential. The current recommended white varieties are all relatively short, uncompetitive, and can suffer badly in weedy conditions. Unless weed-free conditions prevail with high levels of husbandry, they are unlikely to be suited to organic systems.

The differences between winter varieties available in the UK are relatively small. Many winter bean varieties have a good yield potential and reasonable disease resistance, are moderately tall, but can be relatively late maturing, which could present problems on heavier land.

On sites which are well supplied with moisture, the taller varieties will be the most useful. On sites prone to very wet weather, and where there is a risk of Ascochyta, and varieties with good resistance to Ascochyta should be chosen.

There are relatively few varieties of spring beans in the UK. The earlier maturing varieties help reduce the risk of the crop failing to mature until late September or early October in cool seasons. Short, stiff-strawed varieties seem to perform relatively well in the west and north of the UK, and often give a better performance relative to taller varieties on soils that are well supplied with moisture.

Spring beans can be rather late maturing in the north on heavier land, and selecting medium or early maturity dates is important in these situations.

Varieties should be selected for reasonable disease resistance to downy mildew (*Peronospora viciae*), rust (*Uromyces fabae*) and chocolate spot (*Botrytis fabae*) to ensure good leaf retention to maximise photosynthesis and productive potential, whilst at the same time maintaining a thick canopy cover for weed management.

Although many modern varieties can be grown in organic systems, it will be important to continue to encourage plant breeders and NIAB to develop and test varieties with characteristics appropriate to organic production, rather than for high yielding and disease resistance properties alone.

Some of the winter varieties better suited to organic production in 2006 were Clipper, Wizzard and Target.

Some of the spring varieties better suited to organic production in 2006 were Fuego, Compass, Quattro, Syncro, Meli and Maris Bead.

Peas

Peas have both low-growing and vining cultivars. The vining cultivars grow thin tendrils from leaves that coil around any available support, and *could* climb to be 1–2m high with support.

Peas are either grown for harvesting as a green crop, often referred to as 'vining peas' and selling for immediate freezing for human consumption, or for harvesting as a hard, dry pea with a combine harvester.

The majority of combinable field peas are grown for the animal feed market, and are harvested hard and dry. A small quantity is grown for niche markets such as drying, canning, as split peas or the pigeon trade. Growers usually obtain contracts with a buyer prior to planting these varieties.

The main varieties are either white (used for animal feed, canning as 'pease' pudding, and split peas for soups); large blues (used for animal feed compounding and micronizing for human consumption); small blues (used for canning and the pigeon food trade); and marrowfats (used for human consumption in the canned and dried form).

The main types of peas grown in the UK tend to be the spring-sown type. Taller crops with a more structured canopy and fuller leaves tend to assist with weed competition. Shorter, semi-leafless varieties are not so competitive against weeds.

Breeding development has resulted in a change from leafed varieties to semi-leafless varieties, the result being less leaf foliage and ground cover. This does mean that weed management under organic conditions is more challenging with these varieties. Most pea varieties are short-strawed and fully lodged or flat at maturity, leading to difficulties with combining, especially on uneven or stony soils where peas tend to be harvested very close to the soil, and where the front of the combine can easily pick up stones even when fitted with stone traps.

Growers should aim to pick varieties that have more, rather than fewer leaves (for ground cover and weed competition), that have a good tall straw which stands well (for competitiveness and ease of harvesting), and which are resistant to pea wilt (*Fusarium oxysporum f. sp. pisi*) and downy mildew (*Peronospora viciae*).

Experience has shown that sowing a small volume of cereal (barley or oats) with organic peas can help keep the peas standing well to harvest, with the cereal component acting as 'pea stick' supports. This can be cleaned out after harvest.

Some of the varieties (all spring sown) better suited to organic production in 2006 were Bilbo, Nitouche, Cooper and Maro.

Lupins

There are three species of lupin being grown and further developed for UK production: white lupin (*Lupinus albus*), blue lupin (*Lupinus angustifolius*) and yellow lupin (*Lupinus luteus*). Flower colour should not be used to identify the type of lupin, since most whites have a blue flower and many blues have a white flower. These all have very different characteristics and should be regarded as separate from each other.

There are both 'branched' and 'spikelet' varieties of the different lupin types. These refer to their foliage structure, with the branched varieties tending to produce lateral limbs and leaf at a lower level than the spikelet type, which can be useful for weed competition.

There are three types of growth habit found in lupins: determinate, where there is no branching; semi-determinate, where the plant produces a certain number of stems; and indeterminate where the plant requires sun or drought in order to reach maturity.

The question is which type of lupin to grow, and this will depend on whether the grower wishes to grow for silage, crimping or grain. Growers who wish to grow silage would grow a semi-determinate lupin due to the larger biomass, whereas grain growers may prefer less biomass if drilling in wide rows and using an inter-row hoe for weed control.

White Lupins
White lupins can be distinguished by their large, flat-shaped seed. All white lupins currently grown in the British Isles are semi-determinate, so they form a tall, thick canopy and are competitive against weeds. They tend to yield higher than blue and yellow lupins, though they do have a longer growing season as they can be drilled earlier (though this will depend on location), and are normally harvested around the middle of September – unless drilling has been delayed, when harvest will come later.

White lupins have a greater tolerance of compaction, which may be of use to growers on heavy land; however, it should be remembered that if drilling is late, then harvest will be delayed.

Blue Lupins
Blue lupins can be either semi-determinate or determinate. Yield is not as high as white lupins, but then harvest may be up to ten to fourteen days earlier. In some parts of the country blue lupins are preferred because white would give an unacceptably late harvest.

Yellow Lupins
Yellow lupins have a yellow flower, and the seed has a distinctly speck-led appearance. In the south of the country, harvest will be earlier than that of white lupins, though in the north they may be harvested at the same time. New breeding programmes are creating earlier-maturing varieties all the time. Yellow lupins can perform well on light acidic soils, and even those with a pH of below five. Most yellow lupins are semi-determinate and can form relatively dense stands as compared to deter-minate types.

Lupin-Type Characteristics

The UK Processors and Growers Research Organisation (PGRO) has funded trials with spring lupin varieties since the year 2000, but on a relatively low number of sites. Consequently data is limited, and even more so for crops grown under organic conditions. Similarly, grower experience spans only a few years, but new information continues to be generated. Different species are being grown, and these have contrasting characteristics. Yields in trials have ranged from 1–5t/ha. A high protein content (30–45 per cent) and a useful level of oil enhances the feeding value to livestock.

Blue, yellow and white lupin varieties have low alkaloids and tannins. Whites are taller, more competitive, and have better yield potential on good soils. Yellow and blue varieties are shorter, have more open canopies and are better suited to poor/acid soils. Winter-sown varieties are not well suited to organic systems due to their poor ability to compete with weeds.

Variety selection should include rapid development characteristics, a good tall structure and standing ability to assist with harvest and to help compete with weeds, and – most importantly – an early maturity, as ensuring maturity and the ability to harvest (without the aid of desiccants) will be critical to the success of the crop in the UK.

Blue, branching, indeterminate types: These tend to have slightly shorter straw and a lower standing ability, with medium maturity and a yield potential that is well above average. These are better suited to more southerly, drier regions of the UK on acid soils, when grown for dry seed.

Two of the varieties (all spring sown) better suited to organic production are Bora and Bordako.

Blue, spiked, determinate types: These tend to branch a little more and are a little taller, with good standing ability. Yields are reasonable, but the grain size tends to be smaller. Protein and oil contents are similar to the other blue lupin varieties. The spike types tend to mature earlier than the branching varieties.

Some of the varieties (all spring sown) better suited to organic production in 2006 were Sonet, Borwetta and Prima.

Yellow, branching, indeterminate and semi-determinate types: These tend to be freely branching, with a high protein content (42 per cent) but low yield. Straw tends to be long, and standing ability is generally good. These types are really only suited to grain production in the drier, warmer southern parts of the UK.

The main varieties better suited to organic production are Wodjil and Bornal.

White, branching, semi-determinate types: These are generally tall, large leaved and large seeded, with a lower standing ability and later maturity than the blue and yellow lupins. They can have a high seed oil content, with a higher protein content than the blue types but a little lower than yellow types. They are suitable for soils with a pH up to 7.5. They are suitable for grain production in the dryer, warmer south-east of the UK, and can be grown for forage in other areas. The main varieties better suited to organic production are Dieta and Amiga.

Soya

Trials of soya have only been undertaken with a small number of varieties in the UK, and on a relatively low number of sites. Consequently data is limited, and even more so for crops grown under organic conditions. Similarly, grower experience spans only a few years, though new information continues to be generated. Yields in trials have ranged from 1–2.5t/ha. A high protein content (40 per cent) enhances the feeding value to livestock.

Variety selection should include rapid development characteristics, a good tall structure and standing ability to assist with harvest and to help compete with weeds, and – most importantly – an early maturity, as ensuring maturity and the ability to harvest (without the aid of desiccants) will be critical to the success of the crop in the UK.

Some new varieties have had some success in some parts of the country in trial years with warm summers, but more recently, the commercial growing of soya in the UK crop has not performed as well, and consequently the area grown has fallen despite strong market demand.

As it is such a new crop to the UK, there are no specific recommendations for soya varieties at this time. Farmers are recommended to speak with research organizations, advisers and seed suppliers to determine a variety suited to their particular soil type, location and resource situations.

Chapter 9

Crop Establishment and Management

'The ultimate goal of the organic farmer with regard to cultivations should be to do as little as possible, and only as much as is needed.'

Stephen Briggs

CROP ESTABLISHMENT

The main objectives of successful crop establishment are to give the planted seed the best possible chance of germination and to optimize its growth potential within its given parameters. Whether the seed will germinate and grow satisfactorily will depend on a multitude of factors, including the availability of water and nutrients, the correct sowing depth, good soil-to-seed contact, the health of the seed, soil temperature, and the avoidance of predators such as slugs and birds. These factors can be controlled to an extent through good planning and cultivation, though they will always be vulnerable to uncontrollable external factors such as the weather.

Seedbed Requirements

Different crops require different seedbed conditions in order to maximize their germination and growth potential. It is important that the seed has good soil contact when planted so that it can access water and air, and that the young plant has sufficient anchorage in the soil so it remains upright. These factors will influence how successful it will be in the growing season, and will also influence the ability of the roots to expand and explore the soil for water and nutrients.

The plant uses water in order to take up nutrients, and water forms around 85 per cent of the plant structure. This water gives the plant turgidity, hence when water availability becomes restricted, the plant will wilt as the water content within it falls. If it falls too far the plant will reach permanent wilting point and will wilt irrecoverably.

Pockets of air are required by the seed in order that its environment doesn't become anaerobic – meaning 'without oxygen' – when it would not survive. It is important that the plant has sufficient soil contact for germination, and enough anchorage in the soil when growing in order that it will not get blown over in the wind and become lodged. The consolidation of the soil immediately after planting is important to ensure good seed-to-soil contact to allow the seed to obtain water, and minerals for germination and development. If the soil is friable, its surface area for a given volume will be far larger than that of a poorly structured soil of an equal volume, making nutrients far more available to the crop's root system.

In general terms, the smaller the seed, the finer the seedbed should be, and the shallower the drilling depth. For example, the seedbeds for grass and clover should be fine and relatively firm in order that the small seed and their subsequent roots maximize their chance of accessing the nutrients in the soil. If the seedbed contained large clods the roots would not be able to penetrate them, and some seed would be lost under the clods and would end up too deep for germination resulting in a sparse and poor establishment.

When dealing with large seeds such as field beans, these have a much larger surface area and can cope with more variable seedbed conditions. Some organic farmers broadcast them on to the surface and plough them in at around 15cm to 20cm deep. (*See* Chapter 7 for bean management.) The bean seed can emerge from this depth quite successfully, as the large seed contains sufficient energy to produce the long shoot. Planting the seed at this depth also reduces the chance of bird damage and frost kill, and ensures that the plant will be well anchored, which is important when field bean plants can grow over 1.75m high.

It is important that seedbeds are not overworked with mechanical cultivations, as these impair the structure of the soil. This can often happen with powered cultivators such as power harrows or rotovators. As an example, there is little need for a very fine seedbed for winter cereals, as these will develop from reasonable seedbeds, and the soil surface layers will further weather around the seed during the winter months. Creating a fine seedbed would not only be an unnecessary expense, it would reduce water infiltration and soil drainage, potentially leading to soil erosion; and as the crop emerges, seedlings would be more prone to wind damage, especially on sandier soils. Ideal seedbed requirements for a range of crops are shown in Table 37.

Table 37 The ideal seedbed requirement for a variety of crops

Crop	Ideal seedbed conditions
Winter cereals	No clods on the surface larger than 8cm across
Spring cereals	No clods on the surface larger than 5cm across
Winter field beans	No clods larger than 10cm across
Spring field beans	If drilled, no clods larger than 8cm across
Field peas	No clods larger than 5cm
Grassland	No clods larger than 4cm across

In non-organic agriculture, growers have been encouraged to establish crops with ever finer seedbeds, mainly to maximize the efficacy of residual herbicides; this practice reduces the long-term capacity of soils and the workability of the soil.

Some very friable soils may form fine seedbeds even when just ploughed, especially those with a high silt content. This would also be typical of a light or medium soil that has had large quantities of organic matter added to it, and after drilling the seedbed may be very fine.

It is important to note that after drilling, many seedbeds will require rolling, to help consolidate the soil to ensure good soil-to-seed contact and thus good germination, and also to help reduce the effect of a loose, soft seedbed which may cause lodging of the crop later on in its growth. Rolling will also help reduce the range of habitable environments available to slugs and their movement.

Some soils, especially those on a chalk parent material or containing lots of stones and brash, are prone to a condition widely known as 'frost heave'. This condition results in a very puffy upper soil layer in the soil in the late winter or early spring as a result of water freezing and thawing differentially around stones and in the soil. Thus winter-drilled crops can be rolled and consolidated at the time of sowing, only to result in a very loose and puffy form the late winter or early spring. These soils and young crops may need to be rolled again to consolidate the soil to provide good anchorage for the crop, access to water and nutrients, and to reduce soil-borne pest habitats.

Tillage Systems for Establishing the Seedbed

There are many methods available to the farmer to create a seedbed and establish a cereal or pulse crop. One option is to use primary cultivations such as ploughing, and another is to minimize soil disturbance by way of non-inversion stubble cultivations plus light cultivations followed by the drill. Direct drilling, other than to establish undersown covers, is unlikely to work well in organic systems as it relies heavily on the use of herbicide inputs for weed control, which are prohibited under organic standards.

Plough-based systems have the distinct advantage on many organic farms, since they can incorporate crop residues, weeds and manures, whilst leaving a trash-free surface for subsequent cultivations. The key points of the different cultivation systems are summarized below:

1. **Plough-based system:** Typical working depth of 15–25cm, crop establishment cost range £85–£120/ha, work rate to establish crop from 0.25–0.4ha per hour.

Advantages:
- Widely adaptable to almost all soils
- Buries trash and weed seeds
- Creates a fine seedbed on light and medium soils
- Uses widely available equipment

Disadvantages:
- Major disturbance to soil microbial populations
- Oxidation of soil organic matter
- Potentially higher mineral leaching
- Low work rates

Ploughing is the favoured cultivation by many organic farmers.

- Many operations
- Relatively high cost of establishment
- Where a plough pan has been created, sub-soiling may be required

2. **Deep-tined cultivations:** Typical working depth 10–20cm, crop establishment cost range £60–£100/ha, work rate to establish crop from 0.4–0.75ha per hour.

Advantages:
- High output
- Fast work rate
- Few passes required

Disadvantages:
- High horsepower requirement
- Can lead to increased grass weed burden with reduced burial effect

Deep spring-tined cultivations.

Reduced, non-inversion 'stubble' cultivations.

3. Reduced non-inversion 'stubble' cultivations: Typical working depth 5–10cm, crop establishment cost range £40–£70/ha, work rate to establish crop from 0.6–1.0ha per hour.

Advantages:
- High output
- Few passes
- Lower cost
- Moisture and nutrient retention

Disadvantages:
- High horsepower requirement
- Potential increase in grass weeds
- Not suited to all soil types

4. Direct drilling: Works on soil surface, crop establishment cost range £30–£45/ha, work rate to establish crop from 1.5–2.4ha per hour.

Advantages:
- Very high output
- Low establishment cost
- Moderate power requirement compared to cultivator drills
- Less soil structural damage
- Less disturbance to soil microbial populations

Disadvantages:
- A rapid build-up of weeds likely
- A build-up of volunteer weeds, especially grass weeds, which will act as a disease green bridge to cereal crops
- Specialist drill required

Direct drilling has many merits, but opportunities in organic systems are limited.

- Relies mainly on herbicide inputs for success (*prohibited in organic systems*)
- Not suitable for all combinable crops
- Not well suited to most organic systems

CULTIVATION EQUIPMENT

There is a wide array of cultivation equipment available to create a seedbed in order to establish a crop. Traditionally the plough was used first, followed by a secondary cultivator that may have been a spring tine cultivator, or light discs followed by the drill.

During the 1970s power harrows appeared on the market. These were attractive to farmers on heavy land due to the fact that a seedbed could be established with one pass after the plough. However, they were not able to break down very dry seedbeds, and tended to smear the soil when too wet. There is also a school of thought against their use mainly because they tended to *force* the creation of the seedbed by overworking the soil and removing its inherent structure.

Rotovators are also popular with some smaller farmers and horticulturalists, as they allow a seedbed to be created from stubble in one pass. The disadvantage of rotovators is that the soil is often smeared through the action of the rotovator creating an impeding layer at the base of the cultivation, which restricts drainage and impairs crop root development and growth. Rotovators also overwork the soil and remove structure and are very expensive to operate. The use of rotovators subsided in the 1980s, and by the early 1990s they tended to be used for just small areas rather than the whole farm.

Power harrow cultivations.

Discs are another popular cultivator on many farms. They work at high speed by slicing through the upper layers of the soil and mixing in crop residues and weed biomass, and have a very high work rate output. They have many advantages, and can be used on organic farms successfully in some situations. On damp soils, discs can smear the soil through their slicing action, creating an impeding layer similar to a plough pan or rotovator. Discs should also be avoided on an organic farm whenever weeds with a rhizome (couch grass, creeping thistle) or where docks are present, as discs will chop these weeds up and spread their regenerative parts, thus further perpetuating the weed problem.

Rotovator cultivations.

Stubble cultivators have gained popularity on many farms in recent years. These work by loosening the soil with large winged tines, followed by small mixing discs that are then often followed by a crumbler roller for light consolidation. These cultivators have the advantages of high speed and work rate, combined with a loosening and lifting action in the soil, without smearing it. The small discs don't slice the soil, rather they mix crop residues and weeds with soil, and the crumbler roller's light consolidation precipitates the rapid germination of weeds and volunteers prior to the next planting. Without the crumbler roller in place, the cultivator can be used for cultivations, which brings weeds to the surface as a sort of mini fallow. Even with the roller in work, they are reasonably effective at placing weeds at the surface for desiccation.

The main disadvantage of these cultivators is that they do not bury crop residues and weeds as effectively as a plough, and a number of passes may be required to obtain a good quality seedbed. But despite their limitations, experience suggests that they have established a place on many organic arable farms, and can be useful for weed management and cultivations between straw crops.

Also popular in recent years are rollers that incorporate heavy tines, thereby allowing the roller to become a cultivation tool in itself, and also scalloped discs trailed by cast rollers; some discs are preceded by legs operating at depth to loosen the soil. Some manufacturers will make cultivation equipment to order, tailoring it to the farmer's specifications.

The plough remains the principal cultivation method for many organic farms, along with a large range of secondary cultivators and tillage equipment. These include combination harrows (incorporating many individual actions), levelling harrows, and furrow presses.

Discs have a limited use on organic farms.

The main challenge with many plough-based systems is that most ploughs have large, deep, short 'semi-digger' bodies that will only provide adequate inversion and burial of crop residues and weeds if they are operated at 20cm to 30cm depth. This is far too deep for organic systems which rely on biological processes, and where deep ploughing will seriously disrupt these biological processes, will also mineralize and leach too much valuable soil nitrogen, as well as mixing the fertile top soil with lower layers. Ploughs that have longer, slender, more curved bodies are much better suited to organic systems as they can be operated at a depth of 10cm to 15cm while providing adequate inversion and burial of crop residues and weeds, thus reducing any disruption of soil biological processes, minimizing the mineralization of soil nitrogen, and maintaining the fertile topsoil in the top layer (*see* the picture below). The other main benefit is that shallower ploughing needs less power and is quicker, thus saving money at the same time.

There are many older-style 'ley' ploughs available, and a number of companies refurbish these and sell them to farms. They tend to be smaller three-, four- or five-furrow ploughs, but many machinery manufacturers and dealers can fit shallower working bodies to many larger seven- to ten-furrow ploughs if requested. One company has recently started marketing an 'Eco' plough, with bodies capable of shallow inversion and high work rates. This has found favour on some organic farms.

The choice of cultivator used will depend on the soil type, affordability, machinery power, geography, crops grown, weeds present and time available. Producers should consider all these factors when purchasing machinery, and ideally see a demonstration first on a soil type similar to their own.

As with organic crop programming, the best approach with cultivations is to 'mix it up' as much as possible. Avoid using the same cultivation system year after year. An organic rotation is likely to dictate a mixture of

Shallow ploughing is the preferred method on organic farms.

cultivation systems, with a plough required to incorporate a grass and clover ley, and tines, discs or stubble cultivators used between cereal and pulse crops. Aim to use different cultivators at different depths and in different crops to produce different results, because like this the soil will be moved in such a way as to minimize unnecessary disturbance and avoid the build-up of structural problems in the form of impeding layers.

Even if a plough is used year after year, aim to alter the depth slightly to avoid building up a plough pan. Remember, it is not the equipment that causes the problem, but how it is used!

The ultimate goal of the organic farmer with regard to cultivations should be to do as little as possible, and only as much as is needed. The aim should always be to minimize any soil disturbance and thus reduce the detrimental impact on soil microbial life (*See* Chapter 2) by reducing the working depth of cultivations to as shallow as possible, whilst at the same time loosening the soil deeper in the profile to provide good structure, air and water movement, and the free exploration of crop roots.

SUB-SOILING

When a good soil structure exists the soil is easy to work, with a good granular structure with evenly spaced pores and fissures. Plants will find it easy to expand their roots and extract nutrients from the soil, while drainage will be free flowing and the sub-angular blocks that are deeper down will allow roots to penetrate and extract water and nutrients. Some soils, such as heavy clays, can suffer from compaction when heavy machinery drives over them, especially when trafficking occurs on wet or damp soils, and in such soil the fissures tend to run horizontally rather than vertically. Repeat cultivations at the same depth can also create an impeding layer known as the 'plough pan'. A compacted layer in the soil, such as a plough pan, can impede drainage and may severely restrict the ability of the plant roots to explore the soil profile and extract nutrients. Where compaction exists, deep amelioration cultivations, commonly known as subsoiling, may be required.

Subsoilers typically consist of a number of heavy steel legs on a heavy-duty frame, used at a depth of 25–45cm in dry soils. There is a wide range of subsoiler cultivation equipment, which includes fixed deep lifting legs with and without wings at the base, PTO-driven vibrating legs, and deep plough-type machines, such as the Howard paraplough.

When subsoilers were first used they comprised straight legs only. However, research undertaken in the 1980s demonstrated the importance of a 'wing' at the base of the leg. These produce a 'bow wave' effect in the soil as they move, fracturing the soil around them, and above and below the wing. This results in a greater area of soil displaced and means that legs can be further apart, reducing the power requirement, raising the work rate and reducing operational costs. The soil structure at depth is greatly improved, and air, water and plant roots can move more freely.

The moisture of the soil is of paramount importance when considering subsoiling: the soil must be in a brittle condition in order to be effectively fractured by the machine's legs and lower wing. Subsoiling in damp or wet soils that have a 'plasticine' state is at best a waste of time and money, as all it will do is strike channels through the soil, without any relief from compaction. Experience shows that too many farmers undertake subsoiling because they think they should, rather than in soils that need subsoiling, and they often do it in soils that are far too wet. When undertaken in the wrong conditions, rather than doing anything to improve compaction, subsoiling and smearing the soil may actually cause further compaction damage to the soil.

Ideally the soil should have a loose, blocky structure with plenty of pores and roots at depth. If the soil lacks roots at depth, with few pores and a dense angular blocky structure, subsoiling is likely to be beneficial.

In some heavy clay soils, the soil swells and expands in the winter as it takes water into pore spaces. In a very dry summer that follows, they dry out, shrink and crack as water is evaporated. While this 'shrink and swell' characteristic allows for a reasonable level of self-structuring, mechanical subsoiling operations may still be needed to relieve severe compaction, often in localized areas.

The cost of subsoiling operations can be high, so before any subsoiling operation is carried out, it is important to dig a number of 'soil pits' to provide a clear picture of the state of the soil at various depths. What looks dry on the surface can be a wet and plasticine soil at depth, and not at all suitable for subsoiling. It is also sensible to dig pits at different locations in

Sub-soiling cultivations should only be undertaken under the correct soil conditions.

The Howard power plough.

the field, as soil conditions vary considerably across even one field, and it may only be necessary to undertake subsoiling on part of a field.

SEED RATES

A great deal of research work has been carried out concerning seed rates, especially with winter wheat, in non-organic production, showing considerable scope for reductions at early sowing. It is generally accepted that seed rates for organic cereal production should be higher (10 to 15 per cent more) than those used in non-organic production, for a number of reasons:

- Higher seed rates should result in higher plant populations, which will result in better crop competitive advantage and better weed suppression, and will allow for some crop plant loss by predation or during mechanical weeding, or die-back from lower fertility.
- Higher plant populations will give a greater number of primary tillers, which are potentially heavier yielding than secondary tillers.
- In the absence of luxury levels of nitrogen from applied synthetic fertilizers, lower numbers of tillers are typically present in organic crops. Therefore higher plant numbers compensate for lower tiller numbers to ensure better ground cover and weed competition.

Work recently published by HGCA (report number 304) shows the influence of seed rate on yields of organic wheat, as shown in the table.

Table 38 The influence of seed rate on wheat yields

Seed rate/sq m	Range of plants/sq m	Yield (% of control)	Range
200	85 to 116	95.0	85.3 to 103.0
300	92 to 161	98.5	90.6 to 107.3
400	143 to 208	100.0	100
500	187 to 268	102.6	96.0 to 109.7
600 (single trial)	216	105.9	216

The optimum seed rates for different cereal types are shown in the other table.

Table 39 Optimum seed rates for different organic cereals

Cereal type	Seed/sq m	Typical kg/ha
Winter wheat	400 to 450	180 to 220
Spring wheat	500 to 550	225 to 275
Winter barley	350 to 400	160 to 200
Spring barley	375 to 425	180 to 220
Winter oats	500 to 550	175 to 225
Spring oats	650 to 700	220 to 270
Triticale	370 to 510	160 to 220
Rye	460 to 570	160 to 200
Spelt	460 to 570	160 to 200

This is a simple formula for calculating the required seed rate:

$$\text{Seed rate kg/ha} = \frac{\text{target population (plants/sq m)} \times \text{thousand seed weight}}{\text{\% establishment}}$$

A more accurate formula is:

$$\text{Seed rate kg/ha} = \frac{\text{target population thousand seed weight}}{\text{\% germination } 100 - \text{field loss}} \times \frac{100}{100 - \text{field loss}}$$

The table below shows how kg/ha of seed varies with changes in thousand grain weight and target seed per metre.

Table 40 Seed rate in relation to thousand grain weight and target seed per metre Seed numbers/sq m

TGW	100	150	200	250	300	350	400	450	500	550	600	650	700
35	35	53	70	88	105	123	140	158	175	192	210	228	246
40	40	60	80	100	120	140	160	180	200	220	240	260	280
45	45	68	90	113	135	158	180	203	225	247	270	293	316
50	50	75	100	125	150	175	200	225	250	275	300	325	350
55	55	83	110	138	165	193	220	248	275	302	330	358	386
60	60	90	120	150	180	210	240	270	300	330	360	390	420

Values within table are kg/ha

The seed-rate formula shows that the better the seedbed and germinative capacity of the seed lot, the lower the actual seed rate that can be used.

FARM-SAVING SEED

It is quite feasible to farm-save seed from organic crops, provided the fields and stocks are carefully selected. This may only be advisable from the first crop grown from certified seed, but nevertheless can represent a useful cost saving. Where farm saving is being practised, emphasis should be placed on careful drying and cleaning of the crop prior to storage. The crop should not be 'over-dried' as this can reduce germination significantly (*see* Chapter 12). The crop should be tested for seed-borne disease levels, and rejected where these are above threshold. These can be undertaken by organizations such as NIAB, PGRO and other private laboratories and agronomy groups. As a priority the following tests should be undertaken for farm-saving seed:

Cereals

Wheat, Triticale, Rye, Spelt: Diagnostic germination, 1,000 seed weight, seed-rate table, moisture, microdochium, septoria, bunt, tetrazolium viability and vigour.

Barley: Diagnostic germination, 1,000 seed weight, seed-rate table, moisture, microdochium, loose smut, leaf stripe, net blotch.

Oats: Diagnostic germination, 1,000 seed weight, seed-rate table, moisture, microdochium, leaf spot, loose smut, covered smut.

Typically 300g of crop is required for each test, which will take seven to ten days to complete.

Pulses

Field beans: Diagnostic germination, 1,000 seed weight, seed-rate table, moisture, ascochyta and stem nematode.

Field peas: Diagnostic germination, 1,000 seed weight, seed-rate table, moisture, ascochyta and mycosphaerella.

Typically 900g of crop is required for each test, which will take seven to ten days to complete.

There are additional costs associated with saving seed that should be carefully considered; these include plant breeders' royalties, the cost of cleaning and bagging, additional costs associated with roguing the crop, any testing charges (germination, 1,000 seed weight, purity, disease loading), cleaning of the harvester and drier, storage, and segregation of the seed during handling. The table below shows the typical royalty rates for different cereal types.

Percentage establishment should be in the range 70 to 80 per cent, but can be much lower in organic cropping systems (as low as 50 per cent).

Table 41 Royalties payable for farm-saved seed in 2007

Crop type	£ per hectare	£ per tonne
Wheat	5.39	30.60
Winter barley	4.57	26.70
Spring barley	6.39	33.67
Oats	3.36	22.05
Triticale	7.49	44.03

Plant population, and therefore field losses, can be measured at any time after emergence, although it is generally accepted to be most meaningful for winter cereals to determine plants per unit area in the spring after establishment and any winter kill.

SEED TREATMENT AND MANAGEMENT STRATEGIES

In non-organic systems, the seed of many crops is treated prophylactically with agrochemical products regardless of the health status of the seed, and the risks which various levels of infection might pose. Increasingly this practice is becoming less tenable, and a combination of seed health testing and treatment according to test result is becoming more common. Even so, this approach is usually confined to the generation of seed used for crop production, and multiplication generations are still treated. In the case of some cereal diseases, there are sound biological reasons for doing this, since the pathogens involved are highly adapted seed-borne fungi which can increase rapidly with each successive seed generation.

For organic production, the removal of the derogation allowing the use of non-organic seed in 2004 resulted in a requirement that a minimum of two generations of seed could not be treated with conventional products. This raised some concerns that organic seed which is farm-saved or sold through the seed trade could be contaminated with a range of seed-borne diseases that could multiply out of control very quickly.

As a result, a recent Defra-funded research project (2002–2006) evaluated the occurrence of a range of seed-borne cereal diseases in the organic production system, and at the same time undertook research into potential seed treatments that would be allowed in organic production, to reduce seed-borne disease potential.

Seed treatment trials were carried out in 2004 and 2005 and included biological, micronutrient and physical treatments including hot air (steam), milk powder, garlic powder, and *Bacillus subtilis* (*see* Table 42). Results showed that none of the treatments tested suppressed loose smut or leaf stripe on barley. When wheat seed had a high level (30 per cent) of *Microdochium nivale* seedling blight infection, none of the treatments used significantly reduced infection or increased final yield. One of the biological treatments tested (Crompton) did significantly improve plant establishment (145 plants/sq m compared to 113 plants/sq m for untreated plots, and 143 plants/sq m for Sibutol, the conventional control treatment), though effects on yield were non-significant. The biological products (Cerall and the Crompton product) and the hot air treatment suppressed bunt in 2005, as did Radiate (ammonium and zinc ammonium complex), though the latter had no significant effect in 2004. The results for bunt are summarized in the table.

Table 42 Bunted ears per plot (12m) counts in seed treatment trials, 2004/2005

Treatment	Bunted ears/plot 2004	Treatment	Bunted ears/plot 2005
Untreated	36.7	Untreated	28.5
Sibutol	0	Sibutol Secur	0
Radiate	32.7	Radiate 7.3	
NMS	35.3	Cerall 7.5	
EM1	42.0	Crompton	8.5
EM1 + micronutrient	48.7	30sec hot air	9.8
Tricet Micronutrient	43.3	60sec hot air	18.8
Bacillus subtilis	38.7	90sec hot air	14.8
Garlic	35.3		
Hot air (90sec)	27.3		
Lsd ($p = 0.05$)	8.48		10.13

Source: Defra research project OF0330

Though the active mechanisms of some of the effects on bunt that were observed are not well understood, it did appear that some biological treatments reduced the disease. Applied spore loadings (2,000 spores/seed) were high, and though such levels have been observed in organic seed, retrieval by seed treatment should probably not be attempted. At lower spore loadings the effect of biological treatments may be efficient enough to retrieve an infected seed lot. Hot air treatment also appeared to be partially effective, and although no attempt was made to optimize the treatment process within this project, the viability of seed, plant establishment, and final yields were not reduced by the thirties treatment length. Work elsewhere has greatly improved heating processes and achieved a high degree of control of bunt without loss of germination, and of the treatments and processes tested, this would appear to offer the greatest potential for reliable use.

Seed cleaning was investigated as a means of improving establishment in *Microdochium*-infected wheat, and gravity separation was carried out on a seed lot with 30 per cent infection. Though establishment counts and early spring counts were improved slightly in the cleaned seed as compared to the uncleaned, the effects were not significant, and final yields were not improved. Incidence of disease in the cleaned versus uncleaned seed indicated that the process, although removing light and shrivelled seed, did not selectively remove infected seed. However, with higher levels of *Microdochium*, which can occur in some seasons, it is expected that cleaning could reduce disease incidence as the proportion of infected, shrivelled seed also tends to be higher.

As part of the research undertaken as part of the organic seed-testing programme over four years, a wide range of organic crop and seed samples were tested for disease to identify the prevalence of seed-borne diseases in organic crops, and thus to quantify the risks to production. The four-year programme of research concluded that there was no indication of emerging seed-borne disease problems. However, there were sporadic incidences of serious levels of seed-borne disease, and in these cases action was necessary. The most serious problem was the occasional occurrence of bunt in wheat seed. As a result, guidelines were developed to minimize the chances of bunt increase, whilst emphasizing the need to pay attention to other diseases which may occasionally affect the quality of seed.

Guidelines on reducing the risk of bunt in organic cereal crops include the following:

- Always test any untreated 'mother' seed. Effective sampling guidelines have been established, and are published in the HGCA Fact Sheet No. 72 (sent to all HGCA levy payers, or available from the HGCA web site).
- For seed destined for further multiplication, the desired standard for bunt is as close to zero spores/seed as possible, but in any case, below one spore per seed.
- For seed destined for crop production, and not for further seed generations, including farm-saved seed, levels of infection should not be higher than one spore per seed.
- Where it is intended to farm-save seed of wheat, crops should be first wheats if possible, since the most consistent bunt problems are associated with minimum tillage and second wheat production. It is possible that soil-borne bunt could survive in this situation, and perpetuate a low level of bunt in a particular area, which could increase quickly in a crop.
- For farm-saving, always keep grain intended for seed separate from ware seed.
- Some varieties tested are either resistant to bunt infection, or escaped infection. However, in the absence of ongoing information on varieties used for organic production, it is not yet possible to rely on variety resistance as a means of control.
- Some seed treatments and treatment processes show promise for the control of bunt, but in the absence of commercially available systems and recommendations, seed producers and growers farm-saving seed in the UK still need to rely on stringent seed health using the standards recommended.
- Wheat seed sown relatively late (November/December) tends to be more prone to bunt infection as emergence is slowed. In cases where growers wish to use seed just above the threshold level of one spore/seed, early sowing will help to minimize any chances of infection; however, the aim should be to avoid using such seed.

Guidelines for using untreated seed in conventional agriculture are summarized in the HGCA publication *Wheat Seed Health and Seed-borne Diseases* (at the time of writing priced at £10, or free to HGCA levy payers). Though aimed at conventional agriculture, the principles described are also relevant to organic production, and if adhered to, will minimize the risks of seed-borne diseases developing in organic seed stocks.

STALE SEEDBEDS

The potentially substantial soil disturbance such as is required to incorporate crop residues and prepare a seedbed suitable for cereal seeds will always stimulate a number of weed seeds to germinate. The 'stale seedbed' technique uses a prepared seedbed when the subsequent flush or flushes of weeds are removed by light harrowing before the crop is sown; this generally allows the crop to establish with much reduced early competition from weeds. This technique works well in early autumn and late spring, but is not so useful in late autumn or early spring when soil moisture levels and temperatures do not kill off weeds effectively (there is more information on weed management in Chapter 10).

Growers can use a comb harrow on a prepared seedbed to stimulate weed emergence prior to the actual seeding operation. Preparation of a seedbed well in advance of the sowing time, typically ten to fourteen days, causes some of the weed seeds lying near the soil surface to germinate, the disturbance of the soil being all the stimulation that the dormant weed seeds require in order to spring into life. These weed seeds are subsequently cultivated out by taking additional passes with the comb harrow. In situations where the soil weed seed bank is high, and where weather and soil conditions permit, it may be a wise and economically worthwhile precaution to effect two germination and cultivation cycles. In this way a maximum number of weeds will be removed from the seed bank and will have less opportunity to affect the following cereal or pulse crop.

A comb harrow being used to undertake a stale seedbed weed pass.

DRILLING TECHNIQUES

There are several methods that can be used to carry out drilling the crop, but the aim is the same: to provide the best conditions, given the circumstances, for the seed to germinate and develop. As previously discussed, depending on the crop, different seeds will require different sowing depths; the table below shows the ideal sowing depth for a range of arable crops, the measurement representing the depth of consolidated soil the seed should have on it.

Table 43 Ideal sowing depths for a range of arable crops

Crop	Sowing depth
Cereals	3cm/1–1.5in
Field beans	3cm/1–1.5in for spring beans winter beans 7.5cm–15cm/3–6in
Field peas	3cm/1–1.5in
Grassland	1cm may be broadcast and harrowed in or drilled on the surface and covered by the drill tines

Cereals may be drilled up to 4cm, and may even emerge from deeper than this; however, there will be a yield penalty, and seed emerging from depth tends to lose vigour and will struggle throughout the season. Research has also shown that planting wheat too deep results in a reduction in tillering and crown root development, maturity is also later, and establishment time will take longer between the first and last plant appearing.

There are two types of emergence: crops with hypogeous emergence (the cotyledons come above ground, as, for example dry beans, lupins), and crops with epigeous emergence (the cotyledons remain below ground, for example peas, beans, cereals). Although this means that the latter group may emerge more successfully from deeper in the soil, it does not allow the grower to sow the crop deeper than recommended without a cost in yield.

Width of Row

There are three options available to the farmer: either to use a wide-row system based on planting at 20–25cm row widths to facilitate the use of inter-row weeders; or to use a narrow-row system based on planting at 10–14cm row widths; or to not use rows at all, but to broadcast the seed. The production of cereals and pulses on wide rows, typically on 20–25cm spacings (but up to 30cm spacing is workable) is gaining popularity with arable producers on both heavier land and land where grass and tap-rooted weeds are a problem, so that the crop can be hoed between the rows with an inter-row hoe. (*See* Chapter 10 for more information on this system.)

The wider the rows are apart, the higher the plant population will be within the row. Although this helps the plants in competing against weeds, they will still occur within the row, and if a wide-row system is adopted the farm is committing itself to having to undertake mechanical weeding.

Research has shown that crops grown in wide rows (20–25cm apart), such as wheat, barley, oats, peas and beans, can produce yields equal to crops grown on narrow rows in organic systems, and in some cases yield can be improved by as much as 20 per cent over unweeded crops, by reducing weed density. Planting at wider rows is not detrimental to yield, with research demonstrating no appreciable yield penalty up to a row width of 30cm.

It is thought that this is due in part to the fact that weeds are better controlled (*see* Chapter 10), and more nitrogen is mineralized by the passing of the tines of an inter-row weeder, which stimulates more crop growth. However, there is also research indicating that some crops grown on 16cm rows, particularly the more competitive crops such as triticale, oats and rye, have a greater effect on weed suppression than those on narrow or wide rows. The table below shows the advantages and disadvantages of different row widths.

However, specific agronomic, management and economic factors need to be considered. Adaptations are required including matching wider drill rows to hoe widths, the timing of weeding, the potential for crop damage, investment in suitable equipment, and the limitations of slow work rates and manual control. Careful placement of the hoe is required to avoid potentially high levels of crop damage. The developments in automated guidance systems, which aim to solve some of these problems, are examined in Chapter 10.

Table 44 Advantages and disadvantages of different row width seeding techniques

Row width	Advantage	Disadvantage
No rows	Each plant will have its own space reducing competition against itself	Crop spacing will be variable. Damage to the crop will be higher with a comb harrow rather than an inter-row hoe. Weeds will become harder to control earlier on
Narrow rows	Maximizes the ability of the crop to smother out weeds	Too narrow for the use of inter-row cultivators but all right for comb harrows. These can damage the crop more than inter-row hoes
Wide rows	Ability to remove weeds between the rows up to 4in tall with an inter-row hoe	Plant populations within the rows increases crop competition against itself. Weeds can still grow in the rows and avoid being removed. Driver accuracy is of paramount importance

Experience suggests that inter-row hoeing can offer a realistic option for weed control, especially on heavier soils (*see* Chapter 10), and can also be used with lower seed rates.

Broadcasting versus Drilling

Placement of seed can range from almost complete coverage of the ground with broadcasting or 'banded shoe' coulters, through conventional row widths, typically 9.5–12.5cm, to wide rows up to 25cm. These differences are discussed in more detail in the weed control section (Chapter 10).

The use of Suffolk, disc or tine coulters on the cereal drill is determined by the soil type, primary cultivation techniques and trash levels anticipated in the seedbeds. It should be noted that where growers broadcast seed rather than drill it, seed rates should be increased 10–15 per cent, and the field should be broadcast over twice, the second time being 90 degrees to the first pass to ensure an even application of seed.

Coulters

Suffolk coulter drills are best suited to light to medium soils using plough- or deep cultivator-based systems. They can sow most combinable crops effectively, but sometimes require finer seedbeds than other drill types.

Disc coulters can cope with surface trash better than Suffolk coulters. Tine coulters are intermediate in their ability to cope with trash, and much will depend on coulter spacing and layout. Disc and tine cultivator drills both cope well with a range of seedbed preparation techniques but are generally best suited to medium and heavier bodied soils. In light to medium soils they can experience some difficulty with regulating drilling depth unless seedbeds are well consolidated.

Tramlines

It is common for non-organic farms to establish crops with 'tramlines' set out across the field at 12m or 24m spacings. Tramlines are used as location rows for the accurate alignment of equipment in the application of fertilizers (through sprayers and spreaders) and agrochemicals (through sprayers). They also serve to concentrate the areas of compaction, which arise from the many passes of sprayers and spreaders throughout the season, often when soils are damp and susceptible, to localized areas. These areas are subsoiled after harvest to remove the compaction.

In organically farmed arable systems there is little need for tramlines as the frequency with which machinery is used in a growing cereal or pulse crop is far less, if at all. Experience shows that where tramlines are put into organic crops, they fill up with weeds quickly as a result of the lack of crop competition.

If a crop has to be weeded with a tined weeder, it can be safely driven upon without significant damage. Tined weeders are typically 6m or 12m wide. The larger machines help reduce soil damage from trafficking. Often weeding at an angle to the crop row is beneficial, thus tramlines would be of no use. Any soil compaction damage is much reduced when weeding as it can only be undertaken successfully when the soils are dry enough.

One instance where there is justification for tramlines is when the crop is planted on a wide-row system and an inter-row hoe is used. Tramlines for locating the hoe are used and set out depending on hoe width, typically 4m or 6m but up to 12m (*see* Chapter 10 on weed management).

FIELD BEAN ESTABLISHMENT

Traditionally winter field beans are drilled or broadcast on to the surface stubble of the previous crop, and then ploughed into a shallow furrow 6–8in deep. There are several reasons for using the plough in the establishment of field beans: firstly, the seed is placed at a depth where they are out of the way of bird predators; secondly, on some heavy and wet soils establishing a seedbed is not always possible; and lastly, large seed such as winter bean seed doesn't always flow well through the drill unless pea and bean rollers are used. Winter beans should not be sown too early, that is, not before the second week of October. Acceptable crops have been harvested from crops drilled in late December though location can make a significant difference. If winter beans are sown too early and a mild autumn is followed by a very cold winter, the beans may suffer from severe winter damage if they're too far ahead, and higher disease incidence can occur in the spring due to lush growth and the crop being too forward.

Spring bean seed is smaller and will flow through seed drills more easily. It is usually drilled as soon as soil conditions permit in the spring, from the beginning of February to mid March. Growers may drill later though yield is likely to be reduced. When drilling spring beans, ideally they should be drilled to a depth of 3in to reduce predation by birds, but many drills struggle to drill any deeper than 1½–2in deep.

Table 45 Ideal seed rate and plant populations for beans and peas

	Final target population (plants per sq m)	Expected field loss %
Winter-sown beans	16–17	15
Spring-sown beans	40	5

(Source: PGRO)

PEA, LUPIN AND SOYA ESTABLISHMENT

Peas, beans and soya are all similar medium-seeded crops requiring a fine, firm seedbed prior to drilling; but over-cultivation and compaction should be avoided, as these crops do not like soil compaction. Seedbeds should be rolled to ensure good soil:seed contact, or where the seedbed is dry, to conserve moisture and aid germination.

On most soil types it is necessary to roll the field to depress stones to permit weed control in the developing crop, and to allow harvesting close to the ground whilst avoiding damage to the combine. Rolling should be done soon after sowing and well before plant emergence. Leaving the soil surface rough after drilling can prohibit mechanical weeding or lead to unacceptable levels of crop damage from the mechanical weeder 'jumping' about and damaging the young crop.

For all crops a good plant population is essential, since low populations are more difficult to harvest, later maturing, and more prone to bird damage. Sowing into a warm, moist seedbed will assist with germination and rapid early plant development, which is important for competition against weeds. Later sowing necessitates a finer seedbed using spring-tined cultivations, but in many locations, later drilling into warmer soils achieves more rapid and even emergence. This is of benefit for weed competition in these relatively uncompetitive crops.

As with cereals, the production of peas, lupins and soya on wide rows, typically on 20–25cm spacings (but up to 30cm spacing is workable), is gaining popularity with arable producers with grass and tap-rooted weeds, so that the crop can be hoed between the rows with an inter-row hoe. (*See* Chapter 10 for more information about this system.) Drilling on wide-row spacings provides an opportunity for post-emergence inter-row weeding, up until when the crop canopy closes and prevents further weeding. Where maximum competition is desirable and weeds do not present a major problem, rows should be no more than 20cm apart in peas; any wider than this, and weed competition can occur further into the season.

See Chapter 7 (Pulses) for more information on the establishment and suitable seed rates for beans, peas, lupins and soya.

UNDERSOWING

Broadly speaking, undersowing is a method of establishing grass, clover or another catch crop or green manure in the spring underneath another, shorter-term crop such as spring cereals or pulses for grain, whole crop or an arable silage crop.

If undertaken correctly, the undersown cover can help reduce weed competition (*see* Chapter 10) and provide an early, timely and cost-effective establishment of the following green manure cover. The crop already in place provides a protective microclimate (or nurse crop) for the undersown grass, clover or green manure during its establishment period, as well as a useful cash grain crop.

Undersowing offers the organic farmer an option to decrease the number of cultivations required. At the same time, the establishment of the grass and clover ley beneath another crop offers the opportunity to minimize the ingress of weeds on to the land, since increasing the level of plant competition below the crop canopy reduces their chances of establishment.

In principle, undersowing is best suited to areas of reasonably high rainfall in spring and early summer, thus avoiding any potential competition effect between the crops that could be caused by drought.

Choice of Cover Crop

The most popular choice of crops to undersow are spring barley, oats, wheat and triticale. In all cases, selection of a stiff-strawed, tall variety is critical. A crop that lodges (not common in organic systems) will smother the emerging cover beneath, and a crop that is too short may be overtaken by the undersown cover beneath, especially in wetter seasons (*see* the picture below). Peas, lupins, black medic, trefoil and soya are not suitable to be undersown due to their low stature and uncompetitive nature. Beans can be undersown successfully.

The choice of undersown cover is also important. White clover, grass, black medic, trefoil, stubble turnips and forage rape are all suitable for undersowing into most spring cereal crops. Black medick and trefoil grow lower to the ground and are very shade tolerant and are suitable for undersowing in the spring into winter and spring field beans (*see* Chapter 10 for more details). Red clover can be undersown into taller crops (oats, triticale and suchlike) but caution is needed in shorter crops such as barley, because experience has demonstrated that with the very competitive nature of red clover, the clover can grow taller than the barley.

The crop is traditionally combined for grain, but the increasingly popular practice of producing whole crop silage is certainly more beneficial to the undersown ley – the principle being, the earlier the cover crop, the better it can be taken. Early removal of the cover crop improves the

A short crop of spring barley undersown with red clover – the clover overtaking the barley.

Field beans undersown with black medick (Medicago lupulina).

growing conditions for the grass and clover at an earlier stage, and encourages a more robust development at an earlier stage.

Arable silage mixtures (usually combinations of barley or oats with forage peas) can also make a useful cover crop and a bulky, protein-rich food, but sowing rates should not exceed 60kg per hectare, with subsequent lower yield expectations. Arable silage mixtures when not undersown can be drilled at up to 200kg per hectare.

Method of Undersowing

Ideally the cereal crop or arable silage crop should be drilled and then rolled with a Cambridge roller. The grass, clover or other catch crop or green manure should then be broadcast on with the comb harrow, whilst at the same time undertaking a weeding of the crop. The undersow should not be rolled as this will re-plant any disturbed weeds.

Ideally cereals should be undersown in the mid to late spring at GS 30 to 32. Beans can be undersown at the last possible weeding occurrence at 8 to 10 leaves when they are approximately 25–30cm tall.

Subsequent Management

When growing two crops at one time it can be necessary to lower one's expectations of either one crop or the other, and with undersowing, a poorer grain cash crop can lead to an exceptionally good undersow establishment; whereas a very thick, high-yielding grain crop can be too dense for a successful establishment of an undersow. Where a good cereal or pulse cash crop is expected it may pay to reduce seed rates slightly to allow a more open crop and a better establishment of the undersow. If this is undertaken it is important that the undersow is established in a timely fashion to avoid weed ingress.

In the right climate and with the correct management, undersown leys can offer a means of generating an extra crop from a piece of land with little or no detrimental effect on the establishment of a grass, clover or other catch crop or green manure.

When the cover crop is arable silage or whole crop taken reasonably early, the benefits of extra feed combined with a good grass establishment makes this technique very attractive indeed.

GRAZING CEREALS WITH LIVESTOCK

Often termed 'the golden hoof', and a traditional practice in some parts of the UK, sheep can be used to graze cereals during tillering (GS 25 to 30) to help remove weeds, consolidate soils, provide additional fertility (from dung) and stimulate crop tillering. Most cereals are suited to this, but not pulses. Provided this is undertaken when the crop is tillering and the soils are not too wet and the stock levels too high, there is little or no detrimental effect on the developing crop.

This practice is also useful for removing unwanted volunteer stands in the crop. Experience has shown that volunteer beans and oilseed rape can be effectively removed by sheep grazing a following cereal crop in the winter.

Chapter 10

Weed Management

'Weeds – nothing more than plants in the wrong place at the wrong time.'

PRINCIPLES OF ORGANIC WEED MANAGEMENT

Weed management is one of the greatest challenges in the successful production of organic cereals and pulses. Weeds compete for light, nutrients and water, they can spread quickly, seeds can remain in the soil 'seed bank' for many years and, once germinated, can slow or impede harvesting, and they can cause crop storage and marketing problems.

Weed control is primarily achieved by using rotations that contain a ley or green manure phase where a fertility-building green manure is repeatedly grazed or mulched. This prevents a build-up of weeds and further seed shed. Within the crop itself, management options include avoiding weed development opportunities, maximizing the competitive advantage of the crop, cultural techniques, and mechanical intervention in the form of mechanical weeding. The selection of crop types and varieties that have vigorous growth characteristics, especially during their early development stages, has been shown to help out-compete weeds.

Many non-organic farms have become accustomed to very 'clean' fields with the regular use of herbicide inputs, and many farmers like a clean crop for cosmetic reasons. But with weed populations largely removed, many of the beneficial insects and invertebrates also disappear as their feed sources and habitats are removed. However, while it is important to consider weed-seed return and its impact on the weed-seed bank build-up, it is often not cost effective to remove weeds where levels are very low, and a certain level of weeds is tolerable without impacting on yields. It should also be remembered that maintaining a low level of weeds in a field is important, as complete removal can destroy important habitats for natural predators that feed on aphids, which in turn can spread viruses and can encourage fungal spores to grow on the ear.

As in non-organic agriculture, there is an economic threshold beyond which the associated costs and crop damage from weeding simply outweigh the losses caused by weeds. Nevertheless, some weeds are more competitive, and different cultural control methods may therefore be necessary. Having said this, it is important to remember that some weeds are more competitive than others, and the cultural control of them may be quite different. For example, some weeds will germinate immediately after harvest, where others require a period of dormancy before they will germinate.

Previous knowledge of the weed spectrum and incidence of a site is invaluable. Where new land is to be farmed organically, farmers should assess the site carefully for existing weed species and density. Examining past agrochemical input records can be a useful strategy as to the weeds targeted.

Control of annual and perennial grasses and perennial broad-leaved weeds is most difficult. A mixed rotation with a range of drilling dates and crop types will usually present the most opportunities for weeding. The more a rotation is based on similar crop types sown and harvested at similar times, the greater the selection pressure on a specific weed spectrum. Farmers should consider this carefully when planning the rotation and cultivation options.

So we know that weeds at one level can be a threat to successful crop production, but are also key to the viability of the arable ecosystem. Is it therefore possible to achieve a balance?

UNDERSTANDING THE PROBLEM – WEED OR ARABLE PLANT?

Weeds are nothing more than plants in the wrong place. The weed could be an annual or perennial weed, or it could be from a previously grown crop of a different type or species growing in the cash crop – these representatives are often called 'volunteers'. Weeds are normally talked about in a negative context, and are associated with yield loss, moisture and fertility diversion from cropping, as hosts for pests and diseases, and as being a visual impairment on the farming landscape.

Weeds do, however, have many positive values. They can be a cheap green manure, and can be useful for soil protection (provided they do not seed or spread). Weeds can be very useful indicators of soil structural and fertility issues: for example, mayweed is a good indicator of acidifying soil or compaction. Some weeds as soil indicators are summarized in Table 46.

Weeds provide food and shelter for invertebrates, food for birds (seeds) (*see* Table 47), and the invertebrates living on the weeds are in turn food for many birds. The biological importance of a range of weeds is shown in Table 48. Some species, such as mayweed, have relatively little biological value; others, such as creeping thistle or common toadflax, are important for birds, flies, beetles and butterflies. These in turn are key to maintaining biodiversity on the farm. Many of the insects, birds and butterflies are natural predators for pest problems.

Table 46 Weeds as soil indicators

Weed	Alkalinity (high pH)	High nitrogen	Acidity (low pH)	Compaction
Groundsel		X		
Fat hen		X		
Docks		X		
Sowthistle		X		
Nightshade		X		
Field pansy	X			
Cornflower	X			
Poppy	X			
Charlock	X			
Pimpernel	X			
Corn spurry			X	
Mayweed			X	
Chickweed			X	
Speedwell			X	
Sorrell			X	
Buttercup				X
Coltsfoot				X
Mayweed				X

Table 47 The biological importance of weed species

	Birds	Hymenoptera		Beetles		Butterflies
		Adult	Larvae	Adult	Larvae	Adult
Creeping thistle	X		X	X	X	X
Weld		X		X	X	X
Small mallow	X	X				X
Mugwort			X	X	X	X
Common toadflax	X	X		X		X
Ground elder		X			X	X
Common nettle				X	X	X
Shepherds purse	X	X		X	X	
Cleavers	X			X	X	X
Prickly lettuce	X	X	X	X	X	X
Creeping buttercup		X		X	X	X
Vipers bugloss		X		X	X	X
Deadly nightshade		X		X	X	X
Scentless mayweed					X	X
Knotgrass	X			X	X	X
Chickweed		X		X	X	X
Hedge mustard				X	X	X
Dead nettle		X			X	X
Wild carrot	X	X		X	X	X
Hedge bindweed	X	X		X	X	X

Many of the rare arable weeds need protection to retain them as a national 'resource', especially those in arable systems. Some long-established organic farms are one of the few remaining areas where rare arable plants such as corn cockle and corn marigold still exist in small numbers.

Table 48 Number of seed-eating bird species with weed species as important food sources

Plant species	Number of bird species
Knot grass and redshank	12
Chickweed	12
Fat hen and related species	9
Charlock	7
Annual meadow grass	6
Mouse-ear	5
Groundsel	4
Field pansy	3
Wild oats	0
Cleavers	0
Dead nettles	0

Source: P. Lutman Rothamstead Research

So how should organic farmers balance the benefits of some weeds against the detrimental effects of others on production? The secret is to target only those weeds with low biodiversity importance ('bad' weeds), whilst learning to tolerate those weeds which are important ('good' weeds). These are summarized in Table 49.

Table 49 The 'good and bad' weeds in arable systems

Good species	Bad species	Both	? Acceptable
(not damaging + good for biodiversity)	(damaging and little value for biodiversity)	(good for biodiversity but competitive)	(can be damaging but also valuable for biodiversity)
Mouse-eared chickweed	Cleavers	Docks	Chickweed
Groundsel	Wild-oats	Thistles	Meadow-grass
Field pansy	Black-grass		
	Barren brome		

Source: P. Lutman Rothamstead Research

CULTURAL CONTROL OF WEEDS

Crop Rotation

Crop rotation planning is a cornerstone of organic farming practice, and it has important implications for weed management. Rotations can be designed to positively influence weed control and to make a useful contribution to the whole farm management system. Typically rotations extend over several years, often with only an annual change of crop, but the inclusion of cover crops, intercrops and green manures increases the crop diversity in a rotation.

Weed population density may be markedly reduced using crop rotation. The success of a rotation depends on the use of crop sequences that create a diverse pattern of competition, allopathic interference, soil disturbance, and production needs (such as the time of sowing and harvesting). There should be regular changes between spring- and autumn-sown crops, between dense leafy crops and those with an open habit, and between crops that require a long growing season and others that mature quickly. Rotation may also allow the use of a range of cultivations and mechanical weeding methods that may be applicable to the different crops. The aim is to provide an unstable and inhospitable environment that prevents the proliferation of a particular weed species.

Green Manures and Fertility-building Leys

The fertility-building period, or ley, will influence the weed population. If it is well established and managed, it can act as a weed-suppressing phase. Many organic arable farms use clovers, vetches and similar green manures, or a combination of these with grasses (*see* Chapter 3). It is important to choose the right species and ensure they establish quickly, especially for mixed arable and livestock systems, where a ley may last for between two and four years. Establishment of leys can be easier in the autumn period than in the spring because sowing in spring coincides with the main spring flush of weeds. The seedbed needs to be well prepared, firm and fine, so as to achieve a good contact between seed and ideally a moist soil to achieve good establishment.

The choice of fertility-building crop is also important. Rotations with grass and clover leys have been shown to be beneficial in reducing weed seed numbers compared with rotations that do not include a grass phase. Research in Scotland (Table 50 and Table 51) show that the larger the percentage of arable crops in the rotation and the lower the percentage of grass, the greater the weed populations. Grassland systems, which have temporary leys rather than permanent pasture, will provide the opportunity to control perennial weeds during the cultivations between ploughing and reseeding.

Table 50 The effect of percentage of arable crops in preceding years on weed seed bank populations (No. seeds m^{-2} to 20cm depth)

% Arable Crops	Tulloch	Woodside
0	16,800	–
25	32,875	19,200
50	29,360	38,867
75	56,700	44,967

(Younie et al., 1996)

Table 51 *The impact of grass in the rotation on mean numbers of weed seeds m^{-2} to 20cm depth over four seasons*

	Year 1	Year 4
Jamesfield		
Conventional rotation*	5,710	12,167
Rotations with grass	26,092	17,782
Rotations with no grass	25,276	42,141
Woodside		
Conventional rotation*	10,500	16,000
Rotations with grass for <2 years	22,140	45,857
Rotations with grass for >2 years	21,688	40,438
Rotations without grass	29,500	153,999

*Conventional arable rotation, with herbicides
Source: Davies et al., 1997

The Cutting and Mulching of Leys and Green Manures

Cutting and topping weeds will have an impact on the type of weed flora in a field, and can be invaluable in preventing the return of weed seed to the soil seed bank. Cutting and topping are important for weed management in grass, clover leys and in some green manures. Topping can also be used as a remedial measure above some cereal crops to prevent weeds from seeding – *see* Weed Surfing on page 285.

Cutting the grassland ley or green manure components of an arable rotation for hay or silage will have an impact on the weed flora. Silage tends to be cut early in the season when the sward is young and fresh, whilst hay is cut at a later stage. There can be both advantages and disadvantages associated with the timing of cutting, depending on the weed flora and the ultimate requirements of the system. Cutting late may allow weeds in the pasture to grow to maturity and set seed. The ripe seeds may contaminate the hay and remain viable when passed through livestock. Dock seeds should not survive low pH silage, however they will survive in a later cut of hay. This mature seed may also shed on the ley surface and find opportunities to germinate in situ, or be transported by livestock to other locations. In contrast, cutting early for silage in fields with, for example, an infestation of creeping thistle, may encourage the spread and growth of this weed. Hence there has to be a balance between the requirements of the farming system and weed control implications.

Research undertaken by ADAS on the effect of mowing a legume fertility-building crop on shoot numbers of creeping thistle suggests that achieving and maintaining a dense competitive ley has more influence than mowing frequency on thistle survival.

Spring-sown leys and green manures should be cut no later than late August to allow recovery before winter, and a good level of competition against weeds. Summer- and autumn-sown leys and green manures should be left unmown until November or the following spring. Where leys and green manures are grazed but a proportion are not eaten by livestock, the field will need to be topped to prevent seed shed.

Where leys and green manures are not grazed, in stockless arable systems, the management of the biomass may include topping at intervals during the summer to a height of around 10–15 cm. Ideally in fertility-building leys the sward should not be allowed to get higher than 40cm (or knee height). If the vegetation gets higher than this, then topping will create a mat of vegetation that will act like a mulch. This can create dead spots in the ley where clover may be prevented from re-growing by the mulch, or which weeds may colonize. Topping the ley regularly will also ensure that tall weeds that may have germinated will not be able to set seed.

The effect of topping and mulching versus cutting and removal of the biomass on nitrogen fixation, and balancing this with weed management, is discussed at greater length in Chapter 3.

Grazing Regimes for Weed Control

In mixed farming systems with crops and livestock enterprises, where the grass/clover leys are used for fertility building, livestock can make good use of the nutrients and they also produce manure, a resource which can be used around the farm to fertilize cash crops. Most green manure and ley components can be grazed by livestock to some extent or another, including clovers, trefoils, grasses, cereals, peas, vetches and so on. Animals can also be used to consume cut weeds or other plant material such as chaff or screenings that are likely to contain some weed seeds.

Different animals have different grazing habits, and it is well recognized that different breeds or individuals are likely to have different tastes and habits. The species, breed, age and individuality of animals will all affect what they will eat and therefore what effect they will have on both weeds and pasture. Variability within the feeding site (for example vegetation, topography) can also be important, as can other factors such as the weather.

Some general variations to consider include the following:

- **Sheep** are recognized as being useful for weed control as they graze close to the ground and will eat a wide range of plants. Their mouth action is to nibble at forage and weeds. They can be used early, and some breeds are very winter hardy. Sheep can be grazed on crops during the winter and early spring to remove weeds, without any significant damage to the crop.
- **Cattle** can be used for early grazing, but there are a large number of different breeds and types with different grazing requirements, including beef, dairy and traditional breeds. Cattle are only suitable for grazing leys and green manures and not in crops due to their size and weight, and the damage this causes to crops. Grazing strategies appear to be related to plant energy content and digestibility, and this will affect *how* plants are eaten (the leaves or stems or other parts of plants), *which* plants are eaten (species), and *the size* of plants eaten (young or more mature plants). Cattle graze by wrapping their tongue around forage

and pulling it out, and they tend to avoid longer, coarser grass and hairy, spiny or poisonous plants for this reason. The selection of certain plant species and plant components, as well as the location of these plants, is based on the previous experience of the animals, or learned from their mothers when they are calves.

- **Pigs** are good at rooting, and have been recommended at various times for digging out perennial weeds such as dock and couch when fenced within fields. This is best achieved by stocking heavily, but moving them on a frequent basis. As a general rule, sows and boars will dig and root out weeds better than young pigs and fattening pigs.

It is important to get the right grazing balance over the year to obtain the maximum benefit for the animals and also to prevent damage to the sward or soil. For example, stocking more lightly in the winter months and in wet periods prevents poaching and soil damage. Many farmers do not realize that the pressure exerted under the hoof of a sheep is greater than that exerted by a combine harvester, which spreads a heavier weight over a larger 'footprint' area. So it is important to think about the right season to graze, how long to graze, how many animals to graze and how long the grazing area will need to recover. Things to consider include the following:

- To benefit the ley or green manure, and promote competition with weeds, time grazing when weeds are most competitive, during early crop development.
- Allow time for the ley or green manure to recover between grazings.
- Make sure that livestock that have been grazing on weedy land, feed on weed seed-free forage for four to five days before introducing them to any weed-free areas of ley or green manure, or the livestock may infect clean areas with weeds from dunging (some seeds will remain viable after passing through animals, which may take a few days).
- If possible use electric fencing to manage the grazing and move stock across fields in systematic paddocks, rather than letting them have free access to the whole field. This will control grazing pressure and weed management better.

Suggestions for rotating livestock grazing, depending on situation, include the following:

- Alternate the grazing of sheep and cattle from year to year, or use mixed grazing for better weed control. Research at the Scottish Agricultural College (SAC) has shown that mixed grazing in the same field may be detrimental to the health of the cattle.
- Exploit animals' different grazing habits in a rotation. For example, sheep may graze out certain weeds such as ragwort, and have been used to graze out grassy weeds in legumes or bean volunteers in cereals.
- Free-range outdoor pigs forage for roots. This might also help to reduce the perennial weed burden.
- Lightly graze young crops of oats, wheat and other cereals with sheep to encourage tillering and hence improve weed suppression.

Managed grazing of a red clover ley with cattle.

Crop Choice and Sequencing

The length of the rotation, and the choice and sequence of crops will depend upon individual farming circumstances: these will include factors such as soil type, rainfall, topography, enterprises, pests, disease pressure and weeds. However, the aim is to produce an unstable environment in which no single weed species is allowed to adapt, become dominant, and therefore difficult to manage. No one rotation can be recommended, but ideally in terms of weed control, rotations should include:

- alternation of autumn and spring germinating crops;
- alternation of annual and perennial crops (including grass and clover);
- alternation of closed, dense crops such as oats, triticale and rye which shade out weeds, and open crops such as barley or peas which encourage weeds;
- a variety of cultivations and cutting or topping operations that directly affect the weeds.

Grain crops such as rye, oats and triticale that develop rapidly, tiller vigorously and have a tall stature, are significantly more competitive against weeds than the majority of modern wheat varieties that have been bred as dwarf varieties with a better harvest index to produce more grain and less stem. This is shown in the diagram on page 238.

The sequence of cereal crops in the rotation has an important influence on weed competition, as does a mixture of autumn- and spring-sown crops within the rotation, which allows mechanical control strategies to be used on different weeds, or on the same weed at different development stages. This is demonstrated in Table 52, where the weed dry matter at harvest increases with the length of the rotation.

Table 52 Weed dry matter at harvest (g m⁻²) as affected by course of rotation

Course of rotation	Weed dry matter (dm, g m^{-2})
RC/WW	151
RC/WW/WW	178
RC/WW/WW/SO	115
RC/POT/WW	115
RC/POT/WW/WO	79
RC/WW/WBN/WW	129
Significance	

Note: RC = red clover, WW = winter wheat, SO = spring oats, POT = potato, WO = winter oat, WBN = winter beans
Source : Bulson et al., 1996

As a general rule, uncompetitive crops such as wheat or peas should be grown at the start of the rotation after a ley, and competitive crops such as oats, triticale and rye should be grown at the end of the rotation where there is less fertility and more weed pressure. Field beans are a bit of an anomaly, as they are very competitive early on, with a dense leaf structure, but very uncompetitive late on once leaves are lost. Strategies for weed management in beans are discussed later in this chapter (*see* page 288).

Various suggestions and observations for crop choice and sequencing include the following:

- Put sensitive crops after clover leys. Research has shown that in the third cropping year after a grass/clover ley there is twice as much weed emergence as compared to the first.
- Include a row crop such as sugar beet or potatoes in the rotation to permit the use of one or more cultivations to kill emerged weeds, and encourage the germination of others, so reducing the soil seed bank and hence potential weed numbers in future crops. Cultivations may also reduce the problem of perennial weeds by disrupting growth and smothering regeneration in the growing crop.
- Uncultivated leys provide a completely different habitat for weeds, and may be used to reduce or eliminate particular weed species via inducing dormancy. Few studies have been made of the effectiveness of leys for controlling weeds, but trials suggest that there is little advantage for weed management in leaving leys down longer than three years. The species' composition, and the mowing and grazing regimes, are important in weed dynamics. Management of weeds at the time of ley establishment is critical, as is the method of ending the ley to avoid a flush of weeds due to the release of seed dormancy by cultivation. A greater proportion of ley in the rotation usually results in lower seed numbers in the seed-bank in comparison with arable crops. It was a traditional way to deal with land infested with wild oat, but does not eliminate the weed completely.
- Where a long grass/clover break does not form part of the rotation, weed problems are likely to be more severe. The problem will be greater

where less vigorous and therefore less competitive crops are grown. Among the cereals, oats, spelt and winter rye are the most competitive, followed by triticale and then wheat.

- Canopy development and shading are important for weed suppression, and choice of cultivar can influence this (*see* Chapter 8).
- A higher seeding rate and narrower row spacing can increase the level of weed suppression.
- Allopathy may play a part in weed suppression.
- The time of planting will determine the species' composition of the weed flora likely to emerge following seedbed preparation. Spring-emerging weeds will predominate in spring-sown crops, autumn-emerging weeds in autumn-sown crops.
- The slow or delayed release of nitrogen (characteristic of organic systems) is likely to favour the crop over the weed. Weeds will always be more responsive than crops to high levels of nutrients or fertilizer.

Crop and Variety Choice for Weed Control

Varieties that consistently suppress weeds are generally more desirable in organic systems (although this might be outweighed by marketing necessities) as opposed to varieties that do not complete with weeds (and which potentially allow weeds to develop and return seeds to the soil seed bank) – as shown in Figure 24.

Selecting crop types and varieties with vigorous growth characteristics, especially during their early development stages, can help out-compete weeds. Grain crops such as oats and triticale, which develop rapidly, tiller vigorously and have a tall stature, are significantly more competitive against weeds than the majority of modern wheat varieties that have been

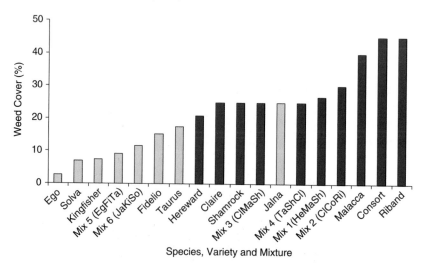

Figure 24 The completeness of different cereal types against weeds

Source: Davies and Walsh 2002

bred as dwarf varieties. Peas, lupins and soya are very uncompetitive and should not be grown where weed levels are likely to be a problem.

When relying on the competitive characteristics of a wheat variety for weed suppression, it is beneficial to choose varieties that develop rapidly, tiller strongly in a prostrate fashion, and provide good ground cover during early development (*see* the photo below). Varieties which stay in their vegetative state longer and tiller more prior to stem extension are suitable, as this provides maximum ground cover and weed competition, as shown in the photos on pages 240 and 241. As crops develop, competition relies on crop height and architecture, with taller varieties having a better competitive advantage (Figure 25), with planofile, wide, flat leaves providing more ground shade than thin, erect leaves, resulting is less weed germination and competition.

Varieties used for organic production should show quick germination and establishment, rapid and vigorous early growth, and the ability to rapidly cover the soil and to shade it (as happens with prostrate or tall varieties) to out-compete weeds at as early a stage in the crop cycle as possible. Varieties are well known to differ in architecture and competitive ability, and whilst those that out-compete weeds are preferred, it should also be borne in mind that those with erect foliage, or that can tolerate some degree of mechanical weeding, are also likely to be useful (*see* notes on variety choice in Chapter 8). Figure 25 shows that a mixture of crop height and architecture increases crop competition, with taller varieties having a better competitive advantage, with planofile, wide, flat leaves providing more ground shade than thin, erect leaves, resulting in less weed germination and competition.

Wheat that tillers strongly in a prostrate fashion provides good ground cover during early development.

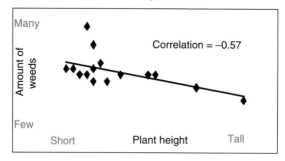

Figure 25 Increased plant height can reduce the competition from weeds

In the grass/clover phase of the rotation, the choice of variety may be dominated by forage value, but if there is opportunity the most vigorous species should be selected, as these will determine the productivity of the whole ley period.

The trend in organic cereal production is to grow the taller-stemmed varieties for their weed-suppressing, shading ability. Some farmers have stayed with the shorter-stemmed varieties widely used in non-organic systems and employed a weed topper/cutter that will remove and, ideally, collect weed seed-heads above the crop. This works as long as there is a difference in height between crop and weed (*see* information about 'weed surfing' later in this chapter). New research in wheat is investigating leaf angle development, and the height and speed of development on weed suppression to aid the farmer in his variety choice.

Prostrate – strong tillering variety.

Erect – weak tillering variety.

Seed Rate and Crop Spacing

There has been much work in cereals on row spacing, pattern, direction of sowing and seed rates (typically 10 to 15 per cent higher in organic cereals). Results are varied, and interactions between the factors are often significant.

The spatial distribution of the young crop, its rooting system and subsequent canopy foliage, is important for weed suppression. In drilled crops the proximity of the plants to one another will determine the competitiveness of the plant stand as a whole. The principle is that the greater the amount of space taken up by the crop in the rows, the less space there is available for the weeds to invade. However, it should be borne in mind that closely spaced crop plants compete with each other and that it is also impotant to allow sufficient space between plants to allow for efficient mechanical weeding should weeds develop and threaten the crop.

Seed rates tend to be higher for organic than conventional crops. There is also an allowance for potentially lower germination rates and loss of the crop by predation and from mechanical weeding operations. High seed rates reduce weed numbers, whereas low seed rates encourage weed development, as shown in figure 26. The challenge is to find the threshold at which the additional cost of seed just outweighs the detrimental impact of weeds in any particular crop.

Most farms grow crops cereal and pulse crops on a traditional crop row spacing of 10–14cm, or occasionally they broadcast the seed and incorporate, as with field beans. The production of cereals and pulses on wide rows, typically on 20–25cm spacings (but up to 30cm spacing is workable) is gaining popularity with arable producers on both heavier land and land where grass and tap-rooted weeds are a problem, so that the crop can be hoed between the rows with an inter-row hoe. (*See* page 279, later in this chapter, for more information on inter-row hoeing systems.)

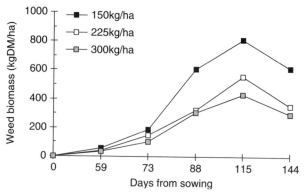

Figure 26 The effect of oat seed rate on weed development

Source: Younie & Taylor, 1995

The use of inter-row weeders facilitates mechanical weed management on larger weeds, on heavier, damper soils, and on weeds that are little affected by spring-tined or comb harrows – for example grass and tap-rooted weeds such as wild oats, brome, charlock, poppy.

When cereals and pulses are grown on wider rows, they grow and behave in very different ways compared to being grown in a narrow row or broadcast system. The same seed rates are typically used per unit area for narrow and wide-row systems. On the wide-row system, twice the seed rate is placed in half the number of crop rows. The closely spaced crop plants then try and compete with each other, growing tall and very competitively. Experience suggests that cereals grown on wide rows tiller less and grow in a more erect fashion. This has a benefit in providing very good competition in the crop row, but leaves the inter-row area more open to weed development. The system will operate well, provided that the inter-row area is hoed on a routine basis with a mechanical inter-row hoe. Where

Cereal grown on a wide-row system showing reduced tillering and erect growth.

	Row Width			
	Narrow	**Medium**	**Wide**	**Mean**
E- W Sowing				
Pegassos	8.4	20.6	16.3	15.1
Eclipse	18.1	19.1	26.9	21.4
Consort	20.3	17.8	41.6	26.6
Rialto	13.4	14.7	23.1	17.1
			Overall	**20.1**
N- S Sowing				
Pegassos	13.0	12.8	6.9	10.9
Eclipse	28.4	24.5	24.5	25.8
Consort	14.6	12.3	19.1	15.3
Rialto	18.0	12.4	16.3	15.6
			Overall	**22.5**

Figure 27 WECOF trial results showing impact of N/S and E/W orientation of cereals on per cent weed ground cover under wheat varieties at GS 49

Source: Younie & Taylor

mechanical weeding is not possible or does not take place as planned, the crop can end up weedier than a narrow-row or broadcast system due to greater light penetration and weed growth in the inter-row area.

A limited volume of work has been undertaken concerning the direction of sowing of cereal crops as part of the EU-funded WECOF trials in three countries, the hypothesis being as follows: that when sowing in a north/south direction, sunlight will always hit all areas of ground under the crop at some point in the day as the sun travels from east to west. But when planting in an east/west direction, an element of shading will result from the crop, which may help with weed suppression. The findings found that in narrow-width rows an east/west sowing was favourable, whereas in wide-row widths a north/south sowing showed better response and fewer weeds. Some of the results of this work are shown in Figure 27.

Seed Size
Crop types and varieties with a larger seed size have been shown to exhibit greater initial vigour of emergence and growth, which may subsequently provide extra competitive ability. If there is a choice available, then the most vigorous species should be selected, as these will be more likely to out-compete weeds and suppress their development.

Organic Seed
Organic certification requires that organic seeds are used for planting and growing organic crops. Experience suggests that organic seeds, which are untreated, tend to have more rapid emergence and good early vigour, but that higher seed losses can be expected (from predation) than in non-organic systems. To compensate for losses, higher seed rates are typically used.

Clean Seed
It is important that the crop seed is free from contamination by weed seeds. Organic farmers are obliged to use organically produced seed, and

this should be clean. It is important if saving seed that it is taken from weed-free crops, and ideally, professionally cleaned. Tolerance levels of contamination should be low, although they are not generally well defined as of yet.

Crop Establishment

The ability of the crop to get off to a good start ahead of the weed flora is critical. Good soil management practices are important to provide the best possible seedbed in which to plant a crop. The impact of a poorly compacted soil or a rough, uneven seedbed can soon be seen on crop establishment and subsequent weed invasion. Rough seedbeds also provide a habitat for soil-dwelling pests such as slugs, which can have a significantly detrimental impact on establishment.

Using a high quality seed drill can also be of benefit, as seed placed evenly spaced at an even depth will emerge uniformly and provide a greater competitive advantage than seed that emerges unevenly. Uneven seed emergence is also more prone to predation by birds, especially rooks and pigeons.

Ensuring that crops are drilled to a high quality is one of the best ways to get the crop off to a good, weed-free start. Growers should regularly check the drill for blockages and keep a keen eye out for missed areas, and if necessary re-drill these areas immediately.

Areas that remain undrilled, or where drill bouts have not been well matched, or where drill misses occur, rapidly fill up with weeds and are difficult to control. Often these become a source of weed infestation in subsequent years and can be highlighted in subsequent crops as weeds areas. Where drill misses occur you should aim to:

- re-drill the missed area as soon as possible;
- drill another variety or crop type in the missed area to smother weeds (sacrificing if required);
- sow clover in the missed area to provide ground cover to smother weeds and build fertility (often this can be done by hand).

Areas where crop drilling is missed and where drill blockages occur, rapidly fill with weeds, and should be avoided wherever possible.

Cover Crops

Cover or break crops in the rotation can be important for weed management. The main purpose of a cover crop is that of nutrient management – absorbing nitrates from the soil to prevent them leaching, and making them available to the subsequent crop. (More information on cover crops is contained in Chapters 3 and 4.)

Many green manures and cover crops are quick to germinate and form a dense ground cover, so they suppress weed emergence. The inclusion of cover crops in the rotation, at a time when land might otherwise lie uncropped, will suppress weed development while maintaining soil fertility and preventing erosion. Some stockless systems rely on short bursts of green manures in addition to shorter-term leys for fertility, thereby losing some of the benefits of a longer weed-suppressing phase. This means that it is essential to achieve good establishment of a dense crop in order to prevent weed problems.

Some cover crops may also exhibit allopathic properties that can have an inhibitory effect on surrounding plants (*see* 'Allopathy' below, page 246). For example, when grazing rye has been incorporated and is breaking down, it has been shown to reduce the emergence of weed seedlings in a subsequent crop. The same is well known regarding the allopathic properties of vetch as a green manure, in that when incorporated, exudates from the roots inhibit seed germination for a period of three to four weeks. This should be borne in mind when using vetch, because planting a cereal straight after incorporation may result in the cereal (or other) seed not germinating. Seeding should be delayed by three to four weeks until this effect has diminished. Horticultural growers use these properties to their advantage by planting pre-germinated plant 'plugs' or 'modules', giving them a few weeks grace ahead of weeds germinating, as the allopathic effect works on weeds as well as planted crops!

Vetch green manures have allopathic properties.

Some other decomposing cover crop and green manure residues release allelochemicals that inhibit the germination and development of crop and weed seeds. Lucerne residue exhibits a strong phytotoxic effect on a range of crops, including the germination of its own seeds; however, it improves the growth of some vegetable crops, though not cereals or pulses. Residues of ryegrass (*Lolium rigidum*) and subterranean clover (*Trifolium subterraneum*) cover crops have been shown to reduce seedling growth of lettuce (*Lactuca sativa*), broccoli (*Brassica oleracea var. italica*), and tomato (*Lycopersicon esculentum*), but thankfully these residues do not affect cereals or pulses. The residues of autumn-sown grazing rye sown as a green manure and incorporated in the spring, have been shown to reduce weed biomass by over 60 per cent as a result of its allopathic properties, in addition to the physical effects of competition.

There are a few disadvantages of using cover crops: they may affect the seedbed preparation for following crops, and could act as a source of infection to those crops where like species are used. For example, avoid using a grazing rye between cereals, as this would act as a green bridge for cereal stem-based diseases such as take-all.

In addition, cover crops that develop a high carbon-to-nitrogen ratio (C:N ratio) could reduce the yield of following crops by diverting microbial activity and soil nitrogen to breaking down the residues, rather than feeding the cash crop (*see* Chapter 4). The response of following crops will depend on the species involved.

Allopathy

Allopathy is a biological control characteristic exhibited by some plants, whereby the allopathic plants are used to reduce the vigour and development of other plants. Allopathy refers to the direct or indirect chemical effects of one plant on the germination, growth or development of neighbouring plants. This can be through the release of allelochemicals while the plant is growing, or from plant residues as it rots down, as shown in the figure below. These chemicals can be released from around the germinating seed, in exudates from plant roots, from leachates in the aerial part of the plant, and in volatile emissions from the growing plant. Both crops and weeds are capable of producing these compounds, and in this case the desired effect is the impaired germination, reduced growth and poor development of weeds.

Potentially allopathy can be used in various ways:

- To manipulate the crop-weed balance by increasing the toxicity of the crop plants to weeds, thereby reducing weed germination in the direct area of the crop, which is the most difficult area to control physically.
- As cover crops to suppress weed germination and development over a whole field in part of a rotation.
- As mulched residues or incorporated residues, which could prevent weed germination and allow transplanted crops to be grown, producing a residual weed control effect.

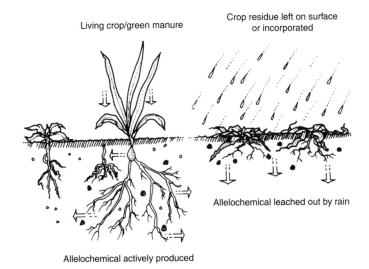

Living crop/green manure

Crop residue left on surface
or incorporated

Allelochemical leached out by rain

Allelochemical actively produced

Figure 28 Allopathic compounds can be produced by some crop and weed species

Source : HDRA

Many crops have been reported as showing allopathic properties at one time or another, and farmers report that some crops such as oats seem to *clean* fields of weeds better than others. The current list of crops that are known to exhibit some level of allopathic property includes wheat, barley, oats, cereal rye, brassicas, red clover, yellow sweet clover, trefoil, vetch, buckwheat, lucerne, rice and sorghum.

Several weed species have also been reported to show allopathic properties: they include couch grass, creeping thistle and chickweed. Where they occur together they may have a synergistic negative effect on crops.

Allopathic effects might also depend on a number of other factors that might be important in any given situation:

- **Varieties:** There can be a great deal of difference in the strength of allopathic effects between different crop varieties.
- **Specificity:** There is a significant degree of specificity in allopathic effects, so a crop that is strongly allopathic against one weed may show little or no effect against another.
- **Autotoxicity:** Allopathic chemicals may not only suppress the growth of other plant species, they can also suppress the germination or growth of seeds and plants of the same species. Lucerne and vetch are particularly well known for this, and have been well researched. The toxic effect of wheat straw or rye residues on following wheat crops is also well known.
- **Crop-on crop effects:** The residues from allopathic crops can hinder the germination and growth of following crops as well as weeds, and a sufficient gap must be left before the following crop is sown. Larger-seeded crops are affected less, and transplants are not affected.
- **Environmental factors:** Several factors impact on the strength of the allopathic effect, including pests and disease, and especially soil fertility.

Low fertility increases the production of allelochemicals. After incorporation the allopathic effect declines fastest in warm, wet conditions, and most slowly in cold, wet conditions.

There have been suggestions that allelochemicals could be isolated and form the basis of 'natural' herbicides. For example, the 'Interceptor' product, which has just received organic certification in New Zealand, is based on an extract from pine needles. Their suitability for use in organic systems in the EU would require discussion and confirmation.

Although crops can be chosen or selected based on their reported ability to produce allopathic chemicals, it is difficult to separate the effects of competition (for light, water and/or nutrients) from allopathic effects in the field. For this reason some researchers doubt the importance of allopathy in practical terms.

For day-to-day crop management it is important to remember that cover crops can provide effective weed control whether allopathic or not. It is of less importance whether a weed-suppressing effect is due to allopathy or competition, as it is the practical outcome that is important. When using cover crops, aim to choose vigorous species and varieties for maximum weed control, but remember that timeliness of establishment is vital, especially for winter cover crops.

Cash cereal and pulse crops vary in their weed-suppressing abilities. Choose strongly suppressive crops such as rye, triticale and oats in the rotation to balance weakly suppressive ones such as peas, lupins and wheat. Row crops such as potatoes can provide a useful opportunity for weed management within the crop. Crop varieties vary slightly in their weed-suppressing ability, so be aware of this and make use of it where appropriate.

When using a crop – especially a green manure, which may be allopathic – leave a gap after incorporation before planting the next, especially with small-seeded crops.

WEED MANAGEMENT BETWEEN CROPS

In arable systems there are often opportunities for some cultural and mechanical weed management between crops. Typically this is between harvest in the mid- to late summer and before sowing the next crop in the autumn or following spring. Where weed problems are known or suspected, it can help to programme an early-maturing crop such as peas, barley, oats, rye or spelt which is harvested in mid-summer, which then leaves a period of four to eight weeks of warm, dry weather where cultivations can be used to help manage weeds. Better still, programming the following crop as a spring crop, preceded by a late-sown green manure, takes the pressure away for crop establishment and allows the farm to focus on weed management. A programme similar to the following could be used:

- **Early August:** harvest oats and undertake stubble cultivations.
- **August and September:** repeat stubble cultivations on a seven to ten-day basis for weed control.

- **Mid September:** sow vetch, phacelia and rye over winter green manure.
- **March:** sow spring wheat or triticale.

When undertaking weed management cultivations between crops, it is important to use the right machinery for the job in hand. All too often farmers use inappropriate tines weeders, which are suboptimal in performance. Experience shows that using cultivator tines with small straight points for weeding cultivations results in many weeds passing between the tines unaffected, with perhaps only 50 or 60 per cent of the ground covered actually being in contact with them.

One improved option is to use a cultivator with large 'A' blade or sweep tines, which are placed to provide an overlap. This may require putting extra legs on to the cultivator, but in this way the maximum area of ground is covered and moved with every pass. The 'A' blades or sweep tines also help to lift and separate the weeds from the soil on to the surface for desiccation.

Two cultivators from continental Europe have recently become available in the UK: the CMN and Kvik-up harrows that use an improved approach for weed management in stubbles. Their similar but slightly different designs combine the use of large sweep tines on heavy legs to first loosen the soil, followed by a powered horizontal rotor fitted with spring tines that can be raised and lowered, and which rotates at twice the forward speed. The spring tines 'flail' the loosened soil to a depth of 4cm to 15cm, throwing soil and weeds into the air. Using gravity, the heavier soil falls first and the weeds fall on top of the soil surface where they desiccate and hopefully die. The machine is relatively expensive and slow to use, but one pass will generally do a better job than two or three passes with a normal stubble cultivator, and is far superior to using a spring tine fitted with straight-tined points.

A tines cultivator with straight points – suboptimal for weed management cultivations.

A cultivator with large sweep tines, useful for weed management cultivations.

Kvik-up cultivator with large sweep tines and powered rotor, very effective for weed management cultivations.

FALLOWING AND BASTARD FALLOWING

It is usually not desirable to have to plan a fallow period into a rotation, but it may be necessary if weeds cannot be controlled during cropping, fertility building or in-between cash crops. Tillage without a crop for a season is sometimes referred to as a 'fallow'. It may not be necessary to stop cropping for a whole year, but instead to employ a *bastard* fallow, meaning no crop for part of the year; this can be as effective as a full fallow, is more suitable for lighter land, and can be fitted into most rotations using similar methods as described in the previous section.

Fallowing is often best during the summer when cultivations can take place and the drier periods allow for root desiccation. This technique is more useful in plough-dominated systems. One aim is to cultivate the soil progressively deeper over time, exposing underground plant parts to desiccation at the soil surface, although in this case dry weather conditions are essential. Ploughing begins in June/July allowing time for an early crop to precede it. A bastard fallow is often used after a ley to reduce perennial weeds before sowing a winter cereal. There is also an opportunity for birds to feed on soil pests such as wireworms and leatherjackets, exposed during soil disturbance.

Fallowing has been shown to reduce perennial weeds within a rotation. The aim is to kill the vegetative organs of the weeds by mechanical damage and desiccation. For a full or bare fallow, heavy land is ploughed in April to give the weeds time to start into growth. It is cultivated or cross-ploughed ten to fourteen days later to produce a cloddy tilth. The soil is then cultivated or ploughed at frequent intervals to move the clods around and dry them out. This is best undertaken by removing the skims from the plough to get full inversion, rather than a mixing of residues and weeds with the inversion. By August the clods should have broken down and the soil is left to allow the weed seeds to germinate. In September/October the weeds are ploughed in and the land prepared for autumn cropping. If a cereal is to follow the fallow, wheat bulb fly may be a problem because it lays eggs on bare ground in July. This can be overcome by sowing a green manure such as mustard to cover the land during this period.

Although there is the benefit of reduced weed control costs in subsequent crops after an effective fallow, the economics of taking land out of production for a full year, together with the undesirable effects on the soil and the environment, make the use of a bare fallow a last resort for weed control in the organic system. There is no financial return during the fallow period, while labour costs accumulate during the fallowing operations. As an alternative to fallowing, the farmer could try using a combination of cultivations between crops (*see* previous section), green manures (*see* Chapters 3 and 4) and cleaning crops such as potatoes.

Tramlines

Organic farmers should avoid using tramlines to aid with weed management. In organic arable systems there is little need for tramlines as the frequency with which machinery is used in a growing cereal or pulse crop is far less, if at all. Experience shows that where tramlines are put into organic crops, they fill up with weeds quickly as a result of the lack of crop competition.

If tramlines are used, the farmer should consider sowing the tramline in the crop with a beneficial green cover such as clover or trefoil. In this way the sown ground cover will prevent weeds establishing, improve soil structure and still allow trafficking by machinery. The use of tramlines is also discussed in Chapter 9.

Non-Crop Area Management

Areas of the farm that are not cropped need to be managed to prevent the spread or invasion of weeds. Weeds growing on wasteland or old manure heaps around the farm should be prevented from seeding, although hygiene may have to be balanced with biodiversity needs, such as encouraging beneficial insects on particular flowering weeds. Many farmers rigorously top or remove weeds like dock from hedgerows and fence lines to prevent seed moving into the field.

Most weed infection sources are from field edges and non-cropped areas, and managing these areas in a positive way can significantly help reduce weed problems on the farm. In its simplest form, non-crop areas should be managed by topping and/or grazing to prevent weeds setting and spreading seeds to adjacent crop land. More positive management can include the establishment of tussocky, competitive grass margins around field edges, which can be managed to prevent the ingress of weeds from hedge bottoms, field edges and non-cropping areas. These margins can provide excellent biodiversity over winter habitats for beneficial insects, especially when wild flowers and herbs are established in mixtures, and can also provide additional income to farms through the various conservation and environmental stewardship programmes operated regionally in the UK.

The other main benefit of these margins is that they can be cut just prior to harvesting a cash crop of cereals or pulses to allow the combine harvester clean access to the crop edge. Where crops are grown right up to field boundaries or hedges, the combine harvester is an excellent and efficient machine at harvesting the crop together with weed seeds and spreading the weed seeds further out in the field, which in turn become a weed problem. The best strategy is to avoid a weed problem altogether.

As with all farm management decisions, it is important to check that any control measures in non-cropped areas conform to any cross-compliance

Established tussocky grass margins around fields can help with weed management.

legislation or conservation or environmental management prescriptions for schemes the farm participates in.

FARM HYGIENE

Weeds are, by their nature, pioneers and almost impossible to eradicate once established. The best form of management is preventing their establishment in the first place. Weeds are easily spread between fields and between farms, and it is worth taking some trouble to try and prevent this with some basic hygiene measures. Ask yourself the following questions:

Have you got a system for detecting weeds early?

- Managing a particular weed will be easier if it is detected early and prevented from spreading.
- Ensure that all people who work on the farm, or who visit it, are alert to the possibility of spreading weeds and weed seed, and ask them to tell you if they notice any particular areas of weeds.
- Keep records of problem weeds and their spread or otherwise. A digital camera can be a useful tool to record the presence of weeds and monitor changes over time.

Is your farm machinery or your contractor's equipment spreading weeds?

- Weed seeds are easily carried in soil, crop residues and on machinery so these should be regularly cleaned down. Toppers, mowers and combine harvesters are particularly effective at spreading weed seeds. It is always best to clean machinery when transferring from field to field or farm to farm.
- If there is a serious weed infestation in a particular field, or machinery is moving through fields where weeds are flowering, then washing down machinery should be a serious consideration.
- Hygiene is particularly important at harvest time. In crops such as cereals, weed seeds may be scattered in the field or caught on the machinery and dislodged later some distance from the original source. It may be necessary to add screens to combines to catch weed seeds at harvest. Older models may already have these features.

Are your inputs spreading weeds?

- Any compost, manure or slurry that is spread on fields must be free from viable weed seeds.
- If you make compost on your farm you will need to turn the heap to ensure the temperature rises high enough to kill any weed seeds that may be present. The temperature needs to remain at 70°C for three days. Ideally compost or manure should be covered to prevent the introduction of wind-borne seeds, or at least kept weeded to ensure no seeds germinate and set seed in it.
- Slurry needs to be adequately aerated to make sure weed seeds are killed.
- Use cleaned seed for planting.

Are livestock causing poaching or allowing weeds to establish?

- Water and feeding points must be rotated to prevent excessive dunging or poaching in one area. Perennial weeds like thistles and docks thrive in high nutrient areas. If animals have been grazing on these plants viable seeds can pass through their systems and germinate when conditions are favourable.
- Try to avoid poaching and damage around areas such as gates.

Techniques to Avoid Weed Competition and Spread

The simplest and most effective way to deal with weeds is to avoid their occurrence in the first place. As with other potential problems, organic agriculture always seeks to avoid problems by developing management tools and avoidance strategies so that the problem remains at a low level of incidence or threshold. The following areas should be considered with regard to avoidance strategies for weed management.

As outlined above, the combine harvester is an excellent and efficient machine at harvesting the crop together with weed seeds. So what options does the organic farmer have to manage this problem?

Return Weed Seeds to the Field
Crops can be harvested with the combine harvester, and the weed seeds and chaff are then spread back on to the fields. Fields should be cultivated as soon as possible in order to germinate the weed seeds. Secondary cultivations are then required to destroy the weeds prior to starting cultivations to establish a subsequent crop. There is always the risk with this system that the weed seeds become a weed problem in the field.

Remove Weed Seeds from the Field
Crops can be harvested with the combine harvester. The combine harvester is set so as to adequately thresh the crop but with low enough concave, drum, sieve and fan settings so as to collect the chaff and weed seed along with the grain. These settings are obviously tricky. The grain, chaff and weed seed can then be transported back to the farm, where the crop has to be cleaned (and dried) and the weed seeds and chaff separated from the crop. The chaff and weed seeds are then normally destroyed (by burning) or they can be fed to livestock.

Chaff Cart System
This system was developed in Australia as a tool to manage herbicide-resistant annual ryegrass in non-organic farming systems. The combine harvester is fitted with a 'slinger', similar to a forage harvester, to collect and feed all the chaff and weed seeds, but not the straw, into the collection cart which is towed behind the combine. The chaff and seeds are either dumped in heaps 'on the go' during harvest and collected later, or are transported back to the farm with a cart changeover. The heaps are later either burned or fed to livestock as feed. In some instances the chaff has been baled using a large square

The combine is fitted with a 'slinger' to collect chaff and weed seed, but not straw, and feed it into the collection cart.

baler and sold for use in dairy cow feed rations. The 'chaff bales' have similar feed value to wheat straw without the need for chopping.

The farms that have adopted the chaff cart system have seen a notable decline in weed problems, especially grass weeds which often mature at the same time as cereal crops. The main advantage is that it removed the need to clean the crop. The main disadvantage is the need to tow a trailer behind the combine, so the system is best suited to large farms with big flat fields.

Since herbicides are never used in organic systems, there is every reason to expect that chaff and weed seed collection could be an effective tool for organic farmers. Seed collected at harvest with the chaff reduces the amount returned to the soil to germinate in the following year. This helps with the production of subsequent cash crops by reducing competition from problem weeds, or at the very least reducing the build-up of the weed seed bank.

The chaff collection system complete with combine 'slinger' and trailer.

FARMYARD MANURE AND COMPOST MANAGEMENT

Mixed farms with livestock often have a good source of farmyard manure or slurry; even stockless arable farms may have access to manures from local livestock farms. The use of farmyard manures and slurry, whilst containing a useful source of nutrients to aid crop production, can be associated with increased weed problems. The problems can arise in various ways, either as a result of weed seeds in the manure, as a result of the way in which it is applied, or due to the stimulatory effect of the nutrients on weeds already present in the soil.

Some manures contain weed seed, either seed that has passed undigested through animals or from bedding materials such as small-grain straw and old hay. High-temperature aerobic composting (recommended under organic standards) can greatly reduce the number of viable weed seeds as long as the temperature is maintained at higher than 60°C for more than three days.

Operationally compost will need to be turned regularly to achieve even heating through the whole heap, and to get material from the outside (where seeds are likely to survive) to the inside (where the highest temperatures are likely to be generated). In a similar way, the aeration of slurry can reduce the number and viability of weed seeds. (*See* information on composting in Chapter 3.)

The Application of Farmyard Manure

When applying manure or slurry, aim to avoid conditions which stimulate weed seeds to germinate (excessive soil disturbance, creating bare patches). Applying slurry to stubble after silage cuts can provide optimum conditions for weed seed germination, since a nutrient-rich bed of cattle slurry will produce a high potassium environment that will favour weeds such as docks rather than grasses. Dock seeds should not survive low pH silage, but will survive in a later cut of hay.

Some research has shown that placing manures and slurry more accurately on crops can benefit the crop rather than the weeds. Crop plants are generally sown fairly deeply, and they germinate from a lower level in the soil profile than weeds, which tend to dwell on the surface and germinate from 0–3cm. Crop plants also root more deeply, and this tendency can be exploited for weed management. In arable systems manure placed 10cm below the soil surface (by injecting) encourages the crop seeds to grow down into the nutrient-rich layer before the surface-dwelling weeds can reach it. This technique can also be used with broadcast-spread slurry, and if it is ploughed in rather than left on the surface it will be available to the crop before the weeds can reach it.

In many cases, the growth of weeds that follows manuring is a result of the stimulating effect manure has on weed seeds already present in the soil. This can be due to the flush of nutrients (for example, the supply of nitrates), enhanced biological activity in the soil, or other changes in the fertility status of the soil. Some work has indicated that excesses of potash and nitrogen in particular can encourage weeds, but in any case it is prudent to monitor the nutrient content of your soil and manure, and to spread manure evenly in order to reduce the incidence of weed problems.

Managing Docks

Applying slurry to crop stubbles or to leys after silage cuts can provide optimum conditions for weed seed germination. A nutrient-rich bed of cattle slurry will produce a high potassium environment that will favour weeds such as docks rather than grasses. Experience suggests that dock seeds survive low pH silage less well, but will survive in a later cut of hay. Don't give the docks an advantage: only apply manures and slurries that have derived from silage/hay cut from land containing docks on to permanent pasture or long-term grass, because in this way any dock seed development can be managed by cutting and/or grazing. Spreading on arable fields is likely to result in a build-up of docks which can undermine the long-term viability of an organic farm.

The author has seen many farms where docks have been spread around the farm as a result of poor slurry and manure management, and as a result have become major problems.

Other Manures and Composts

Poultry litter: The weed seed levels in poultry litter tend to be very low because the feed is mainly grain based. Care should always be taken to ensure that the litter is free from viable weed seeds, but in general, composting is not required for weed management.

Composted manure and mushroom composts: The weed seed levels in properly composted manure (*see* Chapter 3) and spent mushroom compost tend to be very low due to the composting processes used. The end material is generally fairly free of weed seeds, so does not require further composting on farm for weed control.

Green waste compost: The weed seed levels in properly composted green waste compost tend to be fairly low due to the composting processes used; it does not require further composting on farm for weed control. It may, however, contain contaminants such as plastics, which can cause other problems.

Care should always be taken to ensure that organic material such as compost or manure is free from viable weed seeds (normally by ensuring the compost achieves 60°C over a three-day period). Regular soil inspection and analysis will also help to spot potential problems such as nutrient imbalances or areas of compaction that might favour weeds over crop plants.

UNDERSOWING AND BI-CROPPING

Undersowing and bi-cropping (or mixed cropping) involves growing two or more different crop species in the same area. The advantage for weed control is that the *combined* crops cover more ground, so there is less space available for weed emergence. Bi-cropping can involve purely cash crops, or a mixture of cash crops and fertility-building crops.

Undersowing

This process aims to cover the ground with a quick-growing dense layer of vegetation underneath the crop. The undersown species should have a prostrate growth habit so as not to interfere with harvesting the cash crop. The undersown cover is usually leguminous, which adds to, or maintains fertility. It also suppresses weeds. Combining cash cropping with fertility building in this manner potentially produces an economic return, and it may mean there is either no need for an isolated fertility period, or that the length of that phase can be reduced. Short-strawed cereal varieties can be difficult to manage, and straw difficult to save, due to the undersown crop growing up into the cereal. Undersowing cash crops with fertility-building crops has other advantages apart from weed suppression, but so far this technique has mainly found favour in cereal growing, where it helps in re-establishing leys and avoiding bare ground after harvest.

A growing number of organic farms are undersowing cereal and pulse crops on a routine basis and not just to establish leys. Whilst this may seem an expensive option, the legume that has been undersown provides good ground cover after harvest and significantly reduces weed development and a subsequent enlargement of the weed seed bank. At the same time the legume can be used for livestock grazing as additional forage. Most cereals can be undersown in this way, and also field beans. It is particularly beneficial to undersow field beans, as in late season, when the beans defoliate and the crop becomes very open and prone to weed growth, an established undersow can significantly help reduce weed development.

Clover undersown into cereals provides ground cover against weeds after harvest.

Field beans undersown with black medick (Medicago lupulina).

Experience from organic farms employing this technique have found that white clover tends to develop too slowly, red clover grows too tall and can impair harvest, but that black medick (*Medicago lupulina*), often sold by merchants as 'yellow trefoil', performs well in this role. It is leguminous, has a low prostrate growth and is shade tolerant, which helps establishment in the growing crop. Being a different species to clovers (*trifolium sp.*) it can also be used in close rotation with clovers. It is best established by undersowing a cereal or a bean crop in mid- to late spring, ideally at the same time as weeding the crop using a seedbox drill on a comb harrow. Always remember that once the undersow is in place, no further weeding can be undertaken without destroying the undersow. The photo below shows a cereal stubble that has not been undersown, with resultant weeds, and the photo above a cereal stubble undersown with black medick (Medicago lupulina), with a reduced level of weed growth.

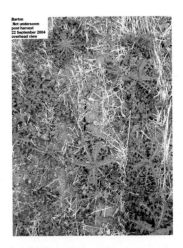

Cereal stubble with weeds, not undersown.

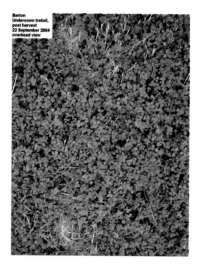

Cereal stubble undersown with black medick (Medicago lupulina), with a reduced level of weed growth.

		Wheat %				
		0	25	50	75	100
Bean %	0	434	302	146	97	124
	25	398	168	148	96	124
	50	346	162	133	80	100
	75	284	138	151	75	36
	100	169	117	72	83	62

Figure 29 Weed biomass (g m-2) as affected by bean and wheat density sown as a bi-crop
Source : Bulson et al., 1991

Bi-cropping

Bi-cropping is widely practised in certain countries, and an enormous variety of systems has been developed. Both component crops may be taken to yield, or one may be left as a living mulch to improve weed control. In successful intercrops weed suppression is usually superior to that of either of the component crops when grown alone. Crop density, crop diversity, crop spatial arrangement, choice of crop species and cultivar will all affect weed growth in intercropping systems. Improved weed control alone is unlikely to justify their use, and there must be other obvious benefits if the change in cropping practice is to prove economic.

In terms of mixed arable cash cropping there have been investigations into organic winter wheat and beans that have resulted in reduced weed growth and gave better yields than sole cropping. (*See* Chapter 6 for information on bi-cropping.) This is shown in the table above. Some farms have also grown peas and barley, or vetch and oats together, and harvested them as grain crops. More farms use this system for growing arable silage mixtures where the crop is not taken to a grain harvest. There is a wide range of combinations that could be designed to suit the farmer's rotations and marketing needs, although some practical experimentation is likely to be required.

IN-CROP CULTIVATIONS FOR WEED CONTROL

Although avoidance strategies, rotations and cultural methods provide the basis for weed management in organic crop production systems, it is likely that some form of direct action will be needed against weeds to prevent crop loss at some time, or to reduce the return of seed to the soil seed bank in a particular crop. Before taking any action it is important to take an overview and assess whether the weeds present are likely to develop to such an extent that they will cause an immediate loss of crop, or will store up potential future problems (for example, by shedding seed and adding to the soil seed bank, so exacerbating future weed problems). If the weed burden is judged to have the potential to cause damage, the cost of this should be offset against the likely costs of any immediate or future direct control measures so that crop weeding is only undertaken when it is economically beneficial to do so.

If action is necessary, a range of mechanical weeding options are available for organic cereal and pulse crops. These are discussed in the following sections. But before deciding *how* to weed, it is always important to consider whether to weed or not.

To Weed or Not to Weed?

The series of questions below were first suggested by Kopf and his colleagues in the year 2000, and can be used as an aid to deciding whether or not a mechanical weeding needs to be undertaken or not:

1. Is weed control needed?
2. When is control needed?
3. Where is control needed?
4. Which method of control?

1. Is Weed Control Needed?

A weed is generally defined as a plant growing where it is not wanted, so the definition of a weed will depend on each individual and the situation. Weeds compete directly with a sown or planted crop for space, light, water and nutrients and are estimated to be responsible for around 10 per cent of all crop losses. To prevent yield losses they need to be removed when they are actively competing with the crop. The following should be considered when deciding if weeds are a problem and if control is needed:

- Will marketable yield be affected by the weeds? Yield reductions occur when weeds compete directly with the crop, so without some form of weed control, will the crop be unharvestable, or yield reduced to an unacceptable level?

- What other negative effects are the weeds having? The physical presence of weeds may impair the harvesting process, shed seeds may affect the quality of the crop if, for example, wild oat seeds contaminate a grain seed crop so that it does not meet seed crop criteria. Where deadly nightshade is present in a lupin crop, this can result in some crops being unharvestable (*see* 'Lupins' Chapter 6).
- What are the long-term consequences of infestation? Lack of control may result in a build-up of the soil weed seed bank or in the spread of perennial weeds, even if there are no short-term reductions in yield, in which case immediate weed control may be justified.
- Are there any benefits associated with the weeds? Some weed covers might deter pests such as aphids, although once again there will be some competition effect on yield.

Two tips emerging from the knowledge gained by a participatory research programme on organic weed management, conducted by the Henry Doubleday Research Association, are as follows:

- Remember that removing all weeds is not only difficult, it is not always desirable, and it does not necessarily make sense economically. Many pieces of research have shown that the cost of removing weeds completely can financially outweigh the resulting yield benefits.
- Thinking long term is crucial in organic farm systems. Weed management should be considered throughout the rotation and as a property of the whole farm system. Weeds may not threaten the current crop, but may have an impact on later crops if allowed to set seed or propagate. The weeding of each crop in the rotation needs to be planned, and the schedule of weeding operations in different crops in one season needs to be co-ordinated.

Experience from a three-year organic crop trials programme during 2003–2006, which split large, one-acre plots of twelve cereal varieties into part unweeded, part weeded (with a comb harrow), showed that no significant yield impacts were observed between the weeded and unweeded areas where low weed competition existed. In fact, undertaking mechanical weeding had a detrimental impact of 0.25t per hectare on yields in two out of three years as a result of increased crop damage when weed competition was low enough not to warrant mechanical weeding intervention.

2. When is Control Needed?
Once you have made the decision to weed, the next logical question is, 'When should it be done?' Before you decide this you need to have the following:

- A good working knowledge of the weed flora and the crop. This will include previous and current observations of weeds present in any particular field and the crop history.
- An understanding of the biology of the weeds on the field. Life cycle, growth habit, potential reproductive capacity will all have a bearing on

the types of management methods that are likely to be the most effective. In particular, a knowledge of whether main weeds are annual, biennial or perennial is important as this will determine their reproductive capacity and influence the method of kill.

- Previous weed management experience on the field. Did measures work in the past, and if not, why not? Are weed levels changing? If so, which weeds are increasing? Which decreasing? Experience is a good start to deciding what measures are necessary and for deciding what other information or knowledge is needed.
- All this information will point to which methods are likely to be the most effective for managing the weeds. In some cases a longer-term preventative approach will be best, and in others a shorter-term control approach will be needed. Further factors that need to be considered are mentioned below.

Timing of weeding: If weeds are left uncontrolled for too long then they may start to compete with the crop. If weeding is too early the weeds may not all have germinated, and you may need to weed again later in the season. Good timing of weeding is most important when the crop is in direct competition with the weeds for resources, but it can depend on many factors, including the type of crop and weed, followed by issues such as soil conditions, weather conditions and time of the year.

Generally, the earlier the emergence of the weeds compared with the crop, the more competitive the weeds are likely to be. Weeds that emerge late tend to have little competitive effect, although they can have an impact on weed seed return and harvesting. There are some factors that can be manipulated by the farmer to give the crop a competitive advantage, for example cultural measures such as stale seedbeds, or using a wide-row drilling approach and inter-row hoeing. These practices tend to widen the 'weeding window' – a period during which weeds need only be removed once – and this makes the absolute timing of weed removal less critical. Widening the weeding window in this way reduces the pressure to carry out a weeding operation within a very narrow time period, which may be difficult or impossible because of weather and resources.

The timing of a weeding operation is influenced by soil/climatic conditions, weed and crop growth stage. Ideally, weeds should be targeted as young as possible either prior to, or just after emergence (see the diagram on page 264 on weed emergence periods). There is little point in targeting mechanical weeding operations at weeds once they have become very well established.

Number of weeding operations needed: Obviously the fewer the number of weeding operations, the lower the immediate weeding costs will be. For cereals and pulses, some, or all, of a carefully timed weed-free period – called the *'critical weed-free period'* – or a single weeding, may be all that

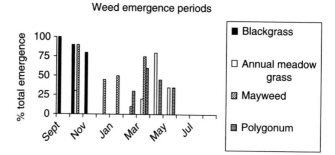

Figure 30 Typical weed emergence periods

is needed to prevent yield loss. (*See* information on critical weed-free period in the next section of this chapter.)

3. Where is Control Needed?

Weeding operations in cereals and pulses can be targeted at different areas of the field depending on the situation:

Broad-spectrum weeding across the entire field area, trying to cover all the ground.

Inter-row weeding is focused between the crop rows.

Patch weeding is where specific patches (or even weed plants) are targeted by mechanical operations or by hand.

Many different tactics can be used in a weed management programme, and for some crops a broad-spectrum weeding approach may be most effective, particularly narrow-drilled cereals and field beans where harrows are used across the whole soil surface.

Inter-row weeding is focused between crop rows, and an increasingly elaborate array of machinery has been designed for this purpose, some of which throws soil back into the crop row to bury intra-row weeds, and some of which is automated guidance (*see* later in the chapter on inter-row hoeing).

It may be possible in some situations just to focus on patches of weeds if there is a particularly problematic local infestation (for example, couch or docks).

4. Which Method of Control?

This obviously depends upon the answers to the previous questions, and especially upon the situation in which control is to be undertaken. Apart from cultural control, there are two main methods of direct control: physical (normally mechanical) and biological.

Cultural control is the basis of any organic weed management programme involving the integration of diverse weed control elements into a rotation. These are discussed in the cultural control section of this chapter.

Physical control forms the majority of post-sowing mechanical weeding methods. It depends upon matching the weeding implement with the weed flora to be managed. Many of these methods are discussed in more details in the following sections of this chapter.

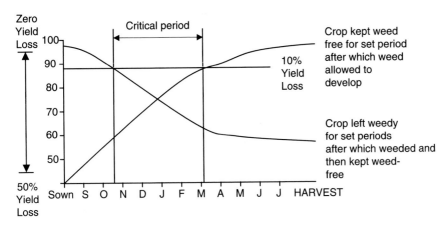

Figure 31 Critical weed-free period

Source : Walsh (1988)

Biological control depends on using other biological organisms (for example fungi, insects, parasitic plants, other plants) to reduce weed populations and prevent population increases. Most of the ideas are speculative and untested at present. One area of promise to organic farmers and growers is allopathy.

Critical Weed-free Period

To keep a crop free from weeds all season may not be economic. Therefore, weeding should be confined to the time that the crop is most vulnerable from weed competition. The 'critical weed-free period' is the period when being free from weeds will make a significant difference to final crop yield.

Assuming that a 10 per cent yield loss due to competition from weeds is acceptable (*see* the graph above), the greatest benefit from weed control operations in a winter wheat crop can be obtained if the crop is kept weed free from the beginning of November to the end of March (*see* graph). Weeds that emerge after this critical period are unlikely to compete significantly with the crop, or the cost of weeding will not be justified by crop yield increase.

MECHANICAL WEED CONTROL

Previous knowledge of the site will be invaluable, otherwise growers should assess the site for weed species and density. The control of annual and perennial grasses and perennial broad-leaved weeds is most difficult. A mixed rotation with a range of drilling dates and crop types will usually present the most opportunities for weeding. The more a rotation is based on similar crop types sown and harvested at similar times, the greater will be the selection pressure on a specific weed spectrum. This should be carefully considered when planning rotation and cultivation options (*see* earlier sections).

Perennial grass and broad-leaved weeds, such as docks, thistles and couch, are best tackled in a short 'fallow' period in dry conditions where the ground can be repeatedly worked to deplete root reserves. Competitive grass weeds such as blackgrass and ryegrass are best tackled by stale seedbeds and ploughing. (*See* cultivations for weed management, below.)

Mechanical weed control may involve weeding the whole crop, or it may be limited to selective inter-row or intra-row weeding. Machines can be used to kill weeds by burying, cutting or uprooting them: these rely on exhausting the energy of the weed by desiccation to cause death, and should be used to best advantage during the 'critical weed-free period'.

Tools without a cutting action are only effective on small weeds. Inter-row implements have been designed that control weeds within the crop row by directing soil along the row to cover small weeds. Mechanical weeders range from basic harrows to sophisticated self-guided devices. These may include cultivating tools such as hoes, harrows, tines or cutting tools like toppers and mowers for leys and green manures. Mechanical methods of control can be used which focus on the use of pre-drilling, pre-emergence and post-emergence mechanical weeding and/or cultivations during the summer months. The choice of implement, and the timing and frequency of its use, may depend on the crop and on the weed population.

The weather and soil conditions under which the operation is carried out will have a major influence on its efficacy. Soil type, surface structure and moisture content affect the choice and efficacy of mechanical weed control implements. The options may be more limited on heavy or stony soils, as most implements work better on light, stone-free soils. Mechanical weeding is less effective when soils are wet during or after weeding operations.

Ideally, soil conditions should be dry and warm, and sunny weather should be taken advantage of to desiccate and kill the weeds on the surface. It is also important to ensure that the weather will be dry for at least twelve hours after weeding to help kill weeds. Rainfall straight after weeding is likely to replant the weeds and keeps them alive.

Weeding should be carried out during the heat of the day, to maximize the desiccation effect on weeds. Care is required during frost periods, because the crop plant tissue may be vulnerable to frosting immediately after weeding. Wind is an organic farmers' assistant. Weeding in windy conditions is ideal, as not only does it dry the soil rapidly, but it draws moisture out of weeds bought to the surface very quickly and aids desiccation and kill. While many non-organic farmers curse wind because it prevents them from spraying agrochemicals, organic farmers welcome it.

In some instances, on silt-based soils where crops have emerged, rainfall can lead to soils 'capping' on the surface, and then it may be prudent to do nothing at all, as the cap prevents weeds penetrating the crust and emerging. A mechanical weeding operation will disturb the cap and allow weeds to flourish.

The optimum timing for mechanical weed control is influenced by the competitive ability of the crop. In organic winter cereals studies have found that corn poppy (*Papaver rhoeas*) was more effectively controlled in the autumn, whilst chickweed (*S. media*) was controlled best in spring when using a spring-tine weeder.

Weed morphology and stage of growth will also influence the selection and efficacy of weeding implement. In experiments to determine the type of physical damage that gave the most effective control of a range of seedling weeds it has been found that burial to 1cm depth is the most effective treatment, closely followed by cutting at the soil surface. Plants need to be buried totally to be killed, but plant size, angle and growth habit influence the depth of covering required.

Experience suggests that if weeds have emerged, you are already late with your weeding operation, and that the best time to kill them is at the white thread stage. At high weed densities, even with the most effective mechanical weeders, sufficient weeds are likely to survive control measures and to profoundly reduce crop yield in cereals, and direct control needs to be linked with long-term preventative measures to maintain the weed population at a manageable level.

With most mechanical weeding implements, operator skill, experience and knowledge are critical to success. Drawbacks to mechanical weed control include low work rates, delays due to wet conditions, and the subsequent risk of weed control failure as weeds become larger.

There may be some disadvantages to the greater use of mechanical weed control. The additional cultivations associated with mechanical weeding could harm soil structure and possibly encourage soil erosion. The increased mineralization of soil nitrogen due to cultivation may be seen by some growers as a problem and by others as an advantage (although this is likely to be limited in effect).

Capped soil with weeds unable to emerge.

There is legitimate concern about the impact of mechanical weeding on ground-nesting birds, and consideration should be given to weeding timing in relation to bird nesting and breeding cycles. Management practices may need some alteration in order to minimize disruption at critical times, although evidence is at times contradictory.

Cultivations for Weed Management

Soil cultivation or tillage in its various forms has long been the mainstay of weed control and is the most effective way to reduce the weed seed bank.

The inversion of land by ploughing can be valuable in the reduction of weeds, particularly some of the more pervasive grass weeds; it can also help overcome shallow compaction of topsoil and surface structure damage. However, ploughing is relatively slow and expensive compared to disc or tine cultivation, and costs more in time, fuel, and wear and tear than other techniques. Nevertheless, non-inversion tillage will only be sustainable in well structured soils and where compaction and damage to structure is avoided, and in the absence of grass weeds that are difficult to control.

The mouldboard plough is the traditional implement for burying weeds and crop residues as ground preparation for establishing a new crop. One piece of research has showed that the annual loss of seeds from a natural soil weed seed bank (with no addition of fresh seed) was 22 per cent with no cultivation. When the soil was cultivated twice a year the annual loss was 30 per cent, and when cultivated four times it was 36 per cent. However, it is not just the cultivations associated with the post-harvest incorporation of crop and weed residue that have weed control benefits: the method, depth, timing and frequency of cultivation may influence the composition, density and long-term persistence of the weed population.

Tillage is often divided into three types: primary, secondary and tertiary. However, there are many operations that do not easily fall into these categories or span them all. The main tillage types and their effect on weeds are discussed below.

Primary Tillage

This is the principal cultivation operation before crop establishment. The main choice is between plough or non-plough systems (reduced or minimal cultivation, conservation tillage, no-till, direct-drilling – *see* below). A range of machinery is available for primary tillage, and some combinations can even work a stubble down to a seedbed in a single pass. Ploughing is seen as a method by which weed seeds can be buried below the depth from which they are capable of germinating, and it is sometimes said that ploughing can be used to 'bury a weed problem'. This is particularly useful for small-seeded or annual grass weeds that are often short-lived and may survive being buried. But this short-term solution to poor weed control in a previous crop often leads to long-term problems due to the persistence of the buried weed seeds in the soil seed bank, as viable seeds may be brought to the surface by ploughing in subsequent years and will germinate if conditions are suitable.

The inversion of land can be an invaluable part of an organic system, particularly in the management and reduction of difficult-to-control grass weeds. The table below shows how different cultivation strategies have been able to reduce the number of blackgrass plants in an arable system.

Table 53 Reductions in blackgrass (Alopecurus myosuroides) plants under different cultivation strategies

Cultivation strategy	% control level
Minimum cultivation	0
September plough and drill	70
October plough and drill	77
Stale seedbed and October drill	80

Reduced Tillage
The concept of direct drilling crops without resorting to ploughing became popular after the development of non-residual herbicides, but recently there has been renewed interest, primarily out of concern for soil conservation. Under reduced tillage there can be better control of soil erosion, conservation of soil moisture, more efficient use of fossil fuel and less disturbance to soil microbial populations. However, weed build-up is a major problem, and volunteer weeds are also likely to be a problem. However, not all soils are suitable for reduced tillage, and less nitrogen may be made available to crops where cultivation is reduced to a minimum. The system is only viable in the long term with the use of agro-chemical inputs, on which it is reliant.

Many organic farmers find it impossible to dispense with the plough routinely. Each method of cultivation has its advantages and disadvantages, and the principle should be to use as many different types of cultivation as possible over the rotation, which prevents any one type of weed getting the upper hand.

Secondary Tillage
This is used to prepare seedbeds and leave a level surface for drilling. Typically it involves disking or harrowing to a depth of 10cm. Rotovators and power harrows are also used, and are able to prepare seedbeds even when ploughing has not been carried out. Implements are available that can combine shallow seedbed preparations with some deeper cultivations in a single pass. Others can loosen the soil below the surface while leaving the preceding crop debris on the soil surface. Secondary tillage operations are useful to germinate and destroy weeds prior to drilling, but as they tend to leave the soil surface in a fine seedbed condition, often result in a secondary flush of weeds shortly after finishing the cultivation. This then requires a further or tertiary cultivation.

Tertiary Tillage

This is the soil cultivation that is used directly as a means of physical weed control. It is dealt with in some detail later in this chapter with regard to comb harrow and inter-row hoeing operations. It is possible to time seedbed cultivations so that they avoid the peak flush of weeds that may be problematic in the crop. For example, if the sowing of organic winter cereals is delayed until after mid-October rather than drilling in late September, it has been shown that it may be possible to significantly reduce the occurrence of problem weeds such as blackgrass (*Alopecurus myosuroides*).

The Stale or False Seedbed Technique

This is a seedbed prepared several days, weeks or even months before sowing a crop. The technique is recognized as a strategy suitable for organic farming and has been widely used for many years. The stale seedbed is based on the principle of flushing out weed seeds ready to germinate prior to the planting of the crop, depleting the seed bank in the surface layer of soil, and reducing subsequent weed seedling emergence.

The soil is cultivated about four weeks before drilling to stimulate germination and encourage the first, and usually biggest, flush of weeds. Delaying sowing and undertaking a second weed-strike cultivation extends the stale seedbed effect further. The seed drill itself will also perform a weed strike when planting the crop seed. If possible, leave the seedbed for two or three days prior to rolling so as to kill some of the weeds disturbed when drilling. This will not adversely affect the drilled crop. Rolling a crop immediately after drilling will re-establish the disturbed weeds.

Darkness cultivation can help reduce weed germination.

Cultivation in Darkness

This uses a novel method of reducing seedling emergence by carrying out seedbed preparations in the dark to avoid stimulating weed seed germination. Daytime cultivation with a mouldboard plough has been shown to stimulate weed seedling emergence 200 per cent above that of night-time ploughing. Weed seeds of certain species need to be exposed to light in order to break weed seed dormancy and stimulate germination, for example chickweed, fat hen and annual meadow grass. If the ground is cultivated in the dark, experimental evidence has shown that when seeds are not exposed to light before being reburied, fewer of these seeds germinate by up to 70 per cent.

Covering soil with opaque material immediately after cultivation or leaving it exposed to light made no difference to weed emergence, suggesting that it is the exposure to light during the cultivation operation that is important. When carried out with a tractor and cultivator at night, even the presence of the tractor working lights was found to be sufficient to break seed dormancy, as demonstrated by the weeder being used with and without a large rubber sheet cover to block out the tractor lights (see figure 32 below).

However, weeding in the dark has only a short-term impact, which may be of benefit during the stages of early crop development. In practice, cultivation in darkness is usually much less effective, as it still leaves enough weeds to reduce crop yield. In spite of the variable results, it seems as if dark cultivation may at least delay weed emergence. In crops where weed control is critical, the practice could widen the window of opportunity. Perhaps with the aid of night-vision goggles organic farmers could find a new night-time occupation?

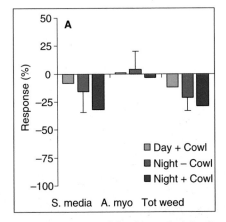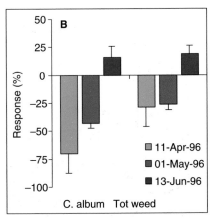

Figure 32 Impact of daytime and night-time cultivations on weed numbers

S. Media (chickweed), A. myo (blackgrass), C. abum (fat hen) Cowl denotes use of rubber cover to block out the effect of the tractor lights

Post-Harvest Tillage

An additional consideration when using tillage to aid weed control is the timing of any form of post-harvest soil cultivation in relation to its effect on the movement and persistence of weed and crop seed shed during or after crop harvest. Cultivations and weed management between crops are discussed in an earlier section of this chapter.

Not all seeds have the same response. The burial of recently shed seeds can induce dormancy when conditions are not appropriate for germination. For example, the burial of winter barley seeds in dry soil can induce dormancy and cause problems in later cropping sequences. Post-harvest cultivation too soon after seed shedding and in sub-optimal conditions for germination can instil a light requirement and as a consequence induce dormancy and persistence in oilseed rape seed shed during crop harvest. Wild oat seed (*Avena sterilis*) requires vernalization to break dormancy, and burial, shallow or deep, will prevent this, allowing the seed to be returned to the surface at a later date to become a weed. It is better to leave dropped wild oat seed on the surface over winter, to vernalize and germinate, before destroying with cultivations. Conversely, barren brome (*A. sterilis*) seeds left on the soil surface persist longer than those buried soon after shedding. In this instance, early cultivation would be more appropriate to ensure control.

Cultivation as soon as practicable after harvest is also recommended for the control of rhizomatous grass weeds such as common couch (*E. repens*) and black bent (*Agrostis gigantea*), and intensive and repeated cultivations are needed to work the soil to the full depth of the shallow rhizome system. The aim is to bring the rhizomes to the soil surface, and this works best in previously undisturbed soil. After the initial cultivation, further passes may only serve to move broken rhizome pieces around, rather than bring them to the surface, and the fragmentation stimulates regrowth of a dormant bud on each rhizome fragment and in the short term may even make the problem seem worse. Cultivations to control regrowth should be repeated every ten to fourteen days or when the grass has leaves 5–10cm long, until no further regeneration occurs. Alternatively, the land may be deep ploughed to bury any regrowth below the depth it will emerge from.

While stubble cleaning may not be appropriate for dealing with the shed seeds of some weed species, it can be an effective way of controlling some important weeds, including charlock (*Sinapis arvensis*), common chickweed (*S. media*), groundsel (*Senecio vulgaris*), wild radish (*Raphanus raphanistrum*), shepherd's purse (*Capsella bursa-pastoris*) and some speedwells (*Veronica spp.*). The surface soil should be cultivated to a depth of not more than 5cm, and this operation repeated at fourteen-day intervals. Some farmers take this a stage further and prepare a seedbed at this time, but leave it unsown until spring. The land is cultivated at regular intervals to deal with seedling weeds, and then ploughed after Christmas in preparation for spring cropping. Weeds controlled by 'autumn cleaning' include blackgrass (*A. myosuroides*) and charlock (*S. arvensis*). Nutrient leaching is likely to be a problem in soil left bare overwinter, and the method may not be allowed under agro-environmental schemes as there are benefits to birds of overwintered stubbles.

EQUIPMENT FOR MECHANICAL WEED CONTROL

Comb harrows are increasingly being purchased by organic farmers and viewed as the definitive method of mechanical weed control in arable crops. However, it should be remembered that the foremost method of weed management should be achieved via avoidance strategies, rotation design, and cultural techniques such as variety selection and stale seedbed techniques. The use of overwinter green manures prior to a spring crop may provide additional opportunities for weed suppression. Mechanical weed control is the last resort for management in the crop itself.

Comb-harrow Weeders

Comb-harrow weeders are very adaptable and can be used in a wide range of crops, including winter and spring cereals, winter and spring pulses, maize, potatoes, brassicas and some field-scale vegetables. They can also be fitted with a seedbox drill and used for undersowing (see 'Undersowing' above, page 258) at the same time as weeding.

Normal crop drill spacing and depth can be used. Increasing cereal seed rates by 10–15 per cent can help out-compete weeds in the row, and can compensate for some damage when weeding. Using the comb harrow to perform weed strikes on false seedbeds prior to drilling can also help crop establishment by limiting competition by weeds.

Reductions in weed density from spring-tine harrowing can range from 0–70 per cent, depending on crop type and growth stage, weed species

Comb-harrow weeder with a seedbox, as used by many organic farmers.

and level of burden, soil conditions and the timing of weeding. The total eradication of weeds is not desirable in organic systems, as an acceptable level of weeds can provide valuable habitats for invertebrates, insects and predators of crop pests. For a positive yield response to weeding, the weed population must be of a competitive level, and the benefit of removing the weeds must outweigh the negative effects of crop damage. Without this, the cost of weeding may outweigh its benefits.

Weeding Programme
The number of passes has little effect on the level of annual weed control. Multiple passes can result in slightly better weed control for more established or tap-rooted weeds, where the first pass loosens the weed and soil and the second pass removes it to the surface for desiccation. However, multiple passes may increase crop damage. If a second weeding is required, allow a seven to fourteen-day period between weeding. This will be sufficient to germinate disturbed weeds.

Timing
Individual weed species differ in their response to the timing of weed control. Weeding can be carried out at three stages during the growing season: pre-crop emergence, early post-crop emergence, and late post-crop emergence. However, the benefits of weeding will be reduced if weeds are removed after they have already had a significant competitive effect.

The most effective way to control many weed species is by using a single weeding as early in the season as conditions will allow, if possible in the autumn, since most weed species are most vulnerable at the seedling stage and the crop has time to recover from any damage. A combined autumn and spring weeding approach achieves the greatest reduction in weed density. Where this is not possible, autumn weeding can provide a good level of reduction. Multiple spring weeding alone will not provide as good a reduction in weed density as autumn plus spring, or autumn weed control.

Target weeds as young as possible, either prior to, or just after emergence. Weeding should be carried out when it is dry and warm, during the heat of the day, to maximize the desiccation effect on weeds. Care is required during frost periods, because the crop plant tissue may be vulnerable to frosting immediately after weeding.

Tap-rooted weeds such as the common poppy, mayweed and volunteer OSR are best controlled at an early growth stage in the autumn when the weed is at a seedling stage, before they can develop a strong tap root. At this stage they can be easily covered with soil and are not securely rooted. Weakly rooted climbing or scrambling weed species such as chickweed, cleavers and field pansy are best controlled by spring weeding at a later growth stage when they can be 'raked' out of the crop, even when they are well established.

On very heavy clay soils, autumn weeding may prove impractical due to wet soil conditions, and access to fields may be restricted at this time of year. Multiple spring weeding passes should be used as an alternative.

However, if an autumn weeding is not carried out in an autumn-sown crop, weed competition will be greater in the spring. On heavy soils an alternative strategy may need to be considered (for instance, mechanized inter-row hoeing in spring).

Soil and Weather Influence

Weed control is better achieved when the soil is in a reasonably dry, friable state when weeds can be more easily uprooted or buried with soil. In wet soils, weed control will be greatly reduced, and soil damage may occur from heavy machinery use. Stones do not adversely affect the efficiency of weeding, but will lead to higher machinery wear rates. Weed control on soils that are dry and hard will be very limited.

For optimal comb harrow use, it is important to leave a fine, flat seedbed after drilling. If winter beans are ploughed in, use a power harrow afterwards to level the soil surface so that the comb harrow can be used efficiently.

The weather conditions for twenty-four to forty-eight hours after the weeding operation are also important so as to ensure good weed kill. Weeding when windy can help dry soils and draw moisture from uprooted weeds for a better kill.

Crop Damage

Crop damage can occur during weeding, commonly at levels between 0 and 20 per cent, especially when performing early autumn or spring weeding, necessitating a compromise between weed control and crop damage. Don't be tempted to weed if the weed burden is low enough so as not to have a detrimental effect on crop yield. Weeding could damage the crop or germinate buried weeds, thereby creating a greater level of weed competition than if the crop had been left unweeded.

Setting up the Comb Harrow

Direction of travel Weeding in line with the drill rows will help minimize crop damage. If multiple passes are used, a second pass at 15–20 degrees to the drill can assist in removing in-row weeds.

Forward speed The fastest practicable speed should be used to comb harrow (10–12+km/hr). If the harrow is set up correctly the tines should vibrate vigorously from side to side as the harrow moves forwards; this produces an effective weeding action. However, if the harrow is not set up correctly, at high speeds larger comb harrows (8m+) can 'yaw', bounce and vibrate, thus limiting effectiveness. If the forward speed is too slow, the tine frames may bounce, with the tines not vibrating or contacting the ground sufficiently, leading to sub-optimal performance. Great care should be taken to set up and adjust the harrow relative to weed growth stage and prevailing soil conditions.

Adjustment/pressure settings The set-up of the harrow is critical for good weed removal. The headstock should be adjusted so that it is at right angles to the ground. With the support wheels set, the tine frames should be adjusted horizontally so that all tines apply equal pressure to the ground, whilst moving forwards. The pressure on the tines (set angle) should be adjusted so that the tines remove weeds without causing excessive crop damage.

The flat adjustment will produce less vigorous weeding than the steeper adjustment. A medium to strong pressure setting should be used for a single pass in cereals (to loosen weeds), followed by second pass at a lower pressure setting. A lower pressure setting should be used for pulses, potatoes and vegetables.

Blind Harrowing

Emphasis should be placed on pre-drilling cultivations using stale seedbed techniques prior to drilling, where a comb harrow can be used to perform weed strikes.

The comb harrow can also be used soon after drilling and prior to crop emergence. This is known as 'blind harrowing', and is used to destroy weeds immediately prior to crop emergence. It is particularly useful in pulse crops such as beans, peas, lupins and soya, but may necessitate a slightly deeper drilling depth. This technique requires very careful timing to avoid damaging the emerging crop, and daily inspections of seed germination and crop development should be undertaken so that weeding is performed when the plant shoot is 1–3cm below the soil surface (depending on the crop and drill depth). Using the comb harrow with a very light pressure setting can disturb weeds immediately prior to emergence, giving the crop the best weed-free start possible.

Experience has shown that placing a sheet of glass (40 x 40cm) on the soil in the field immediately after drilling can advance the germination and emergence of the seed under the glass plate by a few days, the glass acting like a mini greenhouse. This can be used as a guide for blind harrowing, in the knowledge that the rest of the field has slightly delayed crop emergence.

Using Wide-row Systems

Where the crop is drilled on wider rows and an automatic steerage inter-row hoe is used for weeding, the comb harrow may still prove to be a useful piece of equipment. The harrow can be used one or two days after inter-row weeding at an angle across drill lines which will provide a secondary kill on weeds loosened and uprooted by the inter-row hoe, and help control weeds in the crop row itself. Such a technique can achieve high levels of weed control. In some circumstances farmers have mounted an inter-row hoe on the front of a tractor and a comb harrow on the rear, and used them simultaneously. Whilst this is attractive, the slower forward speed required for the inter-row hoe means that the comb harrow is working at a suboptimal forward speed and is not as effective as it could be.

Working in Crops with a Comb Harrow

Winter cereals Use to germinate seeds in stale seedbeds and perform weed strikes pre-drilling. Weed pre-emergence six to seven days after sowing. When soil conditions allow, cereal crops should be weeded at the two- to three-leaf stage (growth stage 13), and if necessary, again at the tillering stage (growth stage 22–30). Weeding to control cleavers or volunteer green manure growth (namely vetch) in cereals can be conducted up until ear emergence (growth stage 51). A medium to strong pressure setting should be used for a single pass in cereals (to loosen weeds), followed by a second pass at a lower pressure setting.

Spring cereals Use to germinate seeds in stale seedbeds and perform weed strikes pre-drilling. Weed pre-emergence six to seven days after sowing. Early spring weed at the two- to three-leaf stage (growth stage 13), followed up with a second weeding at tillering stage (growth stage 22–30). A medium to strong pressure setting should be used for a single pass in cereals (to loosen weeds), followed by second pass at a lower pressure setting.

Field beans Use to germinate seeds in stale seedbeds and perform weed strikes pre-drilling. Early autumn or spring weeding at the second node stage (GS 102), either with or across the rows, and if necessary followed by a second pass along the rows at GS 105 (fifth node), using a light harrow pressure setting to minimize crop damage. Weeding should be kept to a minimum when the plants are approaching early bud.

Peas Use to germinate seeds in stale seedbeds and perform weed strikes pre-drilling in the early spring. Use a pass with the drill rows on a light pressure at an early growth stage (GS 102–105). Later passes can be made along the rows but these may cause damage to pea crops, especially when they have begun flowering. Severe damage will be sustained if passes are made across the rows at later growth stages.

Lupins and soya Mechanical weeding methods should only be used before the lupins reach 15cm in height, or unacceptable damage may be caused to the crop plants. Weeding when the crop plants are around 5cm in height with three to four leaves, and well anchored by the tap root, can be successful, with a first pass at 45 degrees to the rows, followed by a second in the opposite direction a few days later. Weeding in the direction of the rows may result in the lupins being buried, unless a very slow forward speed is used.

Weed Types

Different weed control strategies will be needed for different weeds, with the effectiveness of the comb harrow being dependent on the individual weed species present and their growth stage. Weeding timing, the number of passes and the intensity of weeding needs to be specific to the weed type present, the crop and the soil conditions.

Weeding when the weed burden is low enough so as not to have a detrimental effect on crop yield can have a detrimental effect on crop yield via crop damage. The weeder can also germinate buried weeds, creating a greater level of weed competition than if the crop had been left unweeded.

A wide range of broad-leaved weeds can be well controlled with comb-harrow weeding. The timing of the operation(s) will depend on soil and weather conditions and the size and species of weeds present. The ideal conditions are when the soil is reasonably dry and friable, and the weather is dry, warm and sunny to enable rapid desiccation of the harrowed weed roots.

Tap-rooted species such as mayweed, poppy and volunteer oilseed rape are best controlled at an early growth stage in the autumn when the weed is at a seedling stage, before they can develop a strong tap root. At this stage they can be easily covered with soil and are not securely rooted.

Weakly rooted climbing or scrambling weed species such as chickweed, cleavers and field pansy are best controlled by spring weeding at a later growth stage in the spring when they can be 'raked' out of the crop, even when well established.

Weeds the Comb Harrow is Effective On

The efficacy of comb-harrow weeding is dependent on the growth stage and species of weed present in the crop; thus:

- Broad leaf weeds and polygonums can be readily controlled using a comb harrow, whilst grass weeds are largely tolerant to weeding with a comb harrow.
- Tap-rooted weeds (poppy, mayweed) are best controlled at an early growth stage using an autumn weeding strategy. Mayweed can be controlled as long as it can be targeted at an early growth stage, prior to taproot development. A two-pass, early spring weeding can assist in control.
- Scrambling weeds, for example chickweed and cleavers, are best controlled at later growth stages using a one- or two-pass weeding in the spring. Effective control of chickweed can be achieved by early spring weeding prior to a frost.
- Wild oats, while difficult to control, can be germinated in the stale seedbed prior to drilling the crop.
- Blackgrass is very competitive in the crop from an early stage, and requires control in the autumn to avoid serious yield penalties. Where blackgrass is a problem, it should be controlled as early as possible, during the seedling stage. However, if autumn weeding is undertaken, there is a risk of stimulating more blackgrass to germinate. The germinating blackgrass will be at a competitive disadvantage compared to the crop, but may cause a problem in the spring.
- The most effective control for blackgrass is having a minimum of two years ley in the rotation, as blackgrass has a relatively short persistency in the soil seed bank (about two years). Short ley phases (about one year) in arable rotations may not provide sufficient control.

Weeds the Comb Harrow is Not Effective On

- A comb harrow will not be effective on grass weeds or rhizomatous weeds (annual meadow grass, couch), as the action of the comb harrow will not remove the entire rhizome.

- Comb-harrow weeding is generally not effective for controlling established tap-rooted weeds such as dock, thistle, charlock. However, harrow weeding at pre-emergence, cotyledon stage can help to restrict weed growth during the critical weed-free period.
- Some firmly rooted weeds (namely creeping buttercup) do not respond well to comb harrow weeding with only very limited uprooting achieved. In friable soils, if it is possible to bury the weed with soil at an early growth stage, this can assist with control.

Inter-row Hoe Weeders

The primary method of weed management should be achieved via rotation design, variety selection and pre-sowing cultivations, using stale seedbed techniques. The use of overwinter green manures prior to a spring crop may also provide additional opportunities for weed suppression. Inter-row hoeing, having a wider window of operation, can provide an effective weed management strategy, especially when dealing with grass weeds.

The most common organic approach is to plant combinable crops on 10–12cm rows and select competitive varieties that develop rapidly, using mechanical weeding intervention as a last resort. If mechanical weed control intervention is required, the spring tine or 'comb harrow' is the choice of most farmers. Research has demonstrated that whilst being a useful tool, its effectiveness on tap-rooted and grass weeds is restricted and the window of opportunity for use is limited, especially on heavier soils. When weeding is most effective, soils are often too wet for effective comb-harrow use and when soils are dry enough, weeds have developed and can no longer be adequately controlled with a comb harrow. So what are the alternative options on heavy soils?

The Principles of Inter-row Hoeing in Cereals and Pulses

Inter-row hoe weeding works by planting combinable crops (cereals and pulses) at wider row spacing (typically 20–30cm) and matching drill spacing to the hoe set-up. This can be achieved by blocking up alternate drill coulters, or by adapting drills. Normal seed rates are used, providing a denser planting in the row for competition. The inter-row hoe (normally 4m or 6m wide) is either front- or rear-mounted on a small-/medium-sized tractor. The tool bar is equipped with 'A' or 'L' hoe blades mounted on independently sprung parallelogram legs which follow the contours of the land closely. Baffle plates can be fitted on either side of the hoe blade for crop protection. The hoe is operated at a forward speed of 4–6km/hr, producing a work rate of 3.2ha/hr (compared to 13.5ha/hr for a comb harrow).

When comparing the two principal methods of mechanical weed control available to organic farmers, comb-harrow weeding and inter-row hoeing differ considerably in their effectiveness under different soil conditions and in the control of different weed species and growth stages (Table 54).

Table 54 Response of weeds to weeding methods (reported as a comparison of unweeded plots. Positive percentages mean less weeds)

| | Comb harrow weeding | | Inter row weeding | |
	Mean	Range	Mean	Range
Weed density	–31%	+27% to 74%	–50%	+11% to –93%
Weed density in May	–17%	+33% to –84%	–42%	+6% to –88%
Weed density at crop maturity	+54%	+223% to –73%	–33%	+75% to –99%
Crop density	–8%	+7% to –22%	–11%	+10% to –80%
Crop yield	–13%	+13% to –38%	–6%	+28% to –41%

Source : Welsh (1998)

The efficacy of inter-row hoeing is less dependent on weed species, weed growth stage and soil conditions than comb-harrow weeding, offering greater flexibility for timing, particularly on heavy soils. Moreover the inter-row hoe has been shown to be more effective on a wider range of weed species than the comb harrow, particularly for grass weeds (Table 55).

Table 55 Comparison of weeding strategies

	Inter-row hoeing	Comb-harrow weeding
Timing	Flexible	Autumn/early spring
Soil type/condition	Flexible	Light friable soils
Crop damage	Yes	Yes
Annual broadleaf weed control	✓	✓
Annual grass weed control	✓	x
Perennial weed control	x	x
Workrate	3.2km/hr (4m)	13.5km/hr (12m)

Research has shown that inter-row hoeing can improve yield by 20 per cent over unweeded crops, by reducing weed density. Planting at wider rows is not detrimental to yield, with research demonstrating no appreciable yield penalty up to a row width of 30cm. However, specific agronomic, management and economic factors need to be considered. Adaptations are required including matching wider drill rows to hoe widths, timing of weeding, the potential for crop damage, investment in suitable equipment, and the limitations of slow work rates and manual control. Careful placement of the hoe is required to avoid potentially high levels of crop damage.

There are several challenges to overcome, including operating an inter-row hoe in cereal and pulse crops, locating the crop rows, controlling the hoe in the rows, achieving accurate alignment for long periods, operating at a suitable work rate, dealing with different crop growth stages, gaps in crop rows and minimizing cost. Without careful placement of the hoe, potentially high levels of crop damage can result; however, new developments in automated hoe guidance systems aim to solve some of these problems.

Automated Hoe Guidance System Developments

The biggest challenge is in guiding the hoe to where you want it to go. Manual steerage hoes have been used for many years but are limited by work rate, and the size of hoe that can physically be handled, and are subject to the capability and fatigue of the operator, which will inevitably be variable.

One option is to guide the hoe relative to the furrow. Marker systems including buried cables, furrow marking and following have been used and can work well provided the furrow does not get filled in after drilling, or the reference point is lost. Technological options using global positioning systems, radio beacons and laser systems have been tried, but these are relatively expensive options.

Instead of sensing the furrow, why not look at the crop row? Direct 'tactile' crop sensing is possible in crops such as maize, but is impractical for cereals and pulses. Height differential and spot measurement monitors have been tried, but only offer limited success.

A new crop-sensing alternative was jointly developed by Silsoe Research Institute, IACR-Long Aston, ADAS and Garford farm machinery under a Defra-funded project, and is now commercially available as the 'Robocrop' hoe: it has been adopted by a number of organic and non-organic farmers with great success. The guidance system employs 'video' imaging to locate the crop rows, using mass-produced hardware added to an adapted standard rear-mounted hoe with an additional parallelogram linkage to allow 'side shift' adjustment as the hoe moves forwards. For accuracy, a computer box in the cab is used to image-monitor five crop rows at once, in colour, at a rate of twenty-five images per second. This avoids the problem of locating the crop row in damaged/missing crop

Garford 'Robocrop' inter-row hoe in cereals.

Weed Management

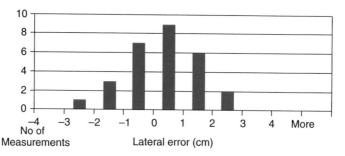

Figure 33 *Small lateral sideways error in locating row with Robydome hoe*

areas, different growth stages, colour changes and shadows. The computer box also 'tracks' the path taken and extrapolates this forwards, comparing it with the image that the small video camera sees in real time. This allows the hoe to stay on course in failed crop areas and at the end of the field when encountering headlands. Night-time work can be carried out by adding lights to the tractor and hoe, or by using an infra-red or night-vision camera filter.

Lateral tolerances are controlled within 3cm, which minimizes potential crop damage from accidental hoeing. Forward work rates are similar to a manual system, with efficient soil movement limiting forward speed. However, the automated system allows hoes wider than 4m to be used without compromising performance. Thus a folding 12m guided hoe has been produced, with a 4m section mounted on the front of the tractor and two 4m wings on the rear (*see* photo below), all operating independently. This significantly increases the work rate in ha/hr, whilst reducing the driver's steering task (*see* Table 56).

12m Garford 'Robocrop' inter-row hoe operating in a cereal crop.

Table 56 Potential cost savings from inter-row hoe operation

Hoe type	Work rate ha/hr	Operational cost £/ha
4m hoe manually guided	0.6	28.0
4m hoe with automated guidance	3.2	10.7
6m hoe with automated guidance	4.8	7.5
12m hoe with automated guidance	9.6	4.0

The capital investment in an inter-row hoe system is high, but once purchased the operational costs are relatively low. With no herbicides, the main wearing components are the hoe blades themselves. Organic farmers have learnt that using new blades at full width in the young crop works best, then once the crop is well tillered, the hoe is switched to part-worn blades to give extra clearance between blade edge and crop. In this way the farm has two sets of hoe blades, but is only ever replacing one set.

Tramlines need to be set up in the crop when operating a wide-row system, to assist with hoe and drill matching and to provide crop access. These can be sown with clover to prevent a weed build-up if necessary. Operating a 4m hoe will require a large number of tramlines, but another benefit of a 12m hoe, as well as a higher work rate, is that it does not have to have so many tramlines.

Research has demonstrated that growing grain crops planted on 20–25cm rows has a wider window of opportunity and achieves more consistent weed kill without detriment to crop performance. Furthermore, recent research developments have resulted in the production of automated guidance systems, making this system increasingly attractive, especially for organic farms on heavy land.

Row spacing may have to be adjusted to allow for the width of the tines. The widths typically range from 15–30cm (6–12in) depending on the drill and the width of the tractor wheels; typically 25cm widths are used.

Wheat grown on a wide-row system.

Oats grown on a wide-row system.

Weeds can be controlled in the crop for a longer period of time than with hoeing where rows are wide apart, because there is less risk of damaging the crop. However, a disadvantage with inter-row hoeing can occur when weeds emerge within the row. When this occurs, some growers can remove them by travelling across the row at a slight angle with a comb-harrow weeder, although this is only attempted when the crop is no more than 3–4in high, or eight to ten true leaves.

Most cereals can be grown on a wide-row system; however, spelt, rye, triticale and oats, due to their tall stature and very competitive nature, are often as competitive on a narrow row, and can out-compete weeds with negligible mechanical intervention. Where specific weed problems exist, inter-row hoeing can still be a useful strategy. Wheat and barley are the cereals that respond best to an inter-row hoeing system.

Peas grown on a wide-row system.

For pulses, it is more difficult to grow winter beans on wide rows unless accurate, deep autumn drilling can be performed. Spring beans, peas, lupins and soya can all be grown on wide rows in an inter-row hoeing system with success.

Limited research has shown that in wide rows, varieties exhibit very different growth characteristics. Thus varieties that normally grow in a prostrate fashion become very erect with the greater intra-row competition. The potential for using lower seed rates on wide rows is also a real possibility. However, careful management of crops is needed, as wheat grown in wide rows has less opportunity for compensation if plant or tiller numbers are poor. A planophile variety will give more shading in a wide-row system, but a tall erectophile variety can achieve a similar effect while allowing later inter-row hoeing with minimal crop damage.

In addition to weed control, the late weeding of grain crops can also increase the mineralization of soil nitrogen during the critical grain development stages, via the disturbance of topsoil and the oxidization of soil organic matter. If matched to critical crop development stages this could be a useful management tool in improving grain protein contents in breadmaking cereals, where protein contents are often low. More research is required in this area, however.

Weed Surfing

A novel approach for crop weed management was developed by Richard and Adrian Steele at Chapel Farm in Worcestershire, England. All the cereal and pulse crops grown on their heavy land farm are grown using a wide-row system. A 4m inter-row hoe, rear-mounted to a small articulated four-wheel-drive forestry tractor is used for weed control with multiple passes during the growing season. The Steeles were the pioneers

Weed mowing machine in action; developed by Richard and Adrian Steele at Chapel Farm in Worcestershire.

CTM weed surfer.

of inter-row hoeing cereals and pulses in the UK, and have been operating the systems successfully for more than fifteen years. They developed the system as a result of having serious blackgrass weeds on their farm, which also prompted their conversion into organic farming as a way forward in controlling the weed which had built up and become herbicide resistant under a non-organic regime. The farm also has some wild oats and charlock as problem weeds, but at a lower level than blackgrass.

Being able to access the crop throughout the growing season, the Steeles developed a weed mowing machine that can be mounted to a tractor front-end loader and used to give the crop a 'haircut', cutting all the weeds above the crop height.

The 'mower' consists of a series of hydraulically driven spinning blades under protection shrouds. The tractor drives through the crop once or twice per season, usually in late season, and cuts the weeds above the crop ears. Weed flowers and seed heads drop to the floor, are eaten by insects, field invertebrates and mammals, or are given far longer to germinate before post-harvest cultivations. Experience suggests that this is an effective method of reducing seed return to the soil weed seed bank. A similar machine called the 'weed surfer' has been developed by a Norfolk-based company, primarily for use in growing sugar beet, but it is also being used by some organic cereal and pulse growers for the same function.

WEED CHALLENGES IN PARTICULAR CROPS

Weed Management in Organic Cereals

A dense stand of cereal may appear relatively competitive against weeds, but at early crop growth stages or under unfavourable conditions, particularly where crop stand is reduced, weeds can take over and cause substantial yield losses. In addition to causing a loss of yield, the weeds may delay ripening, hinder harvesting and reduce grain quality. Weeds such as cleavers, black bindweed and the vetches can reduce crop yields by causing lodging. Weed seeds may contaminate the grain and require additional seed cleaning or make the crop unmarketable for certain purposes.

It is important that only clean cereal seed is drilled; in the past, weed seed contamination was responsible for increasing existing weed problems and introducing new ones. The primary and secondary cultivations used to prepare a seedbed have a considerable influence on the weed population. However, the nature and timing of these cultivations will vary with the previous crop, with soil type, with soil condition at the time of any operation, and with the equipment available.

There is machinery now available that will take stubble down to a seedbed in a single pass. Some farmers feel that a cloddy autumn seedbed gives the emerging crop some protection during the winter, but in general a level crumbly seedbed will give the crop the best start. Unfortunately, the weeds also emerge better and in greater numbers from a fine level seedbed – though in this case, control measures are also likely to be more effective. However, the early sowing of spring and winter cereals to achieve the highest yields gives little time for clearing the land of weeds during seedbed preparation, and this may increase weed populations significantly.

In spring cereals, allowing time for a stale seedbed helps to reduce weed numbers in the growing crop. Spring-tine weeders such as the comb harrow are probably the most widely used form of mechanical weed control in organic cereals. Under the right conditions, harrowing with spring-loaded tines can continue after the cereal crop has emerged and is past the three-leaf stage to control scrambling weeds, but it is only effective against young seedlings of tap-rooted weeds.

Established weeds in general are controlled more effectively by inter-row hoeing, but this requires the crop rows to be drilled at a suitable spacing. Inter-row hoeing may reduce weed density by over 90 per cent, but crop injury can be a problem without the use of a reliable guidance system. Mechanical inter-row weeding is possible up to ear emergence using computerized visual imaging systems that can automatically adjust tines in line with crop planting to eliminate driver error.

The crop may also benefit from increased nitrogen mineralization through the soil disturbance.

Weeding operations should be conducted at an early stage to eliminate crop competition and to avoid disturbing ground-nesting birds.

Hand-rogueing is of value to deal with small numbers of plants of a particular weed to prevent a large population developing. It is also important to pay particular attention to isolated patches of weeds when they occur within a field, to prevent them spreading.

Spelt, rye, triticale and oats, due to their tall stature and very competitive nature, are often able to out-compete weeds with negligible mechanical intervention. Where specific weed problems exist, early mechanical weeding at the three- or four-leaf stage is all that is required. These cereals can be mechanically weeded up until GS 30/31.

Wheat and barley are the cereals that respond best to mechanical weeding. Again, where specific weed problems exist, early mechanical weeding at the three- or four-leaf stage when weeds are small is the best strategy, weeding again if required at GS 30/31. Wheat can be weeded later at GS 35–40 where weed problems are severe, but forward speeds should be much lower to reduce crop damage, and there will inevitably be some crop damage from wheeling when weeding this late. Late weeding is only really viable with wide-row systems and using an inter-row hoe, where tramlines are located within the crop.

Weed Management in Winter Field Beans

Field beans tend to be more effective than peas at competing with weeds due to their foliage density and height, and this is an important factor for the organic farmer. Weeds such as black bindweed and cleavers can cause lodging where present in high enough numbers, and create harvesting difficulties. The options available to the organic farmer are these: using stale seedbed techniques, spring cropping, and either drilling the crop in rows in order to inter-row hoe, or using a comb weeder in the crop up to when the crop is approximately 10cm high.

Autumn-sown field beans are generally more resilient to weed interference than spring beans, which tend to be a more open crop and which are growing at the same time as spring-germinating weeds.

Once the crop is established, a pass with the comb harrow is often effective as early in the season as soil conditions allow. This will help remove weeds before they root too deeply and give the crop time to recover from any damage that may have occurred. Research has shown that the most benefit, in the form of yield gain, is when the crop is kept clean between November and March. Any weeds that emerge after March are unlikely to affect yield. Comb-harrow weeding the crop after this date may cause significant crop damage, and its cost may not be offset by any potential gain in yield. Currently, results indicate the most effective timing is at the second node stage (GS 102) either with or across the rows, and if necessary followed by a second pass along the rows at GS 105 (fifth node).

While field beans are tolerant of mechanical weeding, passes with a weeder should be kept to a minimum when the plants are approaching early bud. Current work is being carried out to evaluate the optimum timings for weed control.

Weed Management in Spring Field Beans

Work to evaluate mechanical weeding in spring field beans was completed in 2006 by the Processors and Growers Research Organisation (PGRO). They found that mechanical weeding can be useful for weed control in spring beans, and two passes tended to give improved weed control over the single pass, with weeding at both GS 102 and again at GS 105 giving the best weed control.

Young bean plant at a suitable stage for mechanical weeding.

Weeding at crop GS 102 is the most effective timing for weed control whilst allowing the crop to recover from any physical damage caused. Weeding across the row is more damaging than weeding in the direction of the rows at GS 102, but there is no difference in damage between directions of weeding when weeded at GS 105. There was no reduction in plant population or height by mechanical weeding.

Populations of fool's parsley (*Aethusa cynapium*) and cleavers (*Galium aparine*) were reduced by the mechanical weeder. However, fat hen (*Chenopodium album*) appeared to have been stimulated by the first pass of the mechanical weeder, with significantly higher numbers compared with the untreated control; but the second pass at GS 105 indicated some control. Mechanical weeding showed good control of field speedwell (*Veronica persica*), scarlet pimpernel (*Anagallis arvensis*) and annual sowthistle (*Sonchus*), with best results achieved by weeding at GS 102/105 across the rows, and with the rows for charlock (*Sinapsis arvensis*).

Farmers should expect that there will be a certain level of damage to the crop; typically this will be around the 0–20 per cent mark, and a compromise is always required between damage to the crop and weed control. However, this can be useful if carried out in very thick bean populations in the spring, and used to thin the crop to an appropriate population (*see* Chapter 7).

It is more difficult to grow winter beans on wide rows unless accurate deep autumn drilling can be performed. Spring beans can be grown on wide rows in an inter-row hoeing system with success.

Weed Management in Peas

Amongst the reasons why field peas are less common than beans in organic rotations are the concerns over peas 'lodging' (falling flat on the floor), and weed control. Good weed control is essential if the pea crop is not to be dominated by weeds. However, a low level of weeds in the crop can actually facilitate combining and reduce moisture by keeping the peas up off the floor.

Field peas tend to be slow to emerge and can be on relatively wide row spacings. Added to this, soil conditions during the early establishment phase are often not suitable for mechanical weed control such as an inter-row hoe. However, using an inter-row hoe later on in the crop's development can be useful up until canopy closure, when no form of weeding is possible thereafter. The use of comb harrows has been shown to be effective in some trials. Increasing seed rates may aid competition with weeds, however this may exacerbate crop lodging later in the season.

Peas are not a particularly competitive crop and if weed incidence is likely to be high, a response to weeding is likely. Weeding when the weed burden is low enough not to affect yield can germinate buried weeds, creating a greater level of weed competition than if the crop had been left unweeded.

A major part of weed control is using stale and false seedbeds. These are cultivations used pre-drilling. A false seedbed is created where a comb harrow can be used to provide weed strikes prior to drilling.

The best control of shallow-rooted annual weeds can be gained using mechanical weeding methods at early growth stages (GS 102–105). This is more effective on lighter soil types, and should be made when the surface is dry. Early weeding passes with a comb-harrow weeder can be made along or across the crop rows without causing significant damage to the crop. Later passes should be made along the rows but may cause damage to pea crops, especially when they have begun flowering. Severe damage will be sustained if passes are made across the rows at later growth stages.

Comb harrow weeding peas along the crop row.

Weed Management in Lupins and Soya

Many farmers use comb harrows, others use inter-row cultivation for weed control in lupins and soya crops. Some find that a competitive variety drilled in a clean seedbed is sufficient; others harrow once with a tined weeder. Where weeds are a particular problem the crop can be harvested as wholecrop silage before weed seed set occurs. This may give a weed control benefit in subsequent crops.

Mechanical weeding methods should only be used before lupins and soya reach 15cm in height, or unacceptable damage may be caused to the crop plants. Weeding when the crop plants are around 5cm in height with three to four leaves, and well anchored by the tap root, can be successful, with a first pass at 45 degrees to the rows, followed by a second in the opposite direction a few days later. Weeding in the direction of the rows may result in the crops being buried, unless a very slow forward speed is used. Forward speed should be reduced to these crops compared to peas and beans as damage is more likely.

Field studies of weeding post-emergence lupins showed that harrowing within the growth stages between the cotyledon and seven to eight leaves shows that lupins are fairly tolerant to soil covering, the primary mode of crop damage in strongly anchored crops. Different forward speeds and number of passes should be used to give as much soil covering of weeds as possible at the cotyledon, three-to-four leaf and seven-to-eight leaf growth stages.

In other field studies and pot tests, white lupin has shown strong suppression of weeds, in particular of corn spurrey (*Spergula arvensis*), fat hen (*Chenopodium album*) and broad-leaved dock (*Rumex obtusifolius*). Both the growing crop and residues incorporated into the surface 2cm of soil inhibited weed growth. Experience in the field suggests that early weed management is critical, as once the crop is developed to 20–30cm height it is impossible to gain access for mechanical weeding without serious crop damage.

Critical Thresholds for Weeding

For all crops it is important to understand the impact that weeds can have on crop yield potential. Competitive weed indices have been developed for non-organic crops, and whilst these may need a little interpretation for organic crops, the principles can be applied in the same way. Table 57 below shows the weed number per metre square needed to reduce potential yield by 5 per cent for a range of weeds; in this way their relative impact can be determined.

For example, in non-organic agriculture, cleavers are the most competitive weed in cereals, seven times more competitive than blackgrass and thirty-seven times more competitive than speedwells. This may not be the case in organic arable systems where cleavers are not able to proliferate where nitrogen fertilizers are not used. An example of some competitive indices and the formula for calculating the value of potential lost yield per hectare is shown below in the Table 57.

Table 57 Competitive Weed Indices

Weed	Competitive Index (% yield loss per weed)	Weeds per m² to cause a 5% yield loss
Cleavers	3.0	1.7
Wild-oat	1.0	5.0
Blackgrass	0.4	12.5
Mayweed	0.4	12.5
Poppy	0.3	16.7
Redshank	0.2	25.0
Common chickweed	0.2	25.0
Knotgrass	0.1	50.0
Crane's-bill	0.08	62.5
Speedwell	0.08	62.5
Field pansy	0.02	250.0

Number of weeds per m² × Competitive Index = % yield loss % yield loss × average yield in tonnes per hectare × average £ per tonne = VALUE OF LOST YIELD PER HECTARE

THE LIFE CYCLE AND CONTROL OF SOME ARABLE WEEDS

Wild Oats

Wild oats can be found as winter or spring germinating. They can grow to 1.8m tall, though some are as short as 50cm. The seeds of the wild oat fall to the ground singly and require vernalization to germinate. Seeds of the winter wild oat can germinate from October to early March, whereas the spring wild oat can germinate in the late winter after vernalization or in the spring, though it is usually in the spring.

One of the strengths of wild oats is the ability of the seed to remain viable in the soil for between ten and twenty years in undisturbed land, or five to ten years in land subject to repeated cultivations. Wild oat infestations are normally associated with continuous cereal production, though they can occur in high numbers in field beans and other crops. Yield losses can be as high as one tonne per hectare. They can spread through contaminated grain, birds and contaminated machinery, and can increase in numbers through poor rotation and weed control.

Control of wild oats will vary upon location, severity and cropping. Where spring wild oats exist, a summer fallow can reduce populations by up to 65 per cent; however, it will have no effect on winter wild oats. Where winter wild oats are a problem, switching to spring cropping can have a significant advantage.

It is important to remember that wild oats contain viable seed even when the panicles are still green, and there will be a proportion of wild oat seed that is non-dormant and can germinate within a few weeks. The majority of seed requires vernalization to initiate germination, therefore leaving seed on or near the surface is important if germinating it forms part of a control strategy: if buried more than 15cm the seeds can remain dormant in the soil once cultivations have started. It is better to leave dropped wild oat seed on the surface overwinter to vernalize and germinate, before destroying with cultivations.

Cultivations prior to spring cropping can effectively control winter wild oats, and a series of spring crops can help to seriously reduce the problem. Spring wild oats can be controlled by fallows or delayed drilling in the spring.

Blackgrass

Blackgrass is a native of the British Isles. Traditionally it has been a problem mainly on heavy land throughout south-east England, though in recent times it has spread on to chalk land and even sands. It spreads into the field from the hedgerows, though is not seen above 1,000ft. Blackgrass is a source of ergot, and flowers between May to August. It can self-pollinate and cross-pollinate, producing 50–6,000 seeds per m^2. These seeds will be shed before wheat is harvested, and 50 per cent before barley has been harvested, so home-saved barley seed should be thoroughly cleaned before drilling. Once the blackgrass seeds have been shed on to the ground a small proportion will germinate straightaway, while the majority will lie dormant for up to four months.

Exposure to light and fluctuating temperature will stimulate germination mainly in October, but with mild winters this can continue through to the new year. Finer seedbeds don't allow as much light to penetrate, and this can aid in reducing blackgrass germination. Blackgrass will germinate in October and November, though there can also be a small flush in the spring; it germinates best at around 8°C. The seeds retain viability for up to four years even in ploughed systems. Straw burning (when it was permitted) was a very effective way of reducing blackgrass seeds in the seed bank and on the soil surface.

An effective method to help keep blackgrass seed numbers down is to move to spring cropping where possible, allowing blackgrass to germinate in the autumn, winter and in the early spring, when it can be cultivated out at the seedling stage prior to establishing a spring-sown crop. Another method is to have a fallow in the rotation; however, these tend only to reduce numbers if they are at least every four years in the rotation. Where delayed drilling is practised blackgrass numbers can be reduced, as stale seedbed techniques can be employed before drilling. It can also be grazed by sheep in the growing crop, and inter-hoe weed management systems are proving effective on some farms. The main control in organic systems is the one- or two-year 'ley' break which does not allow the blackgrass plant to complete its life cycle, and helps induce dormancy in buried seed by not having cultivations for one or two years.

Blackgrass is becoming a more common occurrence in Scotland, possibly a result of mild winters and a move towards reduced tillage cultivations.

Annual Meadow Grass

This weed is common throughout the country. It is mainly seen on loams and clays though rarely on chalk land; it cannot tolerate acidic soils or those that are low in phosphate. It has many ecotypes, and can also be a perennial weed. It has been seen at altitudes up to 3,900ft. Its main strength is its ability to colonize bare patches of soil quickly, which is why it is frequently seen on tracks, poached land and around gateways.

The weed tends to germinate from February to November and to flower from February through to the end of the year, depending on the season. Peak seed production takes place through April and September. Where the soil is dry, germination is significantly slower. The earlier the plant germinates, the larger it will be at the end of the season. Seeds produced in the summer and the early autumn can germinate a month later, while seeds produced from August onwards may overwinter before germination in the spring.

In trials where no seed was allowed to return to the soil it took four years for 99 per cent of the seed produced previously to germinate. This shows that even if the most optimistic farmer thinks they have controlled the weed, there will still be several years of seed in the soil. Seeds consumed by animals tend to lose viability, and those in farmyard manure for a prolonged period also lose their ability to germinate.

The most effective way to control annual meadow grass is through surface tillage. This will encourage germination, and once the seed has germinated the seedlings can then be controlled by further cultivations. Deep ploughing can also aid control, as can inter-row cultivations. However, flame weeding is ineffective as the plant has a basal growing point.

Rough Meadow Grass

Rough meadow grass occurs throughout the country, mainly on grassland, though in recent times its occurrence has increased on arable land. It tends to encroach into the field from field edges and hedge bottoms. It is a perennial grass where some tillers can turn into stolons and allow the plant to spread. Around one third of winter wheat fields contain rough meadow grass.

The plant can shed up to 1,000 seeds between the beginning of June and August, though this will depend on the season. Seeds that mature at the ends of the panicles tend to mature and be more dormant than ones closer to the stem. Once the seeds have fallen they require light to germinate; thus seeds on the surface will germinate, and rainy days followed by dry days will increase germination and decrease dormancy. Seedlings will emerge from March to October up to a depth of 3mm. Seeds can occur from depth by worms depositing them in worm casts.

Rough meadow grass needs a period of low temperatures while in the vegetative stage in order to flower, so it is not a problem in spring crops as it occurs in its vegetative state. However, in the winter crop the plant would experience a cold spell and will flower during the grain-fill period. Though it can occur in substantial numbers, yield tends only to be affected by a small amount, mainly due to the fact that it doesn't take up that much room in the field.

Studies have shown that the best control is to allow the weed to germinate, then to bury the seedlings completely: the key is to bury completely either the seedling that is growing or has been cultivated on the soil surface.

Couch Grass

Couch grass can be found throughout the British Isles. It is found more often on heavier soils, though is being found on light land. It does not grow well on acidic land. It tends to encroach into the field from the field edge and hedge bottoms; however, it is also thought to spread from contaminated seed. Where this has occurred, seed viability can be far higher as the new couch can cross with the cloned one already present, producing more viable seed than the cloned plant itself.

Contrary to an old adage stating that couch cannot be killed, it can. This can happen through a number of means, including rotation, cultivations, fallows and correctly timed grazing.

There are two main forms of spread, firstly by seed, either from within the field or from bought-in or home-saved seed; and secondly through rhizomes, the subsurface shoot organs of reproduction. It is now thought that seed plays a larger role than was first thought. The plant flowers from April to May, then the seeds mature in August to September. These seeds then germinate in the autumn. Germination may be delayed by cold temperatures, lack of moisture or if the seed becomes buried. Seeds may even be viable when they're green. Once buried, seeds may survive for between two to three years, though they may remain viable for up to five. Seeds can survive digestion by livestock except for that of pigs, which when in an infested field will eat the rhizomes (another effective control method).

The second form of spread for couch is through rhizomes. Once a couch plant starts to grow, it will send up shoots during the autumn and produce leaves. It will use these to gain photosynthesis and build energy reserves for the following spring and summer, when it will then spread horizontally by the subterranean rhizomes. The plant will continue to do this throughout the season until the following autumn when horizontal spread will cease.

Traditionally and on organic farms, couch has been controlled by summer fallow cultivations, although if these are mistimed and not carried out correctly in the right weather conditions, the problem can become severe. The ability of the grass to spread is one of its strong points, because as a seed it is not competitive against other grasses; however, it is also one of its weak points, as rhizomes require energy to spread.

There are two methods that can be used to control couch – tined fallow cultivations and ploughing, or both used in combination. It is important that the soil has not been dry for a period of time, as this can force the rhizomes into dormancy and cultivation will be ineffective.

It should be noted that the couch should be actively growing once it has been completed. In early May the field should then be loosened to below the depth of the rhizome (that is, to at least 15cm depth), or deep ploughed (with the plough skims taken out of work to ensure total inversion) so as to bring the rhizome to the soil surface. The field should then be repeatedly cultivated with a tined cultivator with wide sweep tines for maximum efficiency, to stir the soil and bring the rhizomes on to the soil surface for desiccation. Using cultivators with wide 'duck's feet' to ensure total overlap of blades is more effective than straight tines, where the rhizomes can travel between the tines. The Kvik-up and CMN harrows are a great improvement on the traditional cultivator.

The field should then be left for ten to fourteen days before repeat cultivations are undertaken, allowing time for shoots to appear from the growing point on the rhizome. This is the point where success or failure can occur. If it rains continually from this point on, cultivations will be ineffective and the problem will become worse. Ideally there will be three or four days of dry, warm weather that will desiccate the rhizomes, and they will be particularly vulnerable at this point as they will have put their last amounts of energy into the leaf. Once the plant is at the two-leaf stage cultivation should take place again, and provided the soil is dry and no rhizomes are buried, the couch should die out. In trials, this has shown to be 90 per cent effective. It is important that the aerial shoots are not allowed to become longer than 12cm, as this will indicate that energy replenishment is under way.

It is also important that cultivations minimize the chopping and cutting of rhizomes into small segments, as this will only serve to spread the problem around the field. To this end, discs and other machinery that chop up the soil should not be used in fields where couch is a problem.

It is important to note that in order to kill the rhizomes they must be exposed to the air for three or more days. A good test to see when soil conditions are ideal for this activity is when the soil falls loosely from the rhizomes. Rhizomes exposed to frost will also perish. These methods, combined with a rotation containing a competitive grass clover ley and ideally a fallow where required, are extremely effective at controlling couch grass, though having said that, they will always be dependent on the weather.

Creeping Thistles

The creeping thistle is a weed of arable and grassland. Although it prefers light soils, it can still be seen on heavy land. The spread of the plant is usually encroachment from the hedge bottom; although the seeds can float in the air, they often break off and fall very near to the parent plant. A soil containing a high level of potassium will favour thistles, as will a low phosphorus soil. Thistles flower between June to September, and the seeds become viable as quickly as six days after flowering; they tend to emerge from between 5 and 15mm from the soil surface. The plant is a poor competitor early on, though once it has become established it may regrow from the tap root after cutting in as short a time as nineteen days.

Once the plant has established a tap root it will spread lateral subterranean rhizomes after eight to ten weeks. These lateral rhyzomes contain adventitious buds that develop. Once these have formed, the plant is able to regrow if cultivated. The rhizome root system is very brittle, and cultivating through the system can break the roots up into fragments. These fragments, if longer than 2cm, then have the ability to grow into new plants, when the cycle is repeated and the plant spreads. It is thought that the roots are allopathic against crops, reducing germination and vigour. Broken pieces of root may remain viable for many years. It is therefore important that any cultivations minimize the chopping and cutting of rhizomes into small segments, as this will only serve to spread the problem around the field. To this end, discs and other machinery that chop up the soil should not be used in fields where creeping thistle is a problem.

In grassland the aim should be to encourage a dense sward that will prevent infestations. Where thistles have become established, frequent cutting and tight grazing can aid control. How many creeping thistles do you see in frequently mown lawns?

It is important that cutting is low enough to remove all the leaves as this will exhaust the root reserves sooner. Topping should take place at least twice a season, though more frequent cutting will reduce infestations sooner though it may take several seasons. Ideally early season topping should be combined with forage conservation cuts and mid-season and late season topping. It is important that the thistle is cut at the early bud stage when food reserves are low; if it is left it will flower, and from this point until the winter will use the foliage to build up reserves. Where the plant is cut before winter it will not have the opportunity to build these reserves.

A quicker method of control is the manual pulling of thistles, but this tends only to be viable on small patches. Stock will eat thistles when these are young, but as they get older palatability rapidly drops.

In arable situations, a fallow is required for control. First, inspection pits should be dug to determine the depth of the horizontal rhizomes. These normally occur at the bottom of the cultivated layer, the path of least resistance for the horizontal rhizome to spread. Where possible, the field should then be loosened to below the depth of the rhizome in early May – that is, to at least 15cm depth – or deep ploughed (with the plough skims taken out of work to ensure total inversion) so as to bring the rhizome to the soil surface.

The field should then be repeatedly cultivated with a tined cultivator with wide sweep tines for maximum efficiency, to stir the soil and bring the rhizomes on to the soil surface for desiccation. Using cultivators with wide 'duck's feet' to ensure total overlap of the blades is more effective than straight tines, where the rhizomes can travel between the tines. The Kvik-up and CMN harrows are a great improvement on the traditional cultivator.

The field should then be left for ten to fourteen days before repeat cultivations are undertaken, allowing time for shoots to appear from the growing point on the rhizome. This is the point where success or failure can occur. If it rains continually from this point on, cultivations will be ineffective and the problem will become worse. Ideally, there will be three or four days of dry warm weather that will desiccate the rhizomes, and they will be particularly vulnerable at this point as they will have put their last amounts of energy into the leaf. It is important that the shoots and leaves are not allowed to develop, as this will indicate that energy replenishment is under way.

It is also important that cultivations minimize the chopping and cutting of rhizomes into small segments, as this will only serve to spread the problem around the field. To this end, discs and other machinery that chop up the soil should not be used in fields where thistles are a problem.

It is important to note that in order to kill the rhizomes they must be exposed to the air for three or more days. A good test to see when soil conditions are ideal for this activity is when the soil falls loosely from the rhizomes. Rhizomes exposed to frost will also perish. These methods, combined with a rotation containing a competitive grass clover ley and ideally a fallow where required, are extremely effective at controlling thistles, though how successful they are will always depend on the weather.

It should be noted that thistles provide a very important food source and habitat for birds, flies and many insects, and the flowers are attractive to honey bees. In emergence trials, thistles germinate for four years, and at five years there was less than 1 per cent emergence, and in other trials there was no emergence after thirty years. Fallowing can aid control and provide time for germination, and the cultivating and hoeing out of the seedlings.

Sterile Brome

This grass weed is common throughout the country and is often seen on waste ground, roadsides and at the bottom of hedgerows. The use of shared machinery and contractors increased its incidence with the weed often occurring in the middle of fields when usually its spread comes from the hedgerow. The weed can produce between 12,000 and 53,000 seeds.

The majority of seeds germinate very soon after harvest, provided there is moisture in the ground. If the ground is very dry and sunny for a prolonged period the seed may go into dormancy and germinate in the following spring. Most seed is shed between June and August. In trials, all seeds germinated after harvest where it was mixed with soil; where the seed remained on the surface for a prolonged period after harvest it tended to go into dormancy. In field experiments, sterile brome remained viable for only one year when it was buried below 20cm.

Provided that the seed has not gone into dormancy, ploughing has shown to be a very effective way of controlling the weed, provided it is ploughed beyond 20cm depth. The grass will attempt to grow to the surface, and will perish where it has run out of energy. If the seed has gone into dormancy it may germinate in the following spring, though it may not flower because a vernalization period in the vegetative period is required.

Where the weed seed has been mixed with soil through light cultivation after harvest, control can be very affective where a spring crop follows, as this gives the opportunity to create a stale seedbed and to kill off emerging seedlings.

Cleavers

Cleavers are native to the British Isles, being a dicotyledonous, or broad-leaved, annual species. At the early seedling stage cleavers produce distinct, large cotyledons with a characteristic notch at the tip. Until the first whorls begin to appear, these seedlings can be confused with those of ivy-leaved speedwell. Cleavers are traditionally seen in hedge bottoms, where they climb up and smother hedgerows. They can easily encroach into the field, where they can divert nutrients and light from the crop, lodge crops and delay harvest. Cleavers tend to grow on nutrient-rich sites and can be found on various soil types. They thrive on high levels of nitrogen in the soil, and in organic systems can be considered as a good guide to soil nitrogen levels. They are surprisingly resilient to drought due to their extensive rooting. They are a major problem in non-organic systems where they thrive on the nitrogen fertilizers applied. For most farms in organic arable farming systems, they are far less of a problem than many other weeds.

A very important aspect of cleavers' biology that contributes to making this a problem weed is the characteristic emergence pattern observed in autumn-sown crops. In autumn, cleavers' seedlings emerge in a flush shortly after the crop is drilled. This flush is more prolonged than for many other broad-leaved weeds. In addition, unlike most problem arable weeds, in cleavers a second flush in the spring of the following year often occurs. The precise timing of emergence, and the balance between spring- and autumn-emerging cleavers, varies from year to year.

Cleavers tend to germinate between August and May the following year, and to peak in March and April. In summer, typically June, numerous tiny white flowers produce pairs of seeds covered in hooked hairs that can attach themselves to clothing or animal fur, partly because of the size the adult plant can reach, and its ability to climb over the top of the maturing crop.

Cleavers set seed between June and August, and produce around 300 seeds per plant, though this can vary considerably. Once harvest has been completed, some seed may be found in the combine tank, while the majority may have already fallen to the ground. Some cleaver seed will be spread around the field by the combine, and animals will also carry cleavers on to clean fields.

If the seed is left on the stubble surface, light will inhibit germination: it is not until there is sufficient moisture and the seed is covered by soil that germination will take place. Seeds can emerge from 4–15cm of soil.

Management of cleavers should aim to harrow the field after harvest and roll to consolidate moisture; this will encourage germination, and once the seeds have germinated they can either be harrowed on a warm day and killed, or ploughed in to below 20cm. If the surface layer is ploughed and incorporated, any ungerminated seeds will lose viability after two to three years. For buried seed there is an optimum depth of burial of 2.5–5cm, with emergence reducing progressively with increasing depth. Little significant emergence occurs from seeds germinating from a depth greater than 15cm down the soil profile. This is one of the reasons why ploughing has the potential to suppress weed populations. In trials, cleavers were eradicated completely where a four-year fallow was used; periods of the rotation containing grass leys for three or more years will also greatly aid control.

A one-year fallow can reduce cleaver infestations by 85 per cent, and a second year can increase this reduction to 90 per cent. During a fifteen-year trial cleaver infestation was reduced by 80 per cent where a fallow was used every five years. Late spring cropping after a winter crop will also allow a long germination window where the fields are cultivated to cover the cleavers with soil to encourage germination, when they can be harrowed out prior to drilling the spring crop.

Docks

There are various species of dock throughout the British Isles, two with agricultural importance, namely the broad-leaved dock and the curled dock. Both of these can cross with each other and produce hybrid plants.

Often called the 'number one enemy' by some organic farmers, docks are one of the hardest weeds to control in organic systems due to the plant's ability to regenerate by the production of up to 60,000 seeds and also vegetative spread. Germination from seed and rapid development is worse in cultivated ground and in bare patches in grassland. The seeds are particularly resilient, and can survive digestion by animals, and they can even survive for long periods in farmyard manure and non-aerated slurry. Over-application of farmyard manure and slurry can cause patches of pasture to die out, and these will be vulnerable to the establishment of docks. The dock is actually not that competitive when it is at the seedling stage, and this is why it is important not to let it get established.

Once the seedling is forty days old it will have a tap root which can provide a reserve of food and put up another shoot even after it has been mowed or grazed. Poached areas are vulnerable to docks, as are high nitrogen situations. For the arable farmer, docks have also proved to be a problem. Seedlings take the opportunity to emerge whenever conditions are favourable, though this is usually around March, April and late July to October. Established plants can withstand mowing, grazing and trampling, and constant cutting can result in very dense foliage. In trials when the plant was cut five to seven times per year grass yield was not impaired, though where it was only cut three to four times per year, grass yield fell.

This will also ring true for the organic arable farmer, though control is only really possible pre-drilling with the use of stale seedbeds and the use of inter-row hoes. It should be noted that the broad-leaved dock does not usually flower in the first year, so post-harvest cultivations should aid control. The dock's main stem or 'crown' sits around 5cm underground, and it is thought that the plant can regenerate from pieces of the tap root from the top 7cm of curled docks and 4cm of broad-leaved docks – so cutting up the tap root can make the problem worse.

The dock can set 60,000 seeds; these may also be slightly different, and may require slightly different conditions in order to germinate. This unfortunately doesn't allow a huge flush of the weed when germination is encouraged with cultivations. Dock seeds can last for more than fifty years when buried in the soil, and the seed can still be highly viable even from cut-down stems.

The most effective method of control of established dock plants is removal with a weeding tool in both grassland and arable land. These can remove the entire root, which should then be completely removed from the field and disposed of.

The basic principle for reducing dock populations in arable fields is to exhaust the substantial energy reserves held in the plant's root system. After harvest the dock roots should be dislodged on to the surface with a heavy spring tine cultivator. Using cultivators with wide 'duck's feet' to ensure total overlap of blades is more effective than straight tines, where dock root segments can travel between the tines. The Kvik-up and CMN harrows are a great improvement on the traditional cultivator. Cultivations should continue through any period of dry weather to get as many dock roots on to the surface as possible in order that they dry out.

The roots should be removed off the field surface unless growers can be absolutely sure that complete desiccation has been achieved. Some farmers cultivate three to four times in the summer months after harvest, which brings the roots to the surface and dries the soil out throughout the profile, aiding desiccation. Deep ploughing mature docks should be avoided, as the burial of the 'crown' will not kill it, and will encourage

Organic wheat in an area of a field having been managed with repeated dock fallowing cultivations.

development of the plant from depths at which control measures are extremely difficult to employ.

The third method of control is only appropriate for mature, established docks. These are best managed at the end of a two- or three-year clover ley period. During this stable period without cultivations, the crown of the dock moves closer to the soil surface and is therefore more accessible. The method is to break up the ley by rotovating to a depth of 10–15cm (4–6in) in May, when there is sufficient soil moisture to undertake the cultivations. This should remove the crowns on to the soil surface. This should be followed with regular fallowing cultivations on a ten to fourteen-day basis during dry weather to desiccate crowns and the roots.

An alternative method is to rotovate progressively lower and lower each week for three or four weeks. The period between each pass allows the plant to put up new shoots, which are then removed. The idea is that the continual requirement to put up new sprouts exhausts the root's energy reserve. Dry weather is vital during the desiccation period in order that the root dries out.

The fallow cultivations should be repeated through May, June, July and into August for a good level of control. This can then be followed by the re-establishment of clover or a short-term green manure such as forage rape for grazing, or by establishing a competitive cereal such as rye, oats or triticale.

Chapter 11

Pest and Disease Management

THE PRINCIPLES OF PEST AND DISEASE MANAGEMENT

Pest damage and disease incidence has the potential to cause serious damage to crops and financial losses to farmers. Thankfully, in cereal and pulse crops, pest and disease pressure is often greatly reduced in organic systems as a result of the greater crop diversity, higher levels of crop bio-diversity, and the level of natural predators and reduced levels of crop stress.

Not all insects are harmful to crops: indeed, some can be a major bene-fit in keeping aphids and other pest numbers down. It is not just the distinctive *adult* ladybirds that eat aphids: they also eat them whilst in their grub stage. Furthermore lacewings, hoverflies, parasitic wasps and many beetles also contribute to keeping aphid numbers down.

The total elimination of pests and diseases is impossible in practice, and even if it could be achieved within the limits set by organic standards, it would incur the risk of a resurgence of pests and diseases, as natural enemies and competitors to plant pathogens are lost from the production system.

Organic Standards

Organic standards recommend the following methods for controlling pests and diseases:

- Creating fertile soils of high biological activity to provide crops with a balanced supply of nutrients.
- Operating diverse rotation and cropping programmes.
- Choosing resistant crops and varieties that are suited to farm condi-tions.
- Carefully planning planting and harvest dates.
- Encouraging natural predators within and around crops by:
  o mixed cropping and undersowing;
  o retaining hedges, wildlife corridors, field margins and windbreaks as natural non-cropped habitats.
- Using good husbandry and hygiene practices to limit the spread of any pest and disease.

Any pest and disease management programme should be designed according to organic principles, and should be planned so as to maximize the probability that pest and disease problems will be kept in check without excessive intervention by the farmer. In practice this means that pest and disease levels only need reducing where necessary, and to levels where they do not cause significant economic losses, using a range of preventative techniques and avoidance strategies.

Avoidance Strategies

The simplest way to deal with pests and diseases is to avoid their occurrence in the first place. As with other potential problems, organic agriculture always seeks to avoid problems by developing management tools and avoidance strategies so that the problem remains at a low, manageable or acceptable level of incidence or threshold.

The best way to control pests and diseases in an organic arable system is to carefully design and manage the whole farming system to achieve health, diversity and vitality in soils and crops. This then encourages natural growth and a balanced farm ecosystem. The areas described below should be considered with regard to avoidance strategies for pest and disease management.

Natural Control

Rather than seeking to dominate the problem and control it with intervention approaches, nature can often be harnessed to apply its own checks and balances in the constant and ever-changing battle between pest and predator. This ecological cycle occurs in the above- and below-ground crop environment, and is well understood above ground, with interrelationships between pests and predators and diseases and hosts. Below ground is rather different. The soil environment is often called a 'black box' with a lesser understanding of the biological components and their interactions. What *is* understood, however, is that a 'soil food web' exists, with soil animals, anthropoids, fungi and bacteria all interconnected in a complex food chain with interrelationships existing between pests and predators, diseases and hosts, just as in the above-ground environment.

An ecological approach to pest and disease management must recognize that there is no single factor which is responsible for the pest and disease problem, and a range of approaches and strategies may be required for management. The best way to achieve this is through bio-diverse habitats rather than monocultures. With monocultures comes specialization, not only of the production but also of weeds, pests and diseases, which thrive in their own stable monoculture environments. With monoculture also comes higher risk. With diversity, however, comes a greater degree in crop variation, and following on, a myriad of different pests, predators, diseases and hosts, but all at a much lower level, and one often presenting a blockage or deterrent to another. This creates balance, and this is how nature works.

Both the diversity and stability of pest: predator and disease: host relationships are influenced by many factors:

- The diversity of crop and plant species.
- The permanence of crops (i.e. annual or perennial crops).
- The sequence of cropping and the rotation of crop types used.
- Field arrangements, including the size and presence of habitats for beneficial insects.
- The composition, management and permanence of surrounding plants, including wild species.
- The soil, its health and nutrient status.
- The complexity of interrelationships between crop and non-crop plants, pests: predators and diseases: hosts.
- The type and intensity of the farm management system.

What this means in practical terms is that farms have to develop a range of preventative options and strategies to help manage pest and disease problems on the farm.

Rotations
Rotations are discussed in Chapter 5. Rotations are diversity over time, and will remain the main tool in managing pests and diseases on most organic farms, by programming crops and break periods to interrupt pest and disease life cycles, by seeking to establish cropping environments that are as inhospitable as possible to the particular pests and diseases, and in seeking to avoid the problem as much as possible by cultural methods.

Under an organic system, pests and diseases can largely be controlled by using rotations where one crop is not immediately followed by the same crop, and by selecting varieties that are highly tolerant of crop diseases and pest attack, rather than for high-yielding properties. The continued development of predator-friendly environments around field margins will assist in building natural predator numbers. (*See* 'Natural Control', page 308.)

In many cases pests and diseases are unable to survive in the absence of the host plant (or crop), and the absence of the host will cause a decline in population numbers or occurrence. The mechanisms for decline are varied, and can include starvation and being out-competed by another pest or disease. In all cases, rotations should be designed for the longest practicable period between growing crops of the same family. For example, growing beans adjacent to peas in the rotation is not advisable due to the high risk of pest and disease carryover. Some crops are more tolerant than others; these are discussed in Chapter 5.

The main role of the rotation is in controlling soil-borne pests and diseases. Where rotations are likely to be less successful is in the control of the more mobile insects and pathogens that survive in the soil, on plants and in the air (i.e. spores spread by the wind).

Crop sequences   Crop sequences entail diversity over time and space, with different crops grown at different times in the same place, and the same crop grown in different areas. This helps to avoid the build-up of pests and diseases on the farm as a whole, and in specific crops. Crop sequences are discussed in more detail in Chapter 5.

Crop timing   Diversity over time can be employed by growing a mixture of spring- and winter-sown crops, and by sowing, weeding and harvesting at different times. These will all interrupt the life cycles of different pests and diseases and aid control. Work on the principle of depriving the pest of a suitable host at important stages of its lifecycle, and reducing the time available for the pest to exploit the crop. In particular, altering sowing dates can help with the avoidance of egg-laying of certain pests, allowing plants to become well established before any attack occurs.

Resistant Varieties

Resistant varieties afford an opportunity to create diversity in space. Crop types and individual varieties within each crop type differ in their susceptibility to pathogens. Plant resistance is expressed in many different ways, from natural dieback, to immune system reaction and sophisticated chemical defences that hinder or kill the pathogen. Resistance mechanisms to disease can also include morphological characteristics, such as thick waxy layers on plant structures that prevent pathogens gaining access to the plant tissues. These are common in crops such as brassicas, but less so in cereals and pulses. In the case of some pests, such as insects and nematodes, plants can produce a range of chemicals and physical structures such as hairy leaves that deter a pest from feeding. Some plants even produce semi-chemicals that signal that the plant is under attack, and which can serve to attract other predators of the insects.

In practical terms, as a first step the NIAB, HGCA and PGRO recommended lists for cereals, pulses and oilseeds can be used to narrow the choice from varieties produced for non-organic production systems using a 'best fit' approach. These give excellent information on the resistance to a particular disease problem. Where specific diseases are known to be a problem, or local climatic conditions raise the likelihood of a disease occurring, selecting varieties that exhibit resistance for that disease should take a high priority.

As well as the HGCA, PGRO and NIAB recommended lists, growers should also speak to neighbours and local organic farmers who may have experience of a certain variety grown in similar conditions to their own, as well as visiting organic farms and trial sites or programmes.

Varietal Mixtures

Organic farmers often plant mixtures of varieties, working on the principles that greater diversity brings benefits in terms of reduced risk and improved pest and disease control. This is one of the simplest ways in which diversity can be achieved in an arable crop. Modern plant-breeding techniques have resulted in a move away from genetically diverse ancient landraces, to the widespread use of a few high-performing varieties that are far more susceptible to disease. A good example of this is the development of oil-seed rape which is now the third largest crop in the UK after wheat and barley. The varieties used are based on only two highly productive cultivars, and as a result septoria and alternaria have become significant problems.

On-going research into growing cereals with two or more varieties at the same time has demonstrated that the resistance of one variety can be impaired by susceptible varieties grown at the same time, and that mixtures provide a mechanism for reducing the variability in production performance on an annual basis, resulting in a more reliable performance of a crop type over time. Mixtures usually yield at least as well as the mean of their components, often more so, and sometimes exceed the highest yielding component.

More recent work by Professor Martin Wolfe and his team at Elm Farm Research Centre is evaluating the potential of composite cross population of wheat for organic production. This is taking twelve high-quality bread-making varieties and nine high-yielding varieties, and performing crosses in every conceivable configuration to arrive at an F2 composite cross-pop-ulation, rather than a straight-line variety. These are being trialled on a wide range of farms, and early indications are that they are producing robust wheats that could be very useful for organic producers.

The reasons that mixtures can be successful in reducing disease inci-dence are complex and varied. Different varieties have different resistance patterns, some being more susceptible, some more resistant to particular diseases. Certain diseases, such as powdery mildew in cereals, rely on close contact with neighbouring susceptible plants for dispersal. Mixtures help spread out the susceptible and resistant plants, and this affects the rate at which spores can multiply and spread the disease.

Crop mixtures can also be grown using the same principles. Mixtures of cereals, cereals and grain legumes or cereals and legumes such as clover are all realistic and workable options (*see* Chapter 6).

Crop Vigour

Planting crops and varieties that are vigorous in their development and growth characteristic can help combat pest and disease pressure. A fast-developing, vigorous crop will be able to 'grow away' better from

Vigorous crop of triticale.

disease and pest attack than a crop that is struggling to develop. One prerequisite for vigour is a good supply of plant nutrients, water and air to aid development.

Soil Fertility

The importance of maintaining a biologically, chemically and physically healthy soil has been discussed at length in Chapter 2. Generally, a healthy soil is one that will be able to provide a good supply of nutrition to the crop, allow good plant development, a healthy root system to develop, and which permits high levels of biological activity to allow efficient cycling of nutrients.

A healthy soil is important for pest and disease management for a number of reasons:

- Good growing conditions produce healthy plants, which are better able to defend themselves.
- Vigorously growing plants are better able to grow away from pest and disease attack.
- Stressed plants are more likely to be attacked by pests and diseases.

A good example of a healthy soil being biologically active can be demonstrated with fungal associations, where beneficial fungi occupy sites on plant leaves and roots, which inhibits the colonization by detrimental fungi, by literally taking up space. For example, mycorrhizas are symbiotic, beneficial fungi that colonize the roots of many arable plants, and in doing so the fungi assist the plant in its ability to absorb nutrients, in particular phosphate, in exchange for sugar and starches from the plant for its own use. Their presence on plant roots largely excludes damaging or 'bad' fungi.

A healthy soil with a vigorously functioning ecology can directly suppress soil-borne pathogens, including nematodes and insects. Plant disease pathogens only monopolize plants in the absence of stronger soil organisms than themselves, and can be suppressed in biologically active soils. The phenomenon of soils suppressing disease is well documented, although the exact mechanisms are not well understood.

Natural Predator Habitats

A key element for the successful control of pests and diseases on organic farms is to provide a sufficient range and concentration of suitable habitats for beneficial organisms to survive on the farm all year round. For example, providing suitable habitats for hoverflies and ladybirds to overwinter on the farm retains a healthy population all year round. When pest problems such as aphids occur, these natural predators are already in place to assist with control. In turn this can reduce disease transfer occurrences such as barley yellow dwarf virus (BYDV), which is principally spread by aphids.

Field margins containing tussoky grasses and wild flowers are overwinter habitats for many beneficial insects.

Simple measures, such as establishing field margins containing tussocky grasses and wild flowers, ensure the provision of ideal habitats for many beneficial insects. These also have the benefits for weed control (*see* Chapter 10), and can also form part conservation schemes.

Ladybird eating aphids on a crop of field beans.

Ladybird eating aphids.

Beetle banks can also be used to reduce field sizes and ensure that beneficial insects can gain access to all the crop. These provide habitats in the field for crop pest predators (for example, ground beetles) to overwinter and move out into the crop in spring. Research has demonstrated that they can contain up to 1,500 beetles per square metre only two years after sowing.

There is always a background level of pests and predators present on any farm. Population dynamics change with time, season and food source availability; mortality may also be due to natural predation. Some species (for example, *Pterostichus melanarius*) have been shown to react spatially to aphid colonies and have an impact on their population levels.

The use of beetle banks and field margins has been demonstrated to significantly assist in maintaining beneficial organism numbers. The important things to remember when planting and establishing beetle banks and field margins include the following:

- A range of wild flowers, grasses and herbs are likely to attract the greatest number of beneficial organisms, with different insects attracted to different plants (Figure 34).
- Pollen and nectar availability is important for the development of eggs from wildflowers (Figure 35).
- Adults lay their eggs on a wide range of host plants.
- The better the source of pollen/nectar, the better the increase in the number of eggs laid.
- The ideal is to provide a range of plants flowering at different times during late spring and summer to attract different beneficial insects.

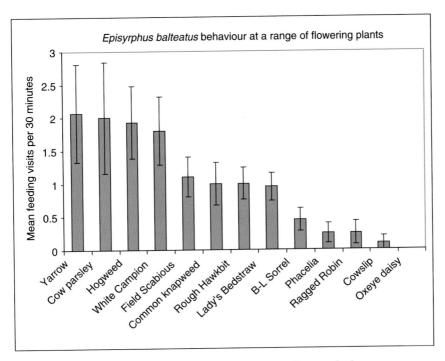

Figure 34 Different species are attracted to different host plants as food sources

Source: P. Northing CSL

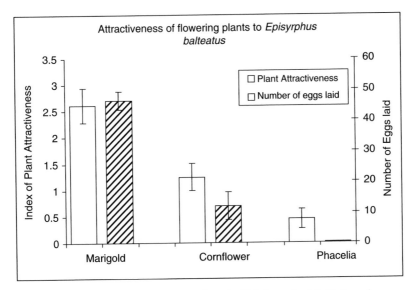

Figure 35 Pollen and nectar availability of marigold is important to Episyrphus balteatus *for egg laying and reproduction of beneficial insects*

Source: P. Northing CSL

Allopathy

The principles of allopathy are set out in Chapter 10. Allopathy also has a role to play in pest and disease management, using the allopathic effects of some plants to deter specific pests and diseases. Little research in this area has been undertaken, and there is great scope to use these naturally occurring effects to greater advantage in organic farming systems.

Brassica plants are known to release glucosinolates from their rooting systems, which are antagonistic to soil-borne pests. This is well known for oil-seed rape and mustard; therefore mustard used as a green manure can assist in controlling soil-borne pest problems.

Recent developments in better exploiting the glucosinolate properties of brasiccas has resulted in the use and marketing of 'Caliente mustards', which have a higher level of glucosinolates than normal mustard. These are being promoted as having the potential to be used as soil biofumigants.

The biofumigant process works on the principle that the unbroken cell of the mustard contains myrosinase and glucosinolate. When destroying the mustard by rapid maceration and incorporation into the soil, the myrosinase and glucosinolate is changed into d-glucose, nitrile and isothiocyanate gas (the latter being the biofumigant). In order for this to work successfully the mustard must be destroyed when mature and incorporated rapidly, ideally to a depth no greater than 15cm in a moist soil, producing a fine tilth, and then rolled immediately to seal in the isothiocyanate gas. Moisture at incorporation is paramount to produce the gas, as it is a reaction that requires water. After incorporation the soil should be left for fourteen days, before drilling as soon after the fourteen days as possible to achieve greatest effect.

Some of the claimed benefits of the high glucosinolate mustards used in rotation are that in addition to their green manorial benefits (*see* Chapter 3), the mustards have a level of nematode suppression, take-all suppression in wheat production, root rot (*Pythium, Fusarium* and *Rhizoctonia*) and *Sclerotinia* suppression in bean production, and the ability to reduce disease pressure from *Aphanomyces, Pythium, Rhizoctonia, Fusarium* and *Sclerotinia* in pea rotations. Whilst this may seem attractive to some farmers, the approach is relatively new, and it is not known if the biofumigant disrupts or destroys beneficial organisms as well as soil-borne pests.

INTERVENTION MANAGEMENT

When the pest: predator and disease: host relationship natural balance is upset, intervention may be required in order to prevent severe damage or loss of a crop. A large range of practices have been developed that can be used to try and manage pests and diseases in organic farming systems: they can be cultural practices, direct physical or chemical controls and, increasingly, biological control agents. These are normally only deployed once a pest or disease has been observed, to stop the problem becoming an epidemic. Many have been developed for the horticultural and fruit-growing market for high value crops. Whilst some may, in theory, be suitable for cereal and pulse production, the practicalities of using them on a field scale basis, and the high costs, are prohibitive.

Pest and disease forecasting is also becoming more popular, and with access to the internet or a mobile phone, it is possible to receive regular updates on pest and disease population dynamics in your own farming area. It has also become possible to predict likely pest and disease outbreaks with the use of modelling programmes, to forewarn of likely population expansions, and to implement the necessary actions or strategies. In the UK, forecasting systems exist for a wide range of pests and diseases, including aphids, orange blossom midge, cereal rusts and septoria and brassica foliar diseases. The main challenge for organic farmers is that there are no miracle cures that can be rolled out: rather, farms have to rely on prevention strategies for pest and disease management.

Permitted Controls

There is a limited range of products that can be used for the intervention treatment of cereal and pulse crops for a range of pests and diseases as shown in the tables 58 and 59. There are many more products available on the market, but most are targeted at the high crop value horticultural and fruit-growing sectors, and the practicalities of using them on a field-scale basis, and the high costs, are prohibitive.

Table 58 Disease control products permitted under organic standards for arable production

Product	Crops	Issue and management
Sulphur (various trade names)	C, P	For the control of mildew – suspension concentrate applied by spraying on to the crop
Wetter's (various trade names)	C, P	Wetting and sticking agents
Biological	C, P	Biological pest control using licensed, naturally occurring predators

C = Cereals P = Pulses
NB it is important to check with your organic certification body as the list of permitted products is subject to change.

It should always be remembered that with one action you may be creating another problem. Many products need to be applied to crops with wetters, sticking agents or adjuncts. Wetting and sticking agents work by removing the waxy protective outer layer on crop leaves and letting the applied product penetrate the leaf, but in doing so they leave the crop more vulnerable to further attack by pests and diseases.

Table 59 Pest control products permitted under organic standards for arable production

Product	Crops	Issue and management
Nemaslug	C, P	*Phasmarhabditid hermaphrodita* nematode, biological control of slugs
Sluggo	C, P	Ferric phosphate slug pellet
Biological	C, P	Biological pest control using licensed, naturally occurring predators
Plant oils	C, P	Plant oils such as mint, pine or caraway, but only as an insecticide, acaricide, fungicide or sprout inhibitor
Garlic	C, P	Plant extract which interrupts the feeding cycles of some pests
Natural plant extracts	C, P	Licensed products based on natural plant extracts
Mechanical traps	C, P	In field traps, barriers or sound deterrents i.e. bird scarers
Agralan	P	Pea and beans weevil monitoring system warning of adult migration

C = Cereals P = Pulses

NB It is important to check with your organic certification body as the list of permitted products is subject to change.

MAIN CEREAL AND PULSE PESTS AND DISEASES

Understanding what conditions predispose crops to infection or invasion by pests and disease is essential in minimizing that risk. The following are the main pest and disease problems most likely to be encountered for UK-grown cereals and pulses. Crops under high levels of stress induced by water or nutrient shortage are prone to disease attack. Equally, luxury growth associated with very high levels of soil nitrogen also predisposes the crop to high levels of disease infection. Varietal selection in conjunction with an understanding of regional differences in likely pest and disease development is vital in minimizing the risk of an epidemic.

Cereals are at risk from numerous diseases due to the level of intensification necessary for profitable production since the 1970s. More recently, varietal diversification and plant breeding have played a prominent part in cereal disease control. Use of break crops and good rotations are also good cultural control measures. The demise of UK straw burning in the 1980s also increased the importance of good disease control.

Ear and Leaf Disease

Research on many farms, and including a recent three-year (2003–2006) study of field-scale plots of wheat, pea and lupin varieties (1 acre of each variety) on five organic farms in Wiltshire, Cambridgeshire, Yorkshire, Lincolnshire and Staffordshire, has monitored cereal leaf and ear disease occurrence in organic cereal crops. Typically there is a wide variation in pest and disease challenge to different crops in different seasons.

The theory of lower pest and disease pressure in organic systems because of greater crop diversity, higher levels of crop biodiversity, more natural predators and reduced crop stress is supported by evidence from research; here, despite experiencing low levels of *Septoria tritici*, eyespot, mildew, brown rust and yellow rust in different seasons, disease levels in organic cereals were generally not significant enough to have a major impact on yields. Higher levels of the foliar diseases, septoria and mildew, are the most often noted, typically at higher fertility sites, but these do not appear to be reflected in reduced yields.

Seed-borne Diseases in Organic Seed

With the exclusion of seed treatments, seed-borne diseases have been seen as a potential threat to organic systems, and emphasize the need for seed testing. Samples from farms and trial programmes over the three-year programme were tested by NIAB for seedling blight (*M. nivale* and *S. nodorum*), ear blight (*Fusarium spp*), bunt (*T. tritici*), loose smut (*U. nuda*) and ergot (*C. purpurea*). Results showed that nearly all samples were below critical levels, but at one site several varieties showed an above-threshold level of bunt in successive years. This suggests a local source of infection, and emphasizes the need for testing! Research results indicate that levels of seed-borne disease in organic crops are no worse, and can often be lower in organic crops compared to conventional crops due to improved diligence in testing.

Main Cereal Diseases

Because many of the cereal diseases affect wheat, rye, triticale, barley and, to a lesser extent, oats, diseases are grouped together with notes on which type of cereal they affect in each case.

Barley Yellow Dwarf Virus (BYDV)

Barley yellow dwarf is a plant disease caused by the barley yellow dwarf virus, and is the most widely distributed viral disease of cereals. It affects the economically important crop species of barley, oats, wheat, rye and triticale in the UK. These notes can therefore be used for all of these cereals, and not just barley.

The symptoms of barley yellow dwarf vary with the affected crop cultivar, the age of the plant at the time of infection, the strain of the virus, and environmental conditions, and can be confused with other disease or physiological disorders. Symptoms appear approximately fourteen days after infection, affected plants show a yellowing or reddening (on oats and some wheats) of leaves, stunting, an upright posture of thickened stiff leaves, reduced root growth, delayed (or no) heading, and a reduction in yield. The heads of affected plants tend to remain erect and become black and discoloured during ripening due to colonization by *saprotrophic* fungi. Young plants are the most susceptible.

Sources and spread   There are two main sources by which a cereal crop might be infected:

1. By non-migrant wingless aphids already present in the field and which colonize newly emerging crops; this is known as 'green-bridge transfer'.
2. By winged aphids migrating into crops from elsewhere; these then reproduce and the offspring spread to neighbouring plants.

Autumn-sown cereal crops usually suffer greater problems with BYDV than spring-sown crops, but mild winters can lead to greater aphid survival and consequently increase the risk of BYDV transmission to spring barley crops in the spring. Aphids can pick up the virus from infected cereal volunteers, winter barley and grasses, and transmit the virus to barley plants in the spring when they migrate. Therefore aim to cultivate out volunteers wherever possible.

Rhopalosiphum padi (L.) The bird-cherry aphid: is usually found on the lower leaves, but if numbers increase they may spread all over the plant. This aphid is one of the important vectors of BYDV, and although they mainly infest winter-sown cereals, numbers can persist after mild winters and they have the potential to move on to spring barley and spread BYDV.

Metopolophium dirhodum (Wlk.) The rose-grain aphid: moves from its winter hosts (for example, roses) in the late spring, and is usually confined to the leaves, seldom moving on to the ears. Large infestations have been known to cause yellowing and premature senescence of leaves in cereals, and this species can transmit BYDV.

Metopolophium festucae (Theob.) The grass aphid: overwinters on cereals and grasses, and after mild winters there is a potential threat of a large increase in numbers in the spring, which can lead to stunting of the whole plant.

Sitobion avenae (F.) The grain aphid: lives all year round on cereals and grasses, with populations becoming noticeable during July when they are the most frequent aphid species found infesting the ears of flowering crops. In some instances, numbers may build up on the flag leaf before flowering, and this can lead to blind or shrunken grains. The grain aphid is the usual cause of direct injury to spring barley crops.

Effect on yield   This is variable, since it depends on viral strain, time of infection and rate of spread. Most severe losses are from early infections and can be as high as 50 per cent.

The role of natural enemies   Because aphids don't usually build up their numbers on spring barley until flowering, natural predators have a good opportunity to keep aphid populations below economically damaging levels. Ground beetles, lacewings, ladybirds, hoverflies and parasitic wasps all feed on aphids, and should be encouraged to exert natural control of aphid populations.

Control 'Green bridge' sources must be ploughed in as early as possible. Drilling dates prior to mid-October favour attacks from winged migrant aphids. However, yield penalties may be experienced from very late drilling.

It is important to distinguish between the two main reasons for controlling aphids on spring cereals: to prevent transmission of BYDV to spring cereal seedlings, and to minimize the effects of direct feeding damage. In most years, when spring cereals are sown, aphid numbers are low because the cool winter temperatures will have taken their toll on aphids overwintering on grasses, cereal volunteers and neighbouring winter cereal crops. However, a mild winter coupled with a warm spring is likely to increase the success of overwintering aphids, and consequently increases the risk of aphids walking and flying from grasses and cereal volunteers on to the spring cereal. To prevent the spread of BYDV, ensure that volunteers are well controlled, and that natural predator numbers are kept as high as possible by ensuring that good overwintering habitats are maintained.

Eyespot (Pseudocercosporella herpitrichoides)
Common eyespot is a fungal disease that attacks the stems of winter cereals. Winter wheat is affected most, but winter barley can also be affected. Spring crops and winter oats are not so badly affected by this disease. It is common in the east and south, despite the drier climate, because of more frequent cereal cropping.

The disease causes lesions around the stem after flowering, and these affect the uptake of water and nutrients into the developing grains; they also weaken the stem. Most yield loss occurs when crops lodge.

Sources and spread The fungus can overwinter in crop residues and volunteers, and is spread mainly by rain-splashed spores, but it can also produce wind-blown spores. Early-sown wheat crops (August and early September) are most at risk. Crops grown under minimum tillage cultivations are at higher risk than under ploughing. Some varieties have good resistance, but there is also a yield penalty when they are grown.

There are two species of eyespot, but they are still known as the 'W' and 'R' types. The R type is now common in the UK, having taken over from the W type, which was common in the 1980s. The R type develops later in the season and is more difficult to control with fungicides. Diagnostics can help determine the presence of both species. The presence of eyespot only forms part of a risk assessment, however, and the presence of disease in the early spring cannot always be linked to disease severity at the end of the season.

Control Spring cereals are less affected than winter-sown cereals, and spring cropping usually escapes the worst effects, apart from crops in the wetter north and west. It generally requires a break of two years or more from cereals to minimize the disease. Reducing grass and cereal volunteers is the main management tool. Using inversion cultivations can help bury grasses and eyespot fungi and spores and helps control spread. Undersowing crops can reduce rain splash from the soil and thus the spread of spores on to the plant.

Some degree of varietal resistance occurs in wheat; most resistant of the currently listed varieties is Wizard, followed by Claire, Consort, Einstein, Madrigal and Tanker.

Sharp Eyespot (Rhizoctonia cerealis)

Sharp eyespot symptoms were first observed on stems of *P. virgatum* in July 2000 in a field experiment at Rothamsted Farm, Harpenden, Hertfordshire, UK. Lesions with a pale centre and sharply defined brown edges, characteristic of those observed on winter wheat, were present at the stem base and upwards to approximately 20cm. The severity of disease increased throughout the summer, the lesions enlarged and coalesced, eventually girdling the stem.

The diseases caused by fungi that affect the stem bases of cereals in the UK and much of continental Europe are: eyespot, the most important of them caused by *Tapesia yallundae* and *T. acuformis;* sharp eyespot, caused by *Rhizoctonia cerealis;* and brown (Fusarium) foot rot caused by *Fusarium spp.* and *Microdochium nivale.* The brown foot-rot fungi also cause ear blight and can be seed borne. *M. nivale* can decrease plant emergence.

Sharp eyespot can cause yield losses in localized patches within crops, but has never been considered a major problem in the UK. The contribution of Fusarium species to stem-base disease has long been recognized, but it came to prominence in the 1990s when research proliferated as concern about eyespot was declining. Losses are unlikely to be large. The main threat from Fusarium is ear blight, a stem-base disease that may be an important inoculum source. So far, mycotoxins that can occur in Fusarium-infected grain are at a low level in the UK, but that situation could change if the climate changes.

Control   For the individual organic farmer, the economic case for controlling eyespot may often be doubtful, as relatively small effects on yield are likely to result from the prevalence of the pathogen *T. acuformis.* Varietal resistance is generally good for most UK varieties. Good rotations with susceptible and non-susceptible 'break crops' help control. Later sowing in the autumn reduces incidence.

Septoria (septoria, septoria nodorum, septoria tritici)

Septoria tritici leaf blotch is an ascomycete fungus that causes a leaf blotch disease of wheat and occasionally other grasses including barley. It is found in all wheat-growing areas of the world, and is the major disease of wheat in the UK. Although more resistant varieties, including Robigus, have recently been added to the HGCA Recommended List, many varieties remain susceptible. *Septoria tritici* can occur at any time during the growing season. It causes the most damage to yield if it attacks the upper leaves, in particular the flag leaf, which reduces photosynthesis and yield.

Sources and spread   Weather conditions dictate the spread of *Septoria tritici* spores from the soil into the bottom of the crop, which then migrate to the upper leaves during wet and windy conditions. Spores ooze from affected lower leaves and are splashed up the plant in heavy rain. Dew and wind can also achieve the same effect. Crop canopy can play a role in spreading the disease. Short crops, or crops where leaf layers overlap, make it easier for the disease to spread. Wet or windy weather conditions at flag leaf emergence are therefore high risk for disease spread.

Control   Selecting varieties with better resistance to *Septoria tritici* can assist with control. Organic crops tend to be thinner and thus spread is more difficult. Undersowing crops with clover or black medick to form an understorey canopy and ground cover has been shown to just about elim-inate septoria, as the undersow prevents the splash inoculation from the soil into the bottom of the crop. Where septoria is present, growing mixtures of cereals can be detrimental, as different cereal node heights provide a very effective ladder for the disease to spread up through the crop to the flag leaf.

Mildew (Erysiphe graminis)
Blumeria graminis is a fungus that causes powdery mildew on grasses, including cereals. It is the only species in the genus Blumeria. The mycelium can cover the plant surface almost completely, especially the upper sides of leaves. Ascocarp are dark brown, globose with filamentous appendages. Ascospores hyaline are ellipsoid, 20–30 × 10–13 µm in size. Anamorph produces on hyaline conidiophores catenate conidia of oblong to cylindrical shape, not including fibrosin bodies, 32–44 × 12–15 µm in size.

Sources and spread   *Blumeria graminis* disperses by scattering conidia and ascospores. It does not grow on synthetic media. It exists in specialized forms, each being restricted to its own particular host. It survives on volunteers and late tillers at harvest. Relatively cool and humid conditions are favourable for its growth. Its relatively great genetic variability enables it often to infect previously resistant plant varieties. Air-borne spores are released following rain, dew or periods of high humidity. Infection and development take place over a wide temperature range, but are retarded by high temperatures (>25°C). Rapidly growing crops are most susceptible.

Control   It favours high fertility situations. Over-application of manures and slurries and luxury growth can result in mildew. Applications of sulphur in suspension with water sprayed on to the crop can keep the problem at bay, but often by the time mildew is identified it is too advanced for sulphur to assist. Varietal diversification can be employed to minimize the risk of spread between adjacent crops. Varietal mixtures in the same field can help to reduce the severity within crop.

Powdery Mildew
Powdery mildew is one of the major diseases affecting barley. There are currently many spring barley varieties that show good varietal resistance, but the popular malting barley variety Optic is very susceptible to this

disease. With so much of the spring barley area using this variety, it has led to an increase in this disease in recent years.

The disease, which affects both winter and spring varieties, produces a white fluffy fungal growth on the leaf surface. This can cause leaves to turn yellow, leading to a loss in green leaf. High levels of disease at an early crop growth stage can kill out tillers. Many varieties show adult resistance, including Optic. This means disease levels are generally lower later in the season.

Control Diversification of varieties is one method of minimizing the spread of disease from one crop to another. Many spring barley varieties have excellent resistance that has been effective in recent years.

Take-all (*Gaeumannomyces graminis*)

Gaeumannomyces graminis var tritici, syn. Ophiobolus graminis var tritici
A disease of cereal roots: all varieties of wheat and barley are susceptible. It is an important disease in winter wheat in the UK, and is favoured by conditions of intensive production, and exacerbated by monoculture. The yield loss can be up to 50 per cent. Take-all is unlikely to be found on organic farms, unless two or more wheat crops are grown in succession.

The fungus attacks the plant roots at any growth stage, early infections causing stunting and yellowing. Affected roots are blackened, sometimes severely. After forming ears in the spring, patches of the crop appear stunted. In severe attacks the worst affected plants are bleached and dead even before flowering. These symptoms give rise to an alternative name for the disease: 'whiteheads'. Yield loss levels of 40 to 50 per cent are often recorded in severe attacks. Modern varieties are stiff and short-strawed, which reduces lodging incidence. This can limit damage from the disease.

There exists a phenomenon known as 'take-all decline'. Experiments performed on the famous 'Broadbalk' field at Rothamstead Experimental Station where continuous monoculture winter wheat is grown, show that take-all build-up occurs in successive crops to reach a peak in the third to fifth cropping year, after which the disease declines, ultimately restoring yields to 80 to 90 per cent of first and second year levels. The decline cycle is destroyed by the introduction of a crop other than wheat or barley as a suitable host plant.

Control The most appropriate control measure is the use of a clean one-year break crop of a non-cereal crop or ley. This reduces the fungus to an acceptably low soil contamination level in about ten months, although the existence of volunteer grasses may nullify any beneficial effects. There is potential that when grasses are included in a ley mixture of grass and clover that the grass may act as a green bridge host. Clover has been shown to reduce the incidence of take-all, especially when undersown in cereals. As regards strategies for minimizing the effects of take-all in second and subsequent wheats, the following should be considered:

- Soil type, with the worst affected soils well aerated with puffy seedbeds; also high organic matter soils, very light soils in wet seasons, and heavy soils with poor drainage and structure.
- pH: extremes either way will aggravate symptoms with 6.5 to 7.0 ideal.
- Soil structure, with well drained, well structured soils suffering less.
- Rotation: avoid back-to-back cereals in the rotation.
- Weed control: a build-up of grass weeds in the rotation will encourage the disease.
- Manganese: reduced availability (loose seedbeds, high pH) favours take-all.
- Phosphate: low soil levels or low availability render the plant more susceptible.
- Sowing date: exaggerated by early and very late sowing.
- Variety: some varieties have been more consistent performers in the second wheat position.
- Undersowing: undersown cereals have been shown to have a reduced incidence.
- Grass-weed populations can act as a 'green bridge' host, and transfer take-all from one cereal to another. Blackgrass is a principal host.

Brown Rust (Puccinia recondite)

Both yellow rust and brown rust can only grow on green plants. Small brown pustules develop on the leaf blades in a random scatter distribution. They may group into patches in serious cases. Infectious spores are transmitted via the soil. Onset of the disease is slow, but accelerated in temperatures above 15°C, making it a disease of the mature cereal plant. Occasionally, more especially in southern and eastern England, the disease can be severe when the weather is hot and dry. The spores survive the harvest period on late tillers and then spread successively to volunteer plants, then to autumn-sown and subsequently to spring-sown cereals. Annual monitoring of the incidence of brown rust and yellow rust races in untreated susceptible varieties is carried out by NIAB. Regular information updates are provided on the current races present, and identify the specific varieties that will be at risk.

Control Select varieties with good resistance if in a susceptible area. Good management of volunteer plants and grass weeds is essential to remove hosts. Ploughing and burial of grass weeds is more effective than non-inversion cultivations.

Yellow Rust

This disease can be very damaging in wheat and barley particularly. New strains soon become widespread once a variety is widely grown. Mild winters and cool moist weather in the spring and early summer favour the disease, but a hot, dry spell will often spectacularly arrest the disease. Typical symptoms are small yellow pustules on the leaves, commonly forming in lines along the leaf. Look out for small pockets of affected plants in the spring. In severe cases, the leaves will appear bleached.

Sources and spread   Yellow rust overwinters on cereal volunteers and grass weeds, either as pustules on the foliage or as tiny threads (mycelium) unseen within the leaf tissues. The disease is favoured by cool, wet weather (10–15°C), but high temperatures will kill the fungus. A spell of warm, dry weather (20–22°C) will check an epidemic. The disease may overwinter on crops, but it rarely flares up until the spring and early summer.

Control   The most effective way to avoid yellow rust is to grow resistant varieties. Note that the popular variety Robigus is susceptible to yellow rust. Farmers should choose a mix of varieties that have different resistance characteristics; thus when yellow rust attacks a variety, you can choose a different one to sow in neighbouring fields, which may be resistant to different races of yellow rust. Good management of volunteer plants and grass weeds is essential to remove hosts. Ploughing and burial of grass weeds is more effective than non-inversion cultivations.

Rhynchosporium
This fungus is frequent in both winter and spring barley, and is often damaging in Wales and south-west England and some coastal regions. It also occurs on rye. It is favoured by cool, wet weather. The destruction of volunteer plants and the wide separation of winter and spring crops will help. It is important to select resistant varieties in susceptible areas.

Net Blotch
This disease can be seed borne and/or carried on debris from the previous season's cereal crop. The ploughing of infected trash is beneficial if cereal follows cereal. Wet weather favours its development and spread. Net blotch is an important seed- and trash-borne disease in barley.

Barley Leaf Stripe
This is a seed-borne disease that is currently rare. It can, however, develop rapidly if levels are high on seed stocks, as happened in the 1990s following resistance in the popular seed treatments at the time.

  The typical symptoms are yellow and brown stripes that appear longitudinally along the leaf. These always start at the base of the leaf. Affected plants can also be stunted. In cases of severe infection, the leaves may die back prematurely.

Control   Test the seed for the presence of the fungus. Where levels are high, do not use for seed.

Oat Crown Rust
Crown rust is one of the main diseases to affect oats, and can significantly reduce yields. It is more prevalent in wetter areas in the west and south of the UK, but it also occurs in the far north. Look for orange-coloured pustules on the leaves.

Control   Some varieties show better resistance to the disease than others.

Seed-borne Diseases

Bunt (Tilletia caries)

Black, fishy-smelling spores colonize the developing grains, turning them into bags of fungus known as 'bunt balls'. All of the grains in the ear and all ears derived from an infected seed are infected. During harvest threshing operations, the 'bunt balls' burst, releasing the spores on to the healthy grain. Infected crops are unfit for milling. Worse, all harvest, transport and storage equipment is contaminated, leading to more contamination of healthy grain. Infection only becomes apparent just before harvest, since an infected ear develops in the same way as a healthy one, although plants may be shorter.

The spores on the outside of the grain infect the young shoot following germination, and from then on, the mycelium of the fungus develops just behind the growing tip of the plant until ear emergence.

Control Although seed treatments provide control for the non-organic farmer, organic farmers cannot use these. Some degree of varietal resistance occurs: for instance, Hereward as a variety would seem to have some resistance. Spring cereals seem to be less affected than winter-sown cereals. Reduced, non-inversion and minimum till cultivations keep spores on the soil surface where they can be transferred to, and infect, subsequent crops. Ploughing can bury spores and induce dormancy. Where bunt is suspected, contaminated crops should not be used for seed; and where bunt-contaminated crops are present, leave to harvest last in the sequence of fields and crops, as the process of combining will spread bunt spores to adjacent or near fields and crops that can become contaminated prior to their harvest.

Ergot (Claviceps purpurea)

Ergot is parasitic on certain grains and grasses. Economically important, *C. purpurea* can affect a number of cereals including rye (its most common host), triticale, wheat and barley. It affects oats only rarely.

An ergot kernel called a sclerotium develops when a floret of flowering grass or cereal is infected by a spore of Claviceps fungus. The infection process mimics a pollen grain growing into an ovary during fertilization. The fungus then destroys the plant ovary and attaches itself to a vascular bundle originally intended for seed nutrition. The first stage of ergot infection manifests itself as a white, soft tissue (known as *sphacelia*) producing sugary honeydew, which often drops out of the grass florets. This honeydew contains millions of asexual spores (*conidia*) which are dispersed to other florets by insects. Later, the sphacelia convert into a hard dry sclerotium inside the husk of the floret. At this stage, alkaloids and lipids accumulate in the sclerotium (*see* the diagram below Figure 36).

When a mature sclerotium drops to the ground, the fungus remains dormant until conditions trigger its fruiting phase (the onset of spring, rain period, and so on). It germinates, forming one or several fruiting bodies with head and stipe, variously coloured (resembling a tiny mushroom). In the head, threadlike sexual spores are formed that are ejected simultaneously, when suitable grass hosts are flowering.

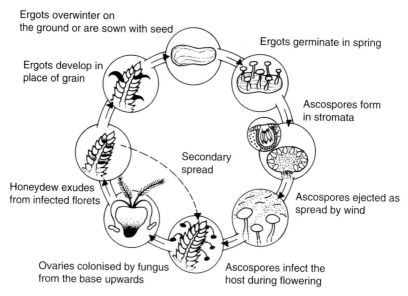

Figure 36 The life cycle of the ergot

Source : HGCA

Black ergot pieces appear in the ear where the grain would have been; these pieces may be up to four times the size of the grain. Ergots can remain in the soil for up to one year, and can germinate to produce more spores. These ascospores then infect the flowering heads in the following summer. The main method of spread is through rain splash and wind from both infected crops and grass weeds from within the field and from field edges. Blackgrass is a principle source of ergot contamination, and carries through from crop to crop. It is important that ergots are not accidentally fed to cattle or humans, as they contain poisonous alkaloids and the grain cannot be used for human consumption.

Ergot infection causes a reduction in the yield and quality of grain produced, and if infected grain (or hay) is fed to livestock it may cause a disease called ergotism.

Control There has been speculation that the increased area of grass margins around the edges of fields as part of conservation schemes has increased the incidence of ergot. However, there is little research that confirms this hypothesis alone.

Planting clean seed is important, and reducing grass volunteers and cereal volunteers is the main management tool. Using inversion cultivations can help bury grasses and ergot, and helps control spread. Burying the ergots to a depth greater than 4cm will stop them germinating the following year and can reduce the chances of seeing the disease. Whatever cultivations you carry out, be aware that mature grass in headlands, set-aside and road verges will act as a source of spores for crops when they are in flower.

Infected grain should be segregated from non-infected grain. The tolerance level for ergots for human consumption is 0.001 per cent by weight (three pieces of ergot per 500g). In practice, grain samples with any ergot may be rejected. It is possible to screen out some ergot, but ergot levels become concentrated in the screenings, and contaminated screenings must be buried or burnt.

Grain cleaners, gravity tables and colour separation equipment can effectively be used to remove ergots from grain prior to sale.

Loose Smut

This disease may infect barley, wheat and oats. Like bunt, the ear emerges as a mass of black sooty spores, and the flag leaf is often discoloured. Again, the infection comes solely from infected seed. Cool, wet conditions that prolong flowering increase the chances of infection. Grain losses of up to 25 per cent can occur, though it is far more common to see 1 to 3 per cent. Where infection is suspected the grain should not be used as seed.

Sources and spread   Loose smut is caused by the fungus *Ustilago nuda*. It is a seed-borne fungus that is deep-seated within the grain. Crops can become infected at flowering from spores that are wind-dispersed from plants showing the typical loose smut symptoms. Grains will develop normally, but the fungus will grow inside the grain. After sowing, plants will germinate and develop normally. The fungus will, however, be growing within the plant. Once the heads have emerged, the heads will be covered in black spores. Affected plants tend to be shorter at this stage, and when the spores have dispersed, it will be apparent that no grains will be produced from the head.

Control   For effective control, test the seed and reject seeds with high levels of contamination: these should not be sown. The maximum permitted infection in certified seed is 0.5 per cent; it is therefore advisable not to sow farm-saved seed where levels exceed 0.5 per cent without an effective seed treatment.

Microdochium nivale

Microdochium nivale can have a big impact on the emergence of crops. It is a seed-borne disease that affects winter-sown cereals. Wheat is affected most, followed by oats, then barley. If seed is heavily contaminated with the fungus, and seed is sown in cold and wet conditions, the impact on germination can be severe. Seed sown in cold and wet seedbeds may be slow to emerge, and these conditions are also ideal for the fungus to germinate and attack the seedlings. This can have a major impact on crop establishment. The disease can contaminate the grain during the period up to harvest when conditions may be cool and wet. At harvest, seed may have high levels of contamination.

Control   Fusarium (*Microdochium nivale*) is not covered by seed certification standards and it can cause severe losses, particularly where seed is sown late in cold wet soils. A seed test will help you decide if a specific weed treatment is required. If you sow early in good conditions all seed treatments will provide effective control.

Air-borne Diseases

One suggestion to reduce the risk of air-borne diseases is to drill the crop so that the rows are in line with the prevailing wind. The spores will still blow into the crop but there is less chance of disease-favourable micro-climates developing.

Cereal Pests

The cultural control of the following pests will be enhanced by understanding their life cycles and the conditions that favour their spread.

Slugs

Slugs are a widespread soil-borne pest with a wide host range. They attack a wide variety of plants and cause havoc in many agricultural crops, feeding on seedlings, leaves, stems, flowers and root structures. Numbers are usually highest on moisture-retentive soils and particularly after crops with dense canopies that keep the soil shaded and moist for long periods, such as winter oilseed rape. The most serious damage to cereals is seed hollowing prior to emergence, although leaf grazing can be severe after emergence as well. Slugs are most active in mild wet weather. During dry weather they are most likely to be found at night, but when weather conditions are ideal they can be found during the day on plants, including the heads of cereals.

Weather conditions play a big role in regulating slug numbers and surface activity. The presence of crop residues and plentiful volunteer seed act as a food source to aid slug survival, and moisture retained in crop residues also enhances slug activity. Fine and firm seedbeds restrict slug movement, and drilling slightly deeper than usual has been shown to reduce grain hollowing. Conditions conducive to rapid establishment favour the crop and give the slugs less time to cause damage.

There are thirty types of slug, but only a few are important to arable producers, principally field slug (*Deroceras spp.*), but garden slug (*Arion spp.*) and keeled slug (*Milax spp.*) can be locally important. The black slug is the largest type but actually does the least damage. The three main predators of the slug are the carabid beetle, the field mouse and the starling, all of whose numbers have severely declined in UK fields in recent years. Providing suitable overwintering habitats for these natural predators can assist in ensuring that there is an active predator population to assist the farmer in his battle with the slug.

Field slugs can cause damage to a wide range of plants, particularly to seedlings, which can be killed out, but also to leaves, stems, grains and tubers.

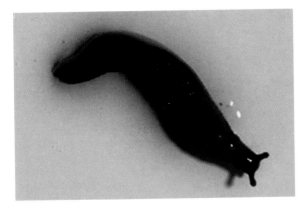

Field slugs (Deroceras spp) can cause serious damage to crops.

Control Cultural control should include well consolidated seedbeds to reduce the habitats and movement of slugs, and establishing natural predator habitats on the farm for carabid beetle, the field mouse and the starling. Where problems exist in damp, shaded fields on heavy soils, drill at higher seeding rates to compensate for losses.

There are two products currently available that can be used in organic farming where problems are especially severe: Sluggo and Nemaslug. Sluggo contains the active ingredient ferric phosphate that occurs naturally in the soil and is attractive to all plant-damaging slugs. This is combined with a cereal-based pellet to form attractive bait for slugs and snails. Ferric phosphate is listed in the EEC list for organic farming (EEC guideline 2092/91). The pellets are applied to the soil surface and attract slugs away from plants. After eating the bait the slugs cease feeding, giving protection to the plants.

The Sluggo bait technology has a novel mode of action in that once ingested, slugs and snails immediately stop feeding. The ferric phosphate causes cellular pathological changes in the feeding (crop) and digestion (hepatopancreas) functions of the slug, and they crawl away to secluded places to die. There is also no strong slime secretion left around plants, and no requirement to collect dead slugs. Vertebrates, including birds, dogs and other large animals, and invertebrates such as earthworms, bees and other beneficials, are not affected.

Nemaslug is a biological approach, whereby a nematode predator is applied to the slugs. It works underground, providing control where the pest is often present. Nemaslug contains nematodes in their vigorously infective stage; applied to the soil, these aggressive organisms actively seek out and attack slugs. When a nematode locates a slug it invades it through the dorsal pore, a small hole at the back of the mantle (the raised area on the slug's back); once inside, it releases bacteria that stop the slug feeding within three days. The mantle swells as the nematodes reproduce within the slug, which will eventually die underground after seven to ten days. The nematodes continue to reproduce as the slug's body is broken down, and further generations of infective nematodes move off into the soil to locate further slugs.

For optimum results, apply Nemaslug to moist soil and when soil temperatures are between 5°C and 25°C (41–77°F). This is a relatively high cost product for small area application only.

Aphids

The main cereal aphids are the grain aphid, the rose-grain aphid, the bird cherry-oat aphid and the cereal leaf aphid. There are two main periods of infestation that are of concern: autumn aphid, which leads to infection with barley yellow dwarf virus (BYDV), which despite its name can infect all cereals; and summer infestation, which can lead to direct feeding damage.

Control Avoid sowing early in autumn (generally before mid-September), since it is these crops that are at most risk from the direct transfer of aphids from 'old' to new 'crop'. Destroy the 'green bridge' of cereal volunteers and grass weeds to disrupt the host availability of the aphids. Encouraging predators with the use of beetle banks and similar areas that harbour aphid predators will reduce the impact of both autumn and summer aphid.

Orange Wheat Blossom Midge

Orange blossom midges (*Sitodiplosis mosellana Gehin*) are potentially damaging to wheat crops. Typical symptoms of damage are shrivelled grain, often with a shallow depression where the larva has been lying. Populations of 6 million larvae/ha are the minimum necessary to pose a risk to wheat crops.

Milling varieties are at most risk from damage, since the effects of the pest include reduction in hagberg falling number and specific weight. There are differences in the flowering and enzyme characteristics of some wheat varieties which affords them some protection from this pest. Open-flowering varieties are thought to be more prone to pest attack. A number of new varieties with a 'resistance' to orange blossom midge attack are arriving on the market, though these are as yet relatively untried in organic production systems. Table 60 below shows the relative infestation of a selection of winter wheat varieties grown on four susceptible sites in 2002.

Table 60 Relative infestation of winter wheat varieties by orange blossom midge on four susceptible sites in 2002

Variety	Infestation
*Welford	3
Soissons	7
Deben	60
Claire	70
Hereward	80
Malacca	85
Savannah	88
Napier	103
Tanker	111
Consort	120

* varieties with 'resistance' to orange blossom midge

Sources and spread The appearance of the midge is dependent on air temperatures during May and June, where mean daily air temperatures exceed 15°C. Crops of winter wheat are only at risk from midge attack between early ear emergence and the beginning of flowering (GS 55 and GS 59), and if mean air temperatures do not reach 15°C during this growth stage of the crop, then the crop will be safe from midge attack. After feeding on the grain, the orange grubs fall to the ground in the summer and spend the winter as cocoons in the soil.

Temperatures above 13°C in May wake up the larvae in the cocoons and they move up towards the soil surface; if the top 10mm of soil is damp, they then form pupae. The emergence of adult midges from the pupae is triggered by mean daily air temperatures exceeding 15°C, and can be spread over several weeks. Once the midges have emerged, they mate immediately and the females begin to search for crops in which to lay their eggs.

Midges can be carried in the wind, or they will fly in a light wind for a few kilometres, attracted to the smell of a wheat crop. Once finding a crop, the midges will not begin to lay eggs on the ear until early evening when the temperature exceeds 15°C. At temperatures below this, they will sit on the lower leaves, often for several days waiting for it to rise. After laying, the eggs can hatch after five to seven days; once the larvae have begun to attack the grain, they cannot be controlled.

To determine whether wheat crops are at risk from wheat blossom midge, growers should keep an eye on crop growth stage and the weather. Midges can only attack crops between early ear emergence and the onset of flowering (GS 55 and GS 59), and if mean daily air temperatures exceed 15°C during dry weather: if these conditions coincide, then growers should look for midges in the crop. The best time for this is on warm, still evenings by standing in the crop and looking for the midges resting or flying around the ears of the crop.

Control Producers should consider site history and the use of 'sticky yellow' pheromone traps to monitor the activity of midges at ear emergence in wheat crops to assess risk. Soil can be tested for the presence of pupae, but their subsequent development depends on sufficient moisture and temperature. Where problems are identified, use resistant varieties – although experience suggests that in organic situations some varieties that have Orange Blossom Midge (OBM) infection may still out-yield some of the current resistant varieties.

Wheat Bulb Fly

Wheat bulb fly is locally important especially in the east of England, parts of Yorkshire and Scotland, and occasionally Warwickshire, Gloucestershire and Wiltshire. The adult appears like a small housefly, on the wing from mid-June to September. The eggs are laid mid-July to early September, with peak egg laying from late July on mineral soils, and mid-August on peat soils. The larvae feed on the central shoot of wheat, causing it to turn yellow, then die. This symptom is known as a 'deadheart' symptom. Sites at high risk from egg laying are fallow (set-aside), land with high grass weed or cereal volunteer populations, early potatoes, vining peas, sugar beet and other roots. Rough tilth and dry soil favour egg laying.

Sources and spread   The female lays eggs during the summer months (July to September), and bare soil is the preferred site. Consequently, wheat sown after potatoes, peas, oilseed rape, field vegetables, fallow and set-aside is particularly at risk. The eggs begin to hatch in January/February and may continue through to mid-March, and the colder it is, the later they will hatch. Severe damage is only caused to crops that have yet to form multiple tillers. As a general rule, crops sown before mid-October will be less affected by this pest. Later sown crops, which may have only a single shoot at egg hatch, may be completely killed by the grub feeding in the single shoot.

A significant proportion of eggs will be predated by beetles, spiders and fungi, so healthy predator populations will help, as will providing suitable overwinter habitats for these. After hatching, larvae die quickly if a host is not available. The producer should aim to produce strong, well tillered plants by early spring, so winter wheat should be drilled before mid-October. Rolling can help attacked crops to recover.

Control   Producers should not grow winter wheat on high risk sites. Soil testing for the presence of eggs is available to help assess risk. Soil samples need to be taken using a standardized method that involves using a shovel 11cm wide, and with sides 3cm high: samples should be taken to a soil depth of 3cm, and to a distance of 7.5cm, which can be marked on the shovel. A total of twenty-four samples should be taken along the longest diagonal of the field.

A wheat bulb fly egg count of 2.5 million eggs/ha (1 million eggs/acre) is the threshold at which damage is likely in the following wheat/spring barley crop. However, lower egg counts may also be of concern, depending how late the winter wheat was sown, and the seed rate to be used. Early sowing (winter wheat) or late sowing (spring barley) should be the primary method for minimizing wheat bulb fly damage if egg counts are high.

Oats are immune, and winter barley is attacked but is usually sown early enough to withstand any damage; nor is it a good host. Spring wheat drilled from mid-March is not affected. Working fallow land to a fine tilth before egg laying will reduce numbers, and deep ploughing will increase egg mortality. A catch crop of mustard or rape after early harvested potatoes or vining peas will provide cover for land during egg laying and will therefore also reduce numbers.

As the wheat bulb fly tends to prefer laying eggs in bare soil, fields planned for non-cereal crops can be cultivated during the summer and used as 'traps' to divert egg laying by wheat bulb fly away from fields destined for susceptible crops.

Frit Fly (*Oscinella frit*)

Frit fly (*Oscinella frit* L.) is a pest of spring oats and winter-sown cereals. Over recent years the prevalence of frit fly has declined, although it is locally common in some areas. There are three generations of frit fly a year, and consequently damage can occur to crops at different times of the year and also to different parts of the plant (shoots and grain).

Frit fly grubs overwinter in the shoots of grasses and cereals, so some damage in the form of 'deadhearts' may be seen in winter wheat, barley or oats during the winter months. This generation of frit fly arises from eggs laid on grasses and volunteer cereals in late summer, and when infested grass or grassy stubble is ploughed in and winter cereals sown, the frit fly grubs can leave the buried grass and invade the shoots of the cereals. Occasionally early-sown winter barley may have eggs laid directly on the young plants. The frit fly grubs feed throughout the winter, and the next generation of flies appears in May/June, targeting their egg laying on to grasses and young spring cereals such as spring oats and occasionally spring barley. The grub burrows into the central shoot and may cause 'deadheart' symptoms in spring oats and barley in June/July. Some spring oats may suffer damage to the ears prior to ear emergence, which can result in blind, withered spikelets.

The grubs pupate within the shoot, and the next generation of adult flies emerges in late July; these target the ears of oats, which are only susceptible for a short period of time after the ears have emerged from the leaf sheath. After flowering is completed, the ears are resistant to any frit fly attack. Frit fly eggs are laid on or beneath the husks of oats, and the grubs burrow into the husk and feed on the oat kernels within. Attacked oat ears are not always evident, and it is only by opening up individual grains that the blackened, thin kernels can be seen. Often a brown frit fly pupa may be present within the grain.

The frit flies arising from this generation lay eggs on grasses and winter cereals, and these become the overwintering generation. By far the best option to minimize any damage from frit fly in winter cereals is to plough in grass and then leave the fallow for as long as possible before sowing the winter crop.

The risk of frit fly damage to winter cereals can be assessed by sampling the grass or grassy stubble for frit fly eggs/larvae before ploughing. Depending on the quantity present, and the date of ploughing and drilling, a recommendation can be made on whether to alter cropping. Frit fly shoot damage to spring cereals is not very common, as most crops will have advanced beyond a stage where significant damage is carried out or is noticeable. However, in crops where sowing was delayed, or the growth of the crop is checked for other reasons, frit fly could cause some problems.

Crops that have reached the four-leaf stage by the time frit flies are actively laying eggs can usually escape any serious damage, as the young side shoots or tillers are preferred by the flies, rather than the main shoot. The next generation's threat to the ears of spring oats in late July is probably of most concern, and is the most difficult to take any control measures against. However, because the ears are only susceptible to attack during early emergence, an early sown or rapidly developing crop can escape this narrow period of risk from frit fly attack.

Control   If grass leys or grassy cereal stubbles are infested with frit fly larvae at the time of ploughing in early autumn, the grubs often leave the rotting turf and migrate upwards to attack the newly sown grass or winter cereal. An interval of at least four weeks between ploughing and sowing will ensure that most of the larvae are predated by birds from the soil surface or die in the rotting turf.

Gout Fly (Chlorops pumilionis Bjerk)

Gout fly is a problem affecting early-sown winter cereals and late-sown spring cereals. Plants attacked by gout fly during early growth become stunted and swollen on 'gouty' stems; the larvae can usually be found within the swollen stems. In spring-sown cereals, late developing small shoots and tillers become stunted and gouty, and fail to produce ears. Advanced crops have larvae feeding down one side of the ear to the uppermost node, checking growth and damaging grain, and causing up to 50 per cent yield loss. Autumn-sown cereals produce stunted, gouty shoots and tillers, similar to affected late spring cereal crops, but autumn attacks are often compensated for by subsequent plant growth. Crops of wheat, barley and rye can be attacked, but not oats or maize.

Sources and spread   There are normally two generations a year. Adult flies are small, about 4–5mm long, with distinctive yellow and black markings, and are present during May and June. The females lay one egg on the upper surface of a leaf, often close to the central shoot. Gout fly eggs are normally 3–4mm long, white and cigar shaped. Eggs will hatch in approximately seven to ten days. The larvae are normally yellowish in colour, and burrow into the centre of the plant where they feed for four weeks; they then pupate – the pupae are brown. Pupation lasts about five weeks before a new adult emerges.

The second generation, occurring in August and September, lays eggs on grasses, volunteer cereals and early-sown winter cereals. These larvae move to the base of the plant stem to feed, causing stunted and gouty shoots in winter and early spring. They pupate in spring, producing adults again in May.

Control   Drilling spring cereals from April onwards, or winter cereals before mid-September, will increase the risk of these crops being attacked by gout fly. Conversely, drilling spring cereals before mid-April, or winter cereals after mid-September, will put crops outside the most susceptible stages of growth at peak times for egg laying.

September-emerging winter cereals in high risk areas should be monitored for egg laying towards the end of September and until late October. There is some evidence that applications of garlic to the infected crop can interrupt the life cycle of the pest by discouraging flies from laying eggs and the grub from feeding.

Leatherjackets (*Tipula spp.* and *Nephratoma spp.*)

Leatherjackets are the common term for larvae of several species of crane fly or 'daddy longlegs', and, like frit fly, tend to be a problem in winter and spring cereals sown after grass. The most common species of leatherjacket is *Tipula paludosa Meig.*, although there are other species. The average leatherjacket density of 2.5 million grubs per hectare is equivalent to 250 grubs per metre square. Crops of wheat sown after grass can be affected by leatherjackets, and spring barley can also be affected when sown in fields with high populations.

The common crane fly, *T. paludosa*, has one generation a year, and the fly is usually seen from September onwards. The females tend to lay eggs in closely grazed grassland rather than unmanaged long grass, although grassy stubbles may also be targeted. The eggs hatch within ten days, and the grubs (2mm in length) feed just below the soil surface on decaying plant material and the roots of plants. As the grubs develop they may feed on the soil surface and attack leaves and stems above ground. The grubs feed throughout the winter, and by the time they finish their feeding cycle in June, can be up to 4cm in length. Consequently, spring-sown crops tend to suffer greater damage than winter-sown crops, as the grubs are much larger when the spring crops are sown. Typical damage symptoms are severed or ragged leaves and damage at the stem bases. Cereals may also suffer damage to the seed, resulting in bare patches within the crop. The leatherjacket grubs can usually be found adjacent to damaged plants just below the soil surface.

Ploughing in grass ahead of establishing an autumn cereal may allow leatherjackets to survive on buried plant material. By far the best option to minimize the threat of leatherjackets is to plough grass early, say in July, before egg laying begins. If this is not a viable option, then grass can be sampled for leatherjackets after egg laying, and the numbers present estimated and the threat to winter cereals and spring cereals determined.

Leatherjacket numbers in grassland exceeding 0.6 million/ha are likely to cause damage to subsequent spring cereals. There are several methods available to estimate the field density of leatherjackets. Growers and advisers are encouraged to carry out 'row scratching' of the crop, whereby several random 30cm lengths of row are removed by trowel to below root depth and hand sorted on a large white tray or other suitable surface. Finding one leatherjacket/30cm of row is equivalent to a leatherjacket population of 0.3 million/ha. As the 0.6 million/ha threshold adopted for leatherjackets in grassland assumes that 50 per cent of the grubs will be killed through ploughing and cultivation, the 0.3 million/ha threshold for treatment is adopted when row scratching in spring cereals.

Additional measures can be taken by following this with an appropriate non-grass/cereal green manure cover crop such as mustard, which will further antagonize the pest. This can be kept in place over the winter and incorporated in the early spring, allowing further pest predation by birds, and subsequent cereal planting in the spring. A green manure should always be used between ploughing the grassland and the crop to absorb and hold soil nutrients until the cereal is sown.

An additional note is that the crane fly only lays on to grass. On stock-less arable farms there is a strong case for planting 100 per cent clover rather than a clover and grass mixture, which will largely eliminate the problem in cropped areas.

Wireworm (*Agriotes spp.*)

Wireworm are click beetle larvae associated with long-term grass (usually in place for longer than four years), particularly where this is on heavier, more moisture-retentive soils; populations of several million per hectare have been recorded. Eggs are laid in grassland or amongst other vegetation, and if this is ploughed, many of the tough wiry larvae can survive the cultivations, and will attack any following crop. Adult beetles are 7–10mm long and brown in colour, and are active from April to July. Eggs hatch in five to six weeks, and the larvae develop slowly, taking up to five years to complete their life cycle. Wireworms are most active in the early spring (March to May) and late summer to early autumn.

They appear to have become more widespread in recent years, also the phenomenon known as 'arable wireworm' populations, that develop in the absence of grass in the rotation, is causing some concern. However, generally the risk of damage is kept low by ensuring grass leys used in the mixed rotation are no longer than four years in length, as well as trying to minimize infestation of soils with grass weed species, particularly couch. Higher percentages of clover and lower percentages of grass in a ley may also limit wireworm infestation.

MAIN DISEASES AND PESTS OF PULSE CROPS

Many factors can affect growth of the pea and bean crop, and the notes below describe the main diseases, pests and disorders which reduce yield and quality. Further information, and colour photographs of symptoms and damage, may be obtained from the publication *Peas and Beans: Pests, Diseases and Disorders* and a list of currently approved pesticides can be obtained from the Processors and Growers Research Organisation (PGRO). (*See* contacts in Chapter 16.)

Flowering field margins with, for example, marigold, ox eye daisy and phacelia, are attractive to aphid predators and can help control numbers.

Spring beans tend to be more susceptible to aphid problems. Beans suffer less from pigeon damage than peas or lupins, they are easier to combine, and growing costs can sometimes be lower.

Chocolate spot can be a problem with winter beans, and brown rust with spring beans, but there are no direct control measures. Downy mildew (*Peronospora viciae*) can sometimes be severe on spring beans. In some conditions, leaf and pod spot (*Ascochyta fabae*) can be severe on winter beans. Varieties differ in their resistance; on sites prone to very wet weather, and where there is a risk of *Ascochyta*, varieties with good resistance to *Ascochyta* should be chosen. The use of occasional foliar sprays such as sulphur for mildew (which can be more of a problem in wetter, warmer parts of the UK) is permitted in organic systems.

Diseases in Field Beans

The control of diseases in organic field beans relies mainly on cultural methods. These include long rotations and avoiding large areas of the farm that are drilled with the same crop (or a crop of the same family, such as peas or vetch), as this can help reduce the spread of disease. The removal of volunteer beans from previous seasons and the burial of trash is also an important tool in reducing the carryover of disease, as is using disease-free seed. The avoidance of cold wet soils and poorly drained and low-lying fields can also reduce disease incidence. Disease incidence within the crop is likely to occur when the crop is tall and dense, largely because the amount of air passing through the crop is restricted; lush crops are more at risk.

Chocolate Spot (Botrytis fabae)
Crops with a high population of tall plants are most at risk from this disease. Winter beans are likely to have higher losses. The disease can occur at first bud or early flowering. Symptoms appear as reddish-brown spots, which eventually enlarge to give a more damaging aggressive phase in cool, wet or damp weather. There is no varietal resistance to chocolate spot.

Although this disease can occur in the organic crop, it is usually with low incidence.

Downy Mildew (Peronospora viciae)
Mildew is now prevalent on spring beans, where it causes greyish-brown, felty growth on the under surface of the leaves. Downy mildew can occasionally affect spring beans to a high degree. Winter beans are not as much at risk. Careful variety selection is important.

Resistance to downy mildew is noted on the National Institute of Agricultural Botany list (NIAB) for spring beans. Some varieties have resistance to the disease, and 1–9 ratings.

Applications of sulphur in suspension with water sprayed on to the crop can keep the problem at bay, but often by the time mildew is identified it is too advanced for sulphur to assist. Varietal diversification can be employed to minimize the risk of spread between adjacent crops. Varietal mixtures in the same field can help to reduce the severity within a crop.

Leaf and Pod Spot (Ascochyta fabae)

Leaf and pod spot is virtually always seed borne, and only clean-tested seed should be drilled in order to minimize risk. Winter beans are more prone to attack, where brown spots appear on the leaf or the pod; these spots contain *pycnidia*, the black fruiting bodies that spread the disease. Once the crop has been harvested it is important to bury the trash in order to minimize disease carryover. The disease prefers wet conditions, and in wet years severe symptoms may occur where disease carryover has occurred from a nearby field from volunteers. The NIAB/PGRO lists varietal resistance on a scale of one to ten.

Some winter varieties that are very susceptible to the disease may develop severe symptoms in wet years, particularly if growing near to previous years' bean fields, where infection can be transmitted from bean volunteers.

Rust (Uromyces fabae)

Rust is characterized by numerous reddish-brown pustules on the leaves. It is more serious on spring beans, and all varieties are susceptible. The disease reaches high levels more quickly on shorter-strawed varieties, but some have a degree of partial resistance. It may occasionally be damaging on winter beans as well if infection begins during flowering and pod set. The NIAB/PGRO lists state varietal resistance on a scale of one to ten, and growers should note these when choosing a winter variety.

Foot and Root Rots (Fusarium species and other fungi)

Foot rot can appear to be quite sporadic in beans. It can occur on seedlings and on more mature plants, causing browning of the stem base and wilting of the leaves. Foot and root rots in beans appear to be more sporadic than those occurring in peas, and the bean crop in general appears less sensitive to root rots than peas.

Over-cropping land with beans is likely to increase the chance of an outbreak.

Sclerotinia (Sclerotinia trifoliorum)

This disease occasionally infects winter beans in damp autumn weather, and infections may be associated with preceding crops containing red clover. Plants develop a watery stem rot, which can spread from plant to plant in dense stands. The related fungus, *Sclerotinia sclerotiorum*, infects spring beans and also peas, rape, linseed, lupins and a range of field vegetables.

Infection in spring beans is, however, very rare, but the risk should be borne in mind when planning rotations with other host crops.

Pests in Field Beans

With all pests, the organic farmer's main defence is cultural control through good crop rotation, and the encouragement of predatory insects through the use of habitat strips and field margins. It is worth noting that increased populations of hover flies, parasitic wasps, ladybird beetle and spiders on organic farms, and the decrease in nitrogen in the crop, create a situation in organic cereals and legumes where pests are rarely a problem.

Black Bean Aphid (Aphis fabae) and Pea Aphid (Acyrthosiphon pisum)

Black bean aphid can be very damaging to field beans if colonies develop just prior to flowering. The pea aphid can transmit several viruses in the crop, which can reduce yield, and the black bean aphid can smother the stems when in high enough numbers. These colonies result from the primary migration from overwintering hosts in mid-May to mid-June. Late-sown crops are generally at greater risk, as the aphid migration is more likely to coincide with an earlier developmental stage of the crop.

Spring-sown crops are usually more likely to suffer damaging attacks than winter beans. As well as forming dense, smothering colonies on the upper part of the stem, these and the less obvious pea aphid are able to transmit several viruses, which add to the yield loss.

Aphids can be controlled by encouraging natural predator numbers (namely lacewings and ladybirds). This can be achieved by ensuring that there are sufficient overwinter habitats for the predators, such as tussocky grass margins and good thick hedge bottoms.

Stem Nematode

The stem nematode can become a problem mainly in wet seasons. Farm-saved seed is at higher risk, and seed should be tested and only clean seed used — although being seed borne it can also infect the soil and affect a wide range of crops such as bulbs and onions. The practice of wide rotations will also aid control.

Bean Seed Beetle

The bean seed beetles emerge from the seed leaving a hole. This will have a serious effect on quality. This may not matter with beans grown for the animal feed market though human consumption market requirements will not be met. Their presence has increased in recent years though the complete burial of trash can aid control as can a long rotation.

Bean Seed Beetle (Bruchus rufimanus)

Bruchid beetle can affect both winter and spring varieties, and is often a serious problem in field bean crops grown for quality or seed. The adults emerge from hibernation as the temperatures increase during May, and oval-shaped, yellowish-green eggs about 0.5mm long are laid on developing pods. These hatch in one to three weeks, and the larvae then bore through the pod and into the seed, where they feed until mature. After two to three months the adult emerges, leaving the characteristic circular exit hole. The beetles do not breed in grain stores.

Flea beetles (Phyllotreta spp)

Small striped flea beetle: 2.5mm long, black with a yellow stripe down each wing case.

Large striped flea beetle: As above, but up to 3mm long.

Turnip flea beetles: Very small and black; one type has a green or blue lustre.

Flea beetles spend the winter as adults, hibernating in tussocky grass, and in debris under hedges and similar situations. In spring they move out on to young plants to start feeding. They may fly for a kilometre or more to find suitable food.

Female beetles lay their eggs in the soil near suitable plants in May/June. The larvae of the large striped flea beetle feed in 'mines' in leaves; other species feed on plant roots. These larvae pupate in the soil, new adults hatching out in the autumn; they feed for a few weeks before hibernating. These adults will survive until the following July or August.

In the spring and summer, adult beetles eat small holes and pits out of young leaves and stems, and this can start even before the seedling leaves appear above ground. A severe attack will check growth and can kill young plants, especially when they are not growing strongly. Small glossy beetles, which jump when disturbed, will be seen on and around plants.

Control   Drilled clover, peas and, to a lesser extent, beans can be affected. Best control is to monitor the beetles' lifecycle emergence with sticky traps, and drill when these are at peak so that the drilled crop plants emerge when the populations are in decline.

Grain aphid (Sitobion avenae)

There are two types of grain aphid: the grain aphid that predominantly feeds on ears, and the rose-grain aphid that prefers leaves. Yield losses of up to 11 per cent have been found in winter wheat, primarily due to direct feeding, with secondary damage from fungi that colonize aphid honey-dew and damaged tissue.

The lacewing (*Chrysopa*) adults and larvae are voracious feeders on aphids. The larvae of some species augment their camouflage colouring by covering themselves with the dried remains of their prey, so that each one looks like a small mass of debris rather than a living insect. Lacewing larvae have unusual sucking mouthparts, with a pair of extremely long, slender and conspicuous mandibles (or jaws) that curve forwards from the front of the head. These mandibles are tubular structures, rather like a pair of hypodermic needles, which are sunk into the victim's body, and the larva then sucks out the body fluids of its prey as if through two tiny drinking straws.

Pea and Bean Weevil (Sitona lineatus)

The pea and bean weevil is a grey-brown weevil with a characteristic snout; it can cause damage to spring beans if large numbers appear when the plants are small. Eggs are laid on the soil and the larvae hatch, burying into the soil to feed on the roots. Pea and bean weevil can dramatically reduce root nodule formation and thus nitrogen fixation. The adult weevils overwinter in peas and beans, either in storage or in spilt crop around the field. They become active in March and will cause considerable damage to young seedlings.

The leaves of attacked plants show characteristic 'U'-shaped notches around the edges, but the main damage occurs as a result of the larvae feeding on the root nodules. Winter beans, although still prone to attack, are usually too advanced in growth for the weevil or the larvae to have any appreciable affect on yield.

Diseases in Peas

Pea Wilt (Fusarium oxysporum f. sp. pisi)

Pea wilt is a soil-borne disease that can occur in any pea-growing area, but is generally confined to parts of Essex, Lincolnshire and South Yorkshire. It can cause substantial reductions in yield, but is effectively controlled by genetic resistance. Race 1 appears to be the most common form, but the majority of varieties are resistant to this race, and growers using land in known high-risk areas should select these. Yields can be severely affected, though genetic resistance effectively controls it. Growers on high-risk land should select resistant varieties.

Infection appears in May to June. A rotation of four or more years between healthy pea crops will control build-up, though where the disease has already occurred the length of the break should be increased.

Downy Mildew (Peronospora viciae)

Once this disease becomes established in the crop there is little that can be done, even by the non-organic farmer. The disease produces resting spores that persist in the soil and initiate primary infections in young pea plants. Though secondary infections can develop, particularly in cool, damp conditions, they are rarely as damaging as primary infections, which can kill plants before flowering. Deep ploughing will aid control, and a variety that has a high resistance to the disease should be selected.

Powdery Mildew (Erysiphe pisi)

Occasionally late maturing marrowfat crops may become covered with a grey-white film of powdery mildew. The disease can delay maturity.

Leaf and Pod Spots

Leaf and pod spots are caused by three fungi: *Ascochyta pisi*, *Mycosphaerella pinodes* and *Phoma medicaginis*, which may be spread by seed infection, or soil or plant debris. The most frequent is *M. pinodes*, which can cause losses in both yield and quality in wet conditions.

The diseases are more common in wet conditions. The organic farmer's best line of defence against this disease is to test seed to ensure it is not infected. There are no varieties at present that show significant resistance to these diseases.

There are no minimum standards specified by the statutory seed certification scheme for *M. pinodes*, but seed, and especially farm-saved seed, should be tested.

Botrytis or Grey Mould

This disease can affect the pods and stems in wet or heavily damp weather, and it occurs when the petals of the flowers fall and land somewhere else on the plant, creating mould growth. Where conditions favour its occurrence there is not a great deal that the organic producer can do.

Foot and Root Rots

Frequent pea crops on heavy land favour the occurrence of root and foot rots, though good drainage and the prevention of compaction will reduce the chances of it occurring. Where it is suspected, a soil-testing kit can be used to predict the chances of serious field losses occurring. If it is thought that rots are going to be a problem, a rotation without legumes should be practised for one rotation cycle. *Scerotinia sclerotiorum* will cause stem rot and can affect peas, spring beans, oilseed rape, linseed and sometimes potatoes.

Bacterial Blight (Pseudomonas syringae pv. pisi)

This blight is a potentially serious seed-borne disease that can occur on all types of peas. Symptoms consist of small brown lesions on the lower leaves, stems and stipules. They become noticeable following periods of heavy rain, hail or frost. The lesions may coalesce and show a fan shape on the leaf, following between the lines of the veins. Some pod spotting may occur. Severe infections have not occurred in spring-sown peas, and the effect on yield has been negligible. This disease has not been noted in spring peas as yet, though it has been noted in isolated incidences in winter peas. As with other seed-borne diseases, clean seed should always be used. A seed test is available from PGRO or NIAB.

Marsh Spot

Marsh spot is a disorder of peas due to a deficiency in, or the unavailability of, manganese. The deficiency causes the formation of a brown spot in the centre of many of the peas produced, and the produce is spoilt for human consumption and for use as seed. It is particularly associated with organic and alkaline soils. When symptoms appear in a crop, 6–11kg/ha of manganese sulphate should be applied at once in a high volume of water with a wetter, or an equivalent application of a manganese spray. Similar treatment must also be carried out when an affected crop is at first pod stage, and repeated ten to fourteen days later, in order to prevent the formation of marsh spot. In some seasons flowering is prolonged, and a third manganese application will be necessary. The amount of manganese in some formulations – for example, chelated manganese – may be too low to be effective at the rate recommended.

Pests of Field Peas

Pea Weevil (Sitona lineatus)
Weevil may cause damage if large numbers appear when plants are small. The leaves of attacked plants show characteristic 'U'-shaped notches around the edges, but the main damage occurs as a result of the larvae feeding on the root nodules. The risk is higher in cloddy seedbeds and conditions of slow growth.

Field Thrips (Thrips angusticeps)
Field thrips feed on the surface of emerging seedlings; the tissue can then appear thickened, and may also become pale. In most cases the peas will outgrow the effects

Pea Aphid (Acyrthosiphon pisum)
Aphids can cause severe yield loss when present in large numbers, and early infestations can result in crops becoming infected with pea enation mosaic virus. The risk increases when colonies are found on 20 per cent of plants. Pea moth larvae feed on the peas within the pod. Yield loss is minimal, but the quality of the peas can be dramatically hit, and growers who are producing peas for human consumption may then find the crop fails quality tests. Ensuring lots of natural predator environments for lacewings and ladybirds is critical, as these are the main predators of aphids (*see* notes earlier in this chapter on aphids).

Pea Cyst Nematode (Heterodera gottingiana)
This is a very persistent soil-borne pest, often causing severe loss. It can build up in the soil where peas and beans are grown too frequently in the rotation. An adequate rotation is essential to minimize the risk of occurrence.

Affected plants are stunted and pale, and the root systems do not develop nitrogen-fixing nodules, but become studded with white, lemon-shaped cysts. It is important that this problem is diagnosed, as subsequent pea crops will be prone to total crop failure unless there is an adequate gap in the rotation.

Pea Moth Larvae (Cydia nigricana)
These larvae feed upon the developing seeds within the pod. Yield loss is minimal, but the effect on quality can be dramatic, and damage to the seed reduces the value of the produce for human consumption and for seed production.

The 'Oecos' pheromone pea moth trapping system should be used to assess population fluctuations, however there is little the organic farmer can do regarding control, other than to encourage high numbers of natural predators. At the time of writing, 'Oecos' pea moth traps are available from 'Oecos' at 11A High Street, Kimpton, Hertfordshire SG4 8RA, and from 'Agralan Ltd' at The Old Brickyard, Ashton Keynes, Swindon, Wiltshire, SN6 6QR, who also produce a pea moth monitoring system.

Pea Midge (Contarinia pisi)

Adult midges emerge from the soil during June. The females emit a pheromone to attract males, and egg laying on shoots and buds occurs soon after. Eggs hatch four to five days after laying, and the larvae feed for about ten days on the buds. The feeding of midge larvae distorts or kills the shoots and flowers, and this can result in so-called 'nettle-heading' in the peas, where the infested shoots become deformed, and attacked flowers abort or produce tiny, misshapen pods. The larvae then drop to the soil and pupate. Some emerge after eleven days as a second generation, whereas others remain in the soil to emerge the following year or in subsequent years. Feeding on the buds causes them to be sterile and encourages the development of botrytis, and hence a reduction in yield.

Stem Nematode (Ditylenchus dipsaci)

The nematode has become a major pest in field beans and can cause severe problems in wet seasons, particularly where farm-saved seed from an infested stock has been multiplied for several generations. The pest is seed borne and can also infest soils, thereby becoming a problem for a wide range of other crops, including bulbs and onions. Seed should be tested for nematode, and only clean stocks should be sown.

This should be of concern to organic farmers, as this pest can also infect red clover, eating out the clover roots and causing dieback. This is often termed 'clover sickness'.

Lupin Diseases

Sclerotinia (*Sclerotinia sclerotiorum*) causes a stem rot, and can occur particularly in warm, wet periods; it can infect all lupins, and the risk should be assessed when planning rotations with other host crops. The worst damage to lupin is caused by anthracnose (*Colletotrichum acutatum*), fusarium wilt, fusarium and other root rot, bacteria and viruses, and seed should be tested prior to planting and not used if infection is found.

Lupins should not be grown in close rotation with other host crops including peas, beans, oilseed rape and linseed due to the potential of disease carryover.

Lupin Pests

There is a major threat to establishment from birds and rabbits, and careful and timely precautions are needed during plant development. The crops are most at threat at emergence. Low-level pea and bean weevil attack is common, but crops are likely to grow away. Lupin aphid (*Macrosiphum albifrons*) can be damaging, and peach potato aphid (*Myzus persicae*) can transmit bean yellow mosaic virus, but infection levels in the UK are rare and at low levels, mainly as a result of the small area of crop grown.

Soya Diseases

Soybean rust, also known as **Asian soybean rust**, is a disease that affects soybeans and other legumes. It is caused by two types of fungi, *Phakopsora pachyrhizi* and *Phakopsora meibomiae*. *P. meibomiae* is the weaker pathogen of the two, and generally does not cause widespread problems. The disease has been reported across Asia, Australia, Africa, South America and the United States, but not in the UK.

Soybean rust thrives on green, growing plants, and requires a climate of high moisture and moderate heat. It is unable to survive the cold winters of northern habitats.

Soybean rust is spread by wind-borne spores, which are released in cycles of seven days to two weeks. It is commonly believed that the disease was carried from Colombia to the United States by Hurricane Ivan!

Soya Pests

The **soybean cyst nematode** (SCN), *Heterodera glycines*, is a plant-parasitic nematode and a devastating pest of the soybean (*Glycine max*) worldwide. The soybean cyst nematode infects the roots of soybean, forming cysts where it lives within the plant tissues. Infection causes various symptoms that may include chlorosis of the leaves and stems, root necrosis, loss in seed yield and suppression of root and shoot growth. It is a significant pest in the soybean growing areas of the USA , South America and Asia, but not in the UK.

The above-ground symptoms of SCN infection are not unique to SCN infection, and could be confused with nutrient deficiency, particularly iron deficiency, stress from drought, herbicide injury or another disease. The first signs of infection are groups of plants with yellowing leaves that have stunted growth. The pathogen may also be difficult to detect on the roots, since stunted roots are also a common symptom of stress or plant disease. Observation of adult females and cysts on the roots is the only accurate way to detect and diagnose SCN infection in the field.

Since SCN is an obligate parasite, a crop rotation involving non-host plants can decrease the population of SCN and has been shown to be an effective management tool. Plants that are already stressed are more susceptible to infection, so good cultural practices, such as maintaining soil fertility, pH and moisture, can reduce the severity of infection.

Chapter 12

Harvesting and Storage

Harvesting and storage are one of the most important aspects of any farming system. Successful cereal and pulse harvesting is subject to a wide range of factors including the range and area of crops grown, machinery available, reliance on contractors, maturity of crops, and ultimately the weather.

ORGANIC STANDARDS FOR HARVESTING

To avoid contamination from non-organic crops and prohibited pesticides, all harvest machinery, trailers, dryers, grain cleaners, conveyors, ancillary equipment, stores, bins and sacks must be cleaned of all crop residues and any other materials that may contaminate the organic grain. Records of all cleaning operations and cleaning products used, movements, treatments and cleaning procedures must be kept in accordance with organic standards and assured combinable crops scheme (ACCS) requirements.

HARVESTING OPTIONS AND CHALLENGES

Most organic cereal and pulse crops are harvested as mature, dry grain. At this time it will ideally be at 14 per cent moisture content, but it can be anywhere between 10 and 20 per cent moisture content, depending on the season. The risks of harvesting before a grain crop is mature include:

- increased drying costs;
- poor threshing and separation of the grain from the rest of the plant;
- reduced germination if the crop is for seed or malting.

The risks of harvesting a grain crop that is over-mature include:

- cereals sprouting in the ear, leading to loss of germinative capacity and quality;
- risk of grain head loss, particularly in barley;
- increased risk of lodging and associated harvesting difficulties;
- loss of pulse grains from split pods.

The optimum harvesting period for any given cereal or pulse crop is not a point in time but a 'window' of a couple of weeks, and it is during this time that the grain should be harvested, provided the weather is dry. Difficulty arises when there is a large area to be harvested with limited resources, and/or when weather conditions make progress slow.

HARVESTING OF CEREALS

There are two options for harvesting cereal crops: as a dry grain, and as whole crop cereals for ensiling or 'crimping' and feeding to livestock.

The use of whole crop silage has gained popularity in recent times, especially in organic systems. The main reason for this is that cereal-based whole crop silage can provide a high dry matter feed, and in seasons where the first cut silage has been adversely affected by rain, it has provided a good quality winter feed. It also allows the crop to be harvested early when many weeds and weed seeds have not reached maturity. Whole crop also complements clover-based silage, which is most likely to be grown on organic farms. Another advantage of whole crop silage is that the grower has the choice, and may opt to cut it for silage, or if the yield of the grass harvest is satisfactory, may leave it to be cut for grain.

The growing of whole crop silage is straightforward. The cereal is drilled in the usual way in the autumn into a fertile site. Once this has taken place, the crop can be left until usually the following May. At this time the crop is then harvested at the cheesy dough stage (typically 25–30 per cent grain moisture content), which is when the starch is being transported to the grain. Before this time the grain will be at the milky ripe stage and too early for harvest. If harvest takes place after the cheesy ripe stage, then it is important that the grain can be cracked open as it passes through the forager: if it does not, it may pass straight through the animal. The table below shows a comparison of grass silage, maize silage and whole crop wheat silage.

Table 61 Comparison of silage and whole crop silage quality

Composition	Grass silage	Maize silage	Fermented whole crop silage
DM (g/kg)	248	275	557
CP (g/kg DM)	132	94	89
ME (MJkgDM)	11.4	10.9	10.7
Ammonia	28	37	33
pH	3.6	3.7	4.0
Starch	29	256	359
Water soluble carbohydrate (g/kgDM) 46		8.2	3.1

Whole crop silage generally results in the highest dry matter but the lowest protein and the lowest energy. Results of experiments have not demonstrated any improvement in yield, fat or protein where the grass silage was of high quality; however, where the quality of the grass was compromised by poor weather, whole crop silage can compensate for the shortfall.

Combine Harvesters for Organic Cereals

Modern harvesting equipment generally has a larger capacity to cope with bulky crops; however, organic crops are likely to have a lower harvest index than non-organic crops, and therefore any harvesting equipment may need to be set up differently. Crops with green material in the bottom such as meadow grass, broad-leaved weeds or undersown, will be cut more effectively with plain knife sections than the more commonly used serrated sections.

There is a trend on many large farms, and with contractors, to have large capacity combine harvesters. Whilst this suits the high input–high output non-organic cereal systems with yields of 8–10t/ha, in organic systems with lower-yielding crops of typically 3.5–6.0 t/ha these larger machines are not always so suitable. Experience suggests that with lower-yielding crops the larger combines are not working at capacity and it is more difficult to adequately load the drum and threshing mechanism; where this is possible, the forward speed required results in other operational problems, and as a result grain losses can be higher.

Organic farmers should perhaps consider either a smaller capacity combine harvester with a larger header, or two older, smaller capacity combines for the task.

Large capacity combine harvester.

Smaller capacity combine harvester.

Straw from organic crops may require more conditioning before baling and removal if there is more green material in the crop at harvest. If there was a large area to be harvested, then it would be sensible to establish a preferred harvesting sequence to maximize grain quality. This will be determined by variety and market, with the priority given to seed, and milling and malting ahead of feed grains.

HARVESTING PULSES

Harvesting of Field Beans

Field beans are usually harvested from September onwards, though this date can vary by a large margin depending on location and season. In some dry years harvesting beans may take place at the end of August, while in wetter, cooler seasons a late September harvest would not be uncommon. The factors influencing harvest date include variety, crop location and season. Beans can cope far better with rain than cereals or field peas due to the crop architecture and thicker pod.

Beans should be harvested when the moisture content is 14 per cent and the beans are hard, and separate from the pod easily. The plant turns black at harvest and is easily threshed when the straw and pods are dry. If there is a chance that pods may shed during the heat of the day, harvesting in the morning or evening when there is more moisture in the air can reduce losses. The bean crop can be harvested at 18 per cent moisture content; however, combine blockages can occur if there is a significant amount of green stems present.

The Harvesting of Peas

The relatively recent efforts to improve the pea harvest through breeding taller and stiffer varieties that are mainly semi-leafless has had varying results throughout the country. In southern and eastern England peas are often erect through to harvest, however in western and northern England the ability of the crop to remain standing through to harvest has proved more difficult. The pea harvest usually takes place before that of winter wheat, with conventional foliage types lodging before harvest. It is important to harvest the peas once they're ready, mainly because they will soon lodge, making the harvest more difficult. Once lodged, weed problems can significantly increase, creating further problems during harvest. Peas should be harvested at 15 per cent moisture for feed. The more specialist human consumption market may have other requirements, which would be detailed in the contracts.

One method favoured by organic farmers is to sow a low percentage of spring cereal with the peas. The cereal acts as 'pea sticks' and encourages the pea canopy to climb, helping to keep it off the ground for harvesting. The grain can be effectively separated using a grain cleaner or dresser.

Whether it is a cereal or pulse crop that has been harvested, the crop should be cleaned and where necessary dried promptly to avoid spoilage of the crop, before placing into storage.

CROP CLEANING

Cleaning is a vital step in the process of harvesting, drying and storing grain. It was once common practice to clean grain on its way into drying and storage facilities, however, this is now much less common in non-organic production, where a combination of herbicide inputs and state-of-the-art combine harvesters are seen as being able to clean out most weed seeds prior to storage. Organic crops are far more likely to contain material other than grain, and although as much of this as possible should be removed in the harvesting process, this needs to be carefully balanced with grain loss in the field. The remainder should be removed before drying, as its presence will slow the drying process and restrict airflow through the crop.

Cleaning is usually carried out to meet the contractual requirements of grain buyers. Generally admixture, such as broken grain, seeds and chaff, should be less than two per cent. Cleaning grain also has numerous other advantages relating to pest control and is therefore strongly recommended for organic grain both pre- and post storage. For example, chaff and debris will reduce the rate of airflow, potentially increasing drying time and efficiency, leaving grain at risk of attack for longer. The productivity of many insect and mite species increases on cracked or broken grain. Cleaners such as aspirated sieves can remove free-living pest stages from infested grain, although hidden stages within the grain, such as larvae and pupae of grain weevils, will survive. In these cases cleaning would have to be integrated with other measures, such as drying, to ensure complete disinfestation.

The use of mobile seed-cleaning plants is permitted, but caution must be used to avoid contamination of organic grains with prohibited treatments. Many mobile cleaners have diversion systems built in to grain and seed treatment plants, which can facilitate this. Even where these are present the mobile plant should be thoroughly cleaned prior to handling the organic crop, and a signed record of cleaning kept.

CROP DRYING

The object of crop drying is to reduce the relative humidity (RH) between the grains to less than 65 per cent, which is below the minimum required by mites and fungi. Accurate drying to recommended levels specific to crops is important. Crops need to be dried sufficiently to maintain their condition and be acceptable to the market. Overdrying of crops can make them difficult to process and will reduce the weight of a given volume, thereby reducing the tonnage sold.

Drying needs to be carried out as soon as possible after harvest to prevent pest growth in the interim. This is particularly important with harvest backlogs, when damp grain may be stored for several days before drying. Backlogs can occur when the hot air dryer cannot cope with the harvested grain, and grain is stored damp in heaps before it is dried. The 'safe' MC of cereals and pulses in equilibrium is about 14.5 per cent. However, this rises with decreasing temperature and varies with different commodities. The safe MC normally lies between the values for 60–70 per cent RH, as given in Table 62.

Table 62 Approximate MC at a range of equilibrium RH at 25C

Crop	Relative Humidity (RH)					
	40	50	60	70	75	80
Wheat	10.7	12.0	13.7	15.6	16.6	17.6
Barley	9.8	11.5	13.2	15.1	16.1	17.3
Beans	9.8	11.3	12.7	14.6	15.9	17.5
Peas	9.5	11.2	13.0	15.8	18.0	20.9
Lupins	8.5	9.8	11.5	13.5	15.0	17.0

Source: McLean

Drying before storage using a hot air dryer can be more effective than using ambient air, as the latter may take too long to get the grain down to a safe moisture content.

Hot air drying

Drying grain using air heated to 40°C or above means that grain can be dried in a matter of hours regardless of the weather. With recirculating batch dryers the amount of drying depends on the running time because the MC continues to fall as long as the grain circulates in the dryer. These

Opico mobile batch dryer.

machines are usually mobile, giving the opportunity to share equipment with other farmers.

In continuous flow dryers, the grain flows through a heated air section and the drying time is determined by the speed of discharge. If using continuous flow dryers when crops are wet, avoid reducing MC too much in one pass. High temperatures and 'cooking' grains can split seeds and this is likely to impair the germination of seed crops. Aim to separate passes by one to two days at moderate temperatures of 38–49°C. In all cases the grain must be cooled after drying. To avoid damaging grain, particularly the germination and gluten quality in bread-making wheat grains, it is normally recommended that a maximum of 65°C at 20 per cent MC is used, reducing by 1°C for every 1 per cent increase in MC. For feed grain a maximum of 120°C for one hour or 100°C for three hours is safe.

Ambient air drying

The UK climate means that ambient air is only sufficient for drying grain to below 14.5 per cent MC in one year out of five, therefore the dryer will need some form of heat or a dehumidification system. Grain can be dried in bin or on floor in layers of 1.5–4m using minimum airflows of about 180m3/hour/tonne with ambient air, or air heated by just 5°C above ambient. As the airflow depends on the grain resistance, the height of the grain heap should be reduced by 0.5m for every 1 per cent increase in grain MC above 20 per cent. During 'in-bin' or 'on-floor' drying, a crust may form on the top of the grain: this can seriously restrict air flow, and may need to be broken by turning or stirring the grain.

Continual flow dryer.

Ambient on-floor drying.

COOLING GRAIN IN STORE

Grain should be cooled to below 15–20°C within three weeks of entering the store. Within 140 days the grain should be cooled to below 10°C, and ideally by the end of December, temperatures should approach 0°C. To reach the temperature target should take only seventy-five to 100 hours, and no more than 300 hours in all.

In practice this can be achieved in the UK using an ambient air ventilation system with a differential thermostat. It is a good idea to run fans in frosty weather: by using this technique, bulk grain can be kept cool right through to the following summer.

Objectives of Cooling

Biological activity halves for each 10°C drop in temperature, so pest increase and quality changes, such as germination loss, will all be retarded by lowering the temperature. Furthermore insects, mites and fungi all have a temperature threshold below which they cannot breed. The most common beetle, the saw-toothed grain beetle (*Oryzaephilus surinamensis*) will not breed below 17°C, while the grain weevil (*Sitophilus granaries*) will breed at 10 to 12°C. These temperatures can be easily achieved by aeration in the UK. Mites and many fungi will increase below 5°C, albeit slowly, so the primary control measure for these organisms is adequate drying (*see* section Drying, above). Finally, cooling also evens out temperature gradients which otherwise might permit condensation when warm air, rising from the grain bulk by convection, meets cool surface grain in the autumn. The speed of cooling is also important to pests: the grain must be cooled to below the insects' breeding thresholds (15°C for saw-toothed grain beetles, 10°C for grain weevils) as it goes into the store and before eggs laid by wandering insects can complete their development into adults. This usually requires an airflow of 10m3/hour/tonne in the UK.

Aeration Systems

The equipment required for cooling by aeration includes fans to deliver the air and ducts to distribute the air, usually beneath the grain and the correct airflow for the tonnage of grain in store. Instead of above-floor ducts, a more convenient but more expensive alternative is to have slots in the floor into which aeration grills are fitted. These allow tractors and lorries to drive over them without the damage often suffered with above-floor ducts.

A cheaper alternative to the horizontal system is a vertical system, sometimes known as 'pedestals'. These typically comprise a 1m upright perforated sections of duct connected by expandable necks to a centrifugal fan which sucks the air – although conversion to blowing is easily arranged. Such systems are basically semi-permanent installations of aeration spears.

Management of aeration, whether using vertical or horizontal aeration systems, is important. Automated control systems, which ensure that fans run at the most suitable ambient temperatures for cooling, are common.

Aeration should start as soon as drying is finished and the store has grain in it. In this context, blowing air through the grain via the ducts is often preferable to sucking it from the surface down, when warm air from later loads of warm grain placed on top of the pile will re-warm previously cooled grain.

MONITORING CROPS IN STORE

For long-term storage it is essential that grain is monitored accurately for pests, temperature, humidity and moisture content. Even where grain is stored for only a few months, things can go wrong. Fungal and pest

Monitoring grain routinely with pitfall traps.

problems can escalate rapidly. Accurate records should be kept of temperatures and insect catches. If temperatures rise or fall unexpectedly, or if insect catches increase, the aeration system may not be performing correctly and the fault will need diagnosing.

Insects and Mites

Research for the Home Grown Cereals Authority (HGCA) reveals that sampling grain for pests and mites in store is too labour intensive and unreliable, and traps are more reliable. Traps should be placed in the surface layers of the grain where the MC and temperature is likely to be higher than the grain bulk; hence insects and mites are more likely to be found in a 4–6m grid.

Check traps every week when the grain first goes into store; later on, every month should be sufficient. Pests need to be identified and recorded accurately (see picture above). Finding a few pests in a trap does not necessarily mean that drastic action needs to be taken. But if the numbers of pests caught increases or more traps have pests in them, indicating that pests are more widely distributed in the grain, then action needs to be taken. Several types of trap are available, including pitfall traps, probe traps, bait traps and bait bags. Specialist advice should be sought as to which trap to use for a particular crop and range of pests.

Dealing with Storage Problems

If grain rises above 15 per cent MC or 20°C, then insects, mites and moulds rapidly become a problem. Grain must be sold and used quickly, or remedial action needs to be taken.

Fungal Growth

Fungal problems may be visible or invisible. Sometimes fungi may become apparent at the grain surface, often associated with sprouting grain. Grain should be cooled as soon as possible, and mouldy and sprouting grain must be removed and destroyed. Fungal growth can also cause damage in a less obvious way, as fungi will grow slowly at moisture contents (above 14.5 per cent) in equilibrium with relative humidities above 65 per cent. Under dry conditions the species occurring may destroy germination or cause heating.

Hot Spots

High numbers of insects can cause heating in bulk grain, but because these heating pockets are so small they may be difficult to detect. These can lead to mould (*see* figure 37 below). Usually the first sign is a cap of mouldy or sprouting grain. Hot spots usually arise from local infestations of grain weevils, *Sitophilus granarius*, which can breed at lower temperatures (10 –12°C) than other species, albeit slowly. If grain is intermittently venti-lated, even for just a few hours each week, the metabolic heat produced by any surviving pests will be removed and hot spots cannot develop. This can be achieved using the existing aeration system, but if this is not possi-ble a portable aeration spear can be used.

Re-drying grain If the grain is badly infested it may be worth using high temperatures to disinfest it. Temperatures of 50°C will kill insects within one to two hours, and at 60°C they can only survive a minute or so. The disad-vantages of this technique are that it is expensive and the grain must be cooled rapidly afterwards or the warm grain can quickly become re-infested.

Conveyors and cleaners If the grain is badly infested it may be worth re-cleaning it, which will remove some pests. Vibration or percussion is another way of killing insects and mites. When combined with cleaning, 75–90 per cent of mites can be killed; however, it is only a temporary

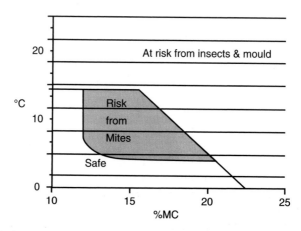

Figure 37 Grain temperature and moisture contents at risk from insects and mould developing

measure for mites, as they can survive inside the grain and populations can quickly build up again.

Cleaning and Drying Cereals

Cleaning grain before it is marketed is an important procedure to avoid rejection of organic grain by millers and feed companies. Buyers require the moisture content of cereals to be 14 per cent, with a maximum of 2 per cent impurities. If the correct sieve sizes are chosen, pests and any beneficial insects contaminating the surface of the grain can be removed, along with any remaining weed seeds, stones or chaff. Using a sieve size of 2mm in an aspirated cleaner can remove 90 per cent of mites.

It is important that cereal moisture content is no higher than 14 per cent for long-term storage, and that cereals are cool as they enter the store. Where the moisture content is high, cereals should not be stored too deeply until the moisture content has been lowered. Radial-ventilated bins, floor-ventilated bins and on-floor ambient systems are all suitable for cereal grain. Spelt cleaning and drying may require a different approach as it is a husked grain.

Home-saving seed is a common management strategy on many organic farms, but careful storage and handling is required, and testing should be undertaken for disease, pest and germination as soon as is practicable.

Cleaning and Drying Field Beans

Merchants require the moisture content of beans to be 14 per cent with a maximum of 2 per cent impurities, though some merchants allow 16 per cent moisture. It is important that the moisture content is no higher than 14 per cent for long-term storage. It is also important that the beans are cool as they enter the store; one hot day spent turning the beans with a loader on an on-floor store at regular intervals can significantly aid the cooling of the grain. Where the moisture content is high, beans should not be stored too deeply until the moisture content has been lowered. If the beans are destined for a premium market it is important that they are not subject to high dryer temperatures as this may cause cracking.

The large seed size of beans makes drying difficult as beans have a low resistance to air flow. It takes time to move moisture from the inside to the outside, and slow, gentle drying with ambient air is best. Mouldy produce is unacceptable for animal feed or other markets.

Radial-ventilated bins allow faster drying than floor-ventilated bins, but care must be taken not to overheat the beans. On-floor drying using ambient or warmed air is also successful, but care must be taken not to load beans too deep if moisture content is high and if lateral ducts are spaced more widely than 1m.

Home-saving seed is a common management strategy on many organic farms, and careful storage and handling is required. Experience suggests that over-drying, excessive cleaning and repeated movement, especially dropping beans from a height on to floors, can cause small internal cracks, which can significantly reduce germination.

Cleaning and Drying Field Peas

Typical standards for pea storage require them to be at 14 per cent moisture, with a maximum of 2 per cent admixture. As a larger seed, peas are slower to dry than cereals. Where the moisture content is above 24 per cent, two dryings may be necessary without the air temperature exceeding 37°C for pea seed and 43°C for peas for human consumption.

High grain dryer temperatures can damage peas for human consumption by making them too tough. During storage, growers should check the peas for any hot spots that may have occurred, and they should also be checked in the upper layers to ensure they are free of moulds. Moving peas from bin to bin can aid in the curing of hot spots, though ventilation and drying to ensure the peas are at 14 per cent moisture content should take place if peas are to be stored for more than four weeks.

The relatively large seed-size of peas makes drying more difficult than with cereals. Whilst damaged peas are still acceptable for compounding, mouldy produce is not. Any type of dryer may be used for peas, but those operating at low temperatures are safer. Floor-ventilated bins are easy and relatively safe to operate. When the initial moisture content is high, the transfer of the peas from bin to bin and the use of warmed air, together with adequate ventilation, may be necessary to avoid mould developing in the upper layers. Radially ventilated bins allow faster drying than floor-ventilated bins, but care must be taken not to overheat the peas. On-floor drying using ambient or warmed air can be used, and provided there is sufficient volume of air and adequate ventilation, peas of relatively high moisture content can be dried using this method. Continuous flow dryers designed to work on a short period/high temperature basis need more careful operation than other systems. The maximum moisture content of peas for safe storage depends upon the method and the length of time they are to be stored. Peas may be safely stored for up to four weeks at 17 per cent MC, but if they are to be stored until the following spring the moisture content should not be above 15 per cent. If the peas are in bulk with forced ventilation or frequently moved, the moisture content can be 1 per cent higher.

CROP STORAGE

It is often said that farmers are more interested in growing their crops than selling them. The gains or losses that can be made to a crop once in the barn, can be more significant than any in the field. A grain buyer once remarked to the author that 'if he was selling his car, he would try and sell it clean and washed, not dirty and full of rubbish, and would expect the same of any grain.' It is therefore prudent to do all that is possible to store the grain to keep it in good condition, and to prepare it for sale in the best way possible to achieve the best price possible, and to avoid rejected consignments and unnecessary costs. This includes drying, cleaning and removing contaminants including weed seeds, chaff and ergots.

Reasons for Storing Crops

Cereal and pulse grain is largely a commodity product which is processed further by users before being turned into end products. There are several reasons for storing crops, some practical and some financial, which include:

- to achieve continuity of supply to customers;
- to reduce post-harvest spoilage;
- to provide a safe and stable environment for large quantities of grain;
- to avoid over-supplying the market at certain times of the year (typically immediately post-harvest);
- providing regular income and avoiding cash flow problems;
- spreading the workload of grain dispatch throughout the year.

For some farmers, especially those with older stores or poor infrastructure, or where there are opportunities to use farm buildings for other commercial activities, it may not be appropriate to store crops at all, rather selling them at harvest for dispatch of the farm in late summer and early autumn. There are advantages (earlier payment) and disadvantages (lower values at harvest) with this system, but it suits some farms.

Organic Standards for Crop Handling and Storage

Organic standards for grain storage are designed to ensure the separation of organic grain from non-organic grain or in-conversion grain, and to prevent contamination of organic grain with non-permitted products.

Storage
Organic grain should be stored observing the following conditions:

- Areas of dedicated storage must be used for organic grain.
- A physical barrier must exist between any organic and in-conversion or non-organic grain; a minimum requirement would be an effective wooden partition, for example.
- Stores must be constructed of suitable materials, maintained in a clean and hygienic state, and covered/sealed to prevent access and contamination by bird droppings and vermin.
- All grain must be clearly labelled as 'organic', 'in-conversion' and 'non-organic'.
- Organic and non-organic grain should be of different varieties so they can be distinguished from each other, if the need arises.
- Storage chemicals, such as fumigants or sprout inhibitors, and ionizing radiation, must not be used in any part of the store where organic crops will be stored on either a short- or long-term basis.

- Organophosphate treatments throughout the grain storage facility are prohibited. The only permitted chemicals are pyrethrum-based, and for application to the store structure only. However, none of these chemicals are currently approved by the Pesticide Safety Directorate.
- Any wood used in the store must not have been treated with organochlorine wood preservatives, such as the organochlorine gamma HCH (Lindane).
- Organic grain must not be allowed to be contaminated with exhaust emissions or fuels used for crop drying.

Transport

Any handling of grain, such as mixing, changing containers, reloading or temporary storage, needs to be carried out on organically certified premises. Whilst hauliers are not required to be organically certified, the rules are designed to ensure traceability and to maintain the integrity of the organic grain. Organic certification bodies can provide guidance on acceptable procedures for grain transport. Most UK buyers are used to these requirements.

STORAGE STRATEGIES FOR CEREALS AND PULSES

Crop storage is important in maintaining continuity and quality of supply to processors and the marketplace throughout the season. Even where crops are to be used on farm as feed, good storage is necessary to avoid harmful toxins developing, to ensure maximum efficiency and to minimize waste. Adequate storage facilities, proper store preparation and monitoring, temperature and moisture monitoring, and record keeping are all essential tasks.

The Home Grown Cereals Authority (HGCA) produces a range of technical information leaflets on grain storage, and the Soil Association produces an excellent technical guide on the 'storage of organic combinable crops', written to provide specific advice on the storing of the same.

The key objective is to store grain in sound condition to maintain its quality prior to sale. The storage strategy elements for bulk grain are:

- store preparation and monitoring;
- low grain moisture content;
- low grain temperature;
- reducing the grain temperature as rapidly as possible;
- ambient aeration system;
- temperature monitoring;
- pest monitoring.

Pre-storage checks cleaning, washing and vacuuming preferably, to remove dust and debris that may harbour pests – are essential. Check compliance with crop production protocols – organic standards, ACCS and so on. Check for the exclusion of vertebrate pests such as birds, rats and mice. Monitor the empty and cleaned store for all pests. Check that equipment such as moisture and temperature metres, fans, ducts and conveyors are all in good working order.

The storage period The two key parameters to managing grain quality in store are temperature and moisture. To minimize the risk of infestation, reduce the grain moisture content to 14 per cent and get the temperature to below 4°C (*see* notes later in this chapter).

Clearly these objectives can be difficult to achieve in some years, with mild or damp autumns or early winters. Grain at the surface of the pile is particularly difficult to control in respect of moisture and temperature, and it is not uncommon for some mite infestation to occur. The use of pitfall cone traps is an essential part of grain store management, and will give early warning of pest infestation.

Grain Storage Options

There is a range of storage options open to organic arable producers, the most common of which are the following:

On farm storage Infrastructure located on the farm capable of handling and storing farm crops for sale and dispatch throughout the year. Different storing infrastructure options are discussed later in this chapter.

Co-operative storage As organic production capacity expands, increasing numbers of group and co-operative stores are developing dedicated facilities for the storage of combinable crops. Organic farmers can pay for allocated space or volume in these stores.

Other organic farms Many arable farms find that they reduce the area and volume of combinable crops that they grow in the process of conversion to organic production. As a result some farms have more storage capacity than they require once organic. There are opportunities for collaboration between organic farms to share storage facilities and costs, and there is even potential for joint marketing activities.

Silos and Bins

Purpose-built steel grain silos or bins are an ideal way of keeping the grain separate. However, they do not provide the same degree of flexibility as on-floor storage as they cannot be used for anything else, and the speed of loading and unloading is limited by the conveyor system. Monitoring for pests is difficult due to lack of access, and they are much more difficult to keep clean and pest free than on-floor systems. If problems do develop they do have some advantages in that only the specific silo will be affected. The construction costs are usually about 25 per cent cheaper than a bulk store.

Temporary Storage

Often farms have feed grain to store for a short period of time – up to two months – with no specialist grain facilities. For small loads grain trailers

Steel grain silos.

can be filled and covered; for larger volumes suitably cleaned buildings can be used, but measures will need to be taken to ensure that pests and vermin do not have access to the grain. Low wooden partitioning (0.3 to 0.6m) can be used to contain the grain, and makeshift bays should be lined with polythene to help prevent moisture ingress. Keep the top of the grain covered with a breathable sheet such as hessian to prevent bird droppings contaminating the grain, and later cover with chicken wire to keep out birds, rats and mice.

More sophisticated are bays made of concrete panels, or round bins of galvanized steel. These are easier to clean, but will still need to be covered with a breathable sheet and wire netting to keep out vermin. Small silo modules can be purchased for around £2,000. These consist of a circular weldmesh frame lined with polythene or bituminized paper, usually in capacities of 17–40 tonnes. For longer-term storage the silos can be fitted with a duct and fan system for cooling. Running costs will be only about 2–5p/t/storage season. A thermostat and an hours' meter increase the cost effectiveness and efficiency of cooling.

One method used by some organic farmers is to purchase used 30-tonne bulk truck trailers. These can be fitted with a cover made of wood or tarpaulin, and used as large 'horizontal' grain silos that can be parked under a weatherproof roof. Mounting a front axle and tow bar with a hydraulic fluid tank on the front converts these into mobile, self-unloading grain stores. Adding a ventilated pipe along the trailer base and a cooling fan can also be useful.

STORE MANAGEMENT AND CLEANING

In the UK the most important grain beetle pests are spread between stores on loads of grain, animal feed, contaminated lorries or equipment. Once infestation has reached a store it can remain in the structure of the building, for many months, or even years, until conditions allow pests to move into a bulk of grain and develop to a noticeable level. Therefore cleaning empty grain stores before refilling with grain is essential in order to prevent infestation by reducing or eliminating pest populations. Cleaning and treatment is relatively easy.

All grain stores, even relatively modern structures, harbour pests in cracks and crevices, dead spaces behind equipment and the inside of conveyers or aeration ducting. In general, most pests avoid light and will remain hidden. Concrete and steel structures are much easier to keep clean than wood. Accumulations of dust, spilt grain and empty bags provide ideal habitats for insects and mites.

Store Cleaning

When cleaning, the aim should be to remove all dirt and debris, so as to remove the food source of pests. Vacuuming with an industrial vacuum cleaner is a more effective cleaning method than sweeping with a brush.

Table 63 Products that can be used when the store is empty

Product	Active ingredient	Form	Treatment
Pyblast	Pyrethrum	Liquid spray for control of pests in empty stores	Spray (no need to wash down). Requires a 24 hour break between application and grain entering the store
Demeter, Diatomaceous earth	Silica (Diatomaceous earth)	Dust for control of pests in grain stores	Apply 30g/m² as dust to the empty store

Steam cleaning and high pressure cleaning with water and hypochlorite solution, followed by rinsing with drinkable quality water, are permitted methods of store cleaning, although this does not suit all systems, especially where buried ducts are present or where exposed steel infrastructures may suffer from corrosion. Alternatively, compressed air can be used to blast clean surfaces, bins or conveyors. Sand blasting helps to remove crop debris, dirt or droppings from inaccessible locations, and can be swept and vacuum collected afterwards.

Store Treatment

Permitted pesticides should only be used as a last resort in any organic system, and permission is required from your organic certification body prior to use.

Pyrethrum

According to organic standards, the only permitted contact pesticides approved for use within organic grain stores are *pyrethrum* or *pyrethrins*, extracted from *Chrysanthemum cinerarifolium*. Currently some pyrethrum and pyrethrin products are approved by the Health and Safety Executive under the Pesticide Safety Regulations for use in agricultural grain storage.

Pyrethrum products should only be applied to the store structure: all grain and other food products must be removed from the store before application, and treated surfaces should not come into contact with grain. Under no circumstances are pesticides to be applied to the organic grain: this is strictly prohibited under organic standards and UK pesticide regulations.

Diatomaceous earth

For organic grain, diatomaceous earths (DEs) may be used as an application to the store fabric or as a means of treating the stored grain where there are infestations. This is a restricted practice, and permission is required from your organic certification body prior to use. Diatomaceous earths (DEs) are formed from the fossils of phytoplankton (diatoms). When insects come into contact with the DEs, the waxy layer is absorbed from their exoskeleton, resulting in water loss, dehydration and death. DEs have extremely low mammalian toxicity.

In the UK, the CSL has run laboratory and commercial-scale tests that indicate applications of a dry dust at 1–2g per kg of surface grain may be needed to inhibit the increase of most insects and mites. It can take a month or more to control existing populations of some pests, particularly grain weevils. The normal cleaning processes of milling grain removes over 90 per cent of the applied dust and, contrary to popular belief, there is no evidence that it is abrasive to, or damages, machinery.

Table 64 Products that can be used when grain in store

Products	Active ingredient	Form	Treatment
Demeter Diatoma- ceous earth Silico-Sec	Silica Diatomaceous earth	Dust for control of pests in stored grain	Prevention 200g/ m^2 in top 30cm Infestation 600g/ m^2 in top 30cm Floor storage as above rake in

ASSURED COMBINABLE CROPS SCHEME

Organic grain that is marketed needs to comply with the assured combinable crops scheme (ACCS), in addition to organic standards. Most UK buyers require evidence of this prior to purchase. This scheme is designed to provide quality assurance for farm-stored grain. It expects the user to comply with certain standards with regard to production, harvesting and storage, and allows individual parcels of grain to be traced from seed to the load leaving the farm in a lorry. A cornerstone of the scheme is the keeping of records.

Within a grain store the following points must be checked, and records kept to confirm that points have been checked, faults noted, and actions taken to rectify faults:

- Condition of the store: soundness of structure, roof, walls and floor
- Cleanliness
- Previous uses
- Infestation – insects, rodents and birds
- Any treatments
- Condition and quality of fittings.
- Conditions of equipment – trailer
- Harvester
- Conveying equipment
- Measurement equipment
- Drying equipment
- Conditions of the grain at harvest, segregation
- Drying
- Details of the grain during storage
- Temperature/moisture records
- Pest monitoring and any control records
- Presence of any rodent or birds – actions taken

This list is not exclusive, and reference must be made to the appropriate documents issued by the scheme organizers. In the above and other cases, records must be made to show that the appropriate actions have been taken. These records must be dated and attributed to an appropriate person.

Chapter 13

Operating Organic Arable Systems

HOW ORGANIC FARMS ARE 'DIFFERENT'

The International Federation of Organic Agriculture Movements (IFOAM) defines organic agriculture as:

> a whole system approach based upon a set of processes resulting in a sustainable ecosystem, safe food, good nutrition, animal welfare and social justice. Organic production is therefore more than a system of production that includes or excludes certain inputs.

- Organic farming aims to create an economically and environmentally sustainable agriculture.
- The emphasis is placed upon self-sustaining biological systems, rather than reliance on external inputs.
- Organic farming is much more than simply replacing synthetic inputs with natural ones, though it is often described as this.

The main features of most organic arable farms embrace the following:

- The concept of recycling, with less reliance on external inputs.
- Operating at lower levels of nutrients, with the challenge of making nutrients available when the crop needs them.
- Greater emphasis on the whole-farm system (rotations, matching crops to differing levels of fertility through the rotation).
- More mixed systems (but not necessarily on all farms).
- Evidence or otherwise of differences in biological activity/functioning; harmful effects of fertilizers or pesticides in conventional agriculture.
- Legislation to define its outputs (the only farming system that is 'legally defined'), minimizing pollution from agriculture management.

ORGANIC LAND CONVERSION STRATEGIES

Organic Conversion Requirements

To become organic, a producer must be registered with an approval certification body, and the land has to be 'converted'. Conversion is typically a two-year process where the land is farmed under organic principles and standards, but any produce from the land is not certified as organic and so cannot be sold as organic, although it may be identified as 'in conversion' (*see* later sections in this chapter). It is the responsibility of the organic certification body to inspect the producer on a regular basis to ensure that he is complying with organic standards.

The Conversion of Land

To be classified as organic, all land must undergo a monitored twenty-four-month conversion period, during which the land must be managed under a conversion agreement that specifically excludes the use of pesticides and artificial fertilizers prohibited under organic standards.

In some circumstances it may be possible to reduce the conversion period by up to twelve months, based on historic land management and inputs. This is normally based on the land having been managed under a formal management agreement, where prohibited inputs have not been permitted. Evidence needs to be provided to the certification body and ACOS prior to the start of conversion if this is being considered.

In the absence of a formal management agreement, if the producer can demonstrate to the certification body that prohibited inputs have not been used on the land for the previous twelve months, it may be possible to reduce the conversion period by up to four months. All land must be managed to full organic standards throughout the conversion period.

The Conversion of Crops

The conversion period for arable, root and horticultural crops is twenty-four months from the last use of any prohibited materials before an organic crop can be sown or planted. It is recommended that when land commences conversion, it is rested from crop production during the twenty-four-month conversion period to help manage pests, diseases and weed populations.

Legumes such as clover are often planted during the twenty-four-month period to build soil fertility and nitrogen reserves, improve soil structure and workability, and help provide a cropping break to manage pests, diseases and weed populations. However, in some circumstances, where large areas of cropping land are being converted and the farm is striving to develop a balanced rotation, it may be necessary to undertake some cropping during the conversion period as part of the conversion programme agreed with the certification body (*see* In-conversion Cropping below, page 370).

In this situation, where a crop is harvested during the first twelve months of conversion, it can only be sold as non-organic. If a crop is harvested in the second year of conversion then it may be sold as 'in-conversion'. A crop can only be sold as organic where the seed has been drilled into land certified as organic.

The certification body may decide in certain cases to extend or reduce the conversion period having regard for the previous use of the land, but in all cases there must be at least twelve months of monitored conversion that is subject to inspection. Land and brown field sites contaminated by environmental pollution or residues – such as heavy traffic, factories, sewage sludge, landfill, residual pesticides – may render the holding ineligible for organic status, or the conversion period may be extended at the discretion of the certification body.

Grass for grazing or cutting for conservation forage has 'in-conversion' status during the second year of conversion, and gains full organic status once the conversion period has completed. Silage and hay crops, including whole crop cereals, that are harvested after the full conversion period, have organic status. Dry-harvested crops for livestock feed or sale have full organic status if they are planted into fully organic land. They have 'in-conversion' status if they are either planted or harvested in the final twelve months of conversion.

Within the definition of 'organic' livestock feed there is an accepted allowance for in-conversion crops that can be fed to livestock. Up to 30 per cent of the average annual dry matter intake (DMI) may comprise in-conversion feedstuffs. When the in-conversion feedstuffs come from a unit within the own holding, this percentage can be increased to 60 per cent (with the minimum percentage of fully organic feedstuffs being no less than 40 per cent).

Use of Organic Seed

Organic seed must be used where it is available. If the grower has a specific variety that they wish to use and this seed is not available organically, the certification body will require an explanation as to why the variety which is only available as non-organic seed is preferred over a similar one that is available organically. The website www.organicxseeds.co.uk gives up-to-date information on the varieties and suppliers of organic seed, and is also used by inspection bodies when considering derogation requests. All seed used for organic production (organic or non-organic seed) must be free from seed treatments containing prohibited materials, as defined in the organic standards.

Part or Whole Farm Conversion

In the UK it is possible to convert part of any farm to organic production and operate non-organic enterprises alongside organic enterprises. For instance, some farms have converted their livestock production to organic status whilst retaining their arable production as non-organic, as they perceive organic arable production as 'too difficult' or 'too risky'. This can be a short-term solution, as many organic livestock enterprises now recognize that it is easier, more secure and more cost-effective to produce organic feed grains on farm and feed them to livestock, rather than importing organic grain on to the farm at a far higher cost.

It is also possible to convert a stockless arable farm to part organic production. In both cases, areas of the organic and non-organic land areas will need to be separated to minimize the risk of prohibited products (fertilizers, sprays and so on) contaminating organic land and crops. Where farms have lots of roads, farm tracks and hedges, or already have conservation grass margins around fields, this should not present a large problem. In some cases an area will need to be established as a buffer zone between organic land and non-organic land. These will also be required between land on the farm and any neighbours' fields that are non-organic. For some farms, part-farm conversion is the only realistic option, because they grow particular crops where there is no current organic market. This is currently the case for the crop known as 'sugar beet' in the UK.

Whilst part-farm conversion may seem an attractive option, it brings additional complications with regard to record keeping and segregation, in particular to prohibited products and use, machinery cleaning and grain storage separation. Farms that are part organic and part non-organic are also required to set aside land under EU rules at the appropriate rate (8 per cent in 2006) across the entire area eligible for the single payment scheme. However, many farms choose this route into organic production.

When converting the whole farm to organic production, many of the additional complexities are removed. Buffer areas within the farm are not required, record keeping is far simpler, and grain stores are easier to manage without separation. Farms that have their whole acreage under organic production are also exempt from having to set aside land under EU rules under the single payment scheme. This then simplifies farm management. However, farms are still required to comply with all appropriate national legislation and cross-compliance rules.

Part or whole farm conversions should also be considered in the context of participation in any agri-environment scheme agreements and their structures.

Parallel Production

Where farms are part organic and part non-organic, parallel production is prohibited. For livestock this means that the same species of animal cannot be kept as an organic animal and a non-organic animal on the same holding (that is, organic cattle and non-organic cattle on the same holding is prohibited). For cereal and pulse production (as with other crops), this means that different varieties of crops must be grown as organic and non-organic crops, with clear and demonstrable separation.

Phased Conversion of Land

This approach is used when converting part of the farm to organic production. Organic standards require that a minimum area is converted which permits the operation of a sustainable organic rotation. For practical reasons, on many arable farms this means a block of 20ha to 40ha as a minimum to fit with field sizes and to produce volumes of crop of a large enough volume to fit transport requirements (typically 10t or 30t loads). Phased conversions can be programmed over two to four years (converting 25–50 per cent of the farm per year), or far longer. Some farms have been known to take ten years to convert all of the farm land to organic production.

Phased conversion has the benefit of 'buffering' the farm to change, allowing the farmer and farm staff to learn new skills, and a gradual realignment of cropping and workload. On some farms it may be a more financially advantageous option, with less risk, although not for all; furthermore, it may allow any capital investment requirements to be spread over a number of years.

On the other hand, phased conversion has the disadvantage of slowing down the volume of crop to market over time, as the supply of organic crop cannot be just turned on and off, as with non-organic production. Phased or fast-track conversions should also be considered in the context of participation in any agri-environment scheme agreements and their structures. Careful planning and financial planning is required whatever route is chosen.

Fast-track Conversion of Land

This approach is used when converting the whole farm to organic production in one step. For practical and policy reasons, on some farms large and small, this is the only sensible option, and has the benefit of gaining access to organic markets in the shortest time possible. However, it also presents its own challenges: it requires the farmer and farm staff to learn new skills quickly, and may require capital investments to be made over a shorter period. It involves a massive re-alignment of cropping and workload on any farm, which entails very careful planning. Certainly on some farms it may be a more financially advantageous option with less risk – although not for all.

With the change in the CAP and introduction of the SPS, fast-track conversion is now the preferred option for many farms. Fast-track or phased conversions should also be considered in context of participation in any agri-environment scheme agreements and their structures. Careful planning and financial planning is required whatever route is chosen.

Conversion Timing

Planning the conversion start is of critical importance, since getting conversion start dates wrong can be very damaging financially, given that organic crops can only be sown into land that has achieved full organic status (that is, has undergone twenty-four months organic conversion – *see* Conversion of Land section above).

For most farms, conversion start and finish dates are programmed in accordance with the cropping calendar: sometimes this is in the spring, but typically conversion start dates are set in the late summer or early

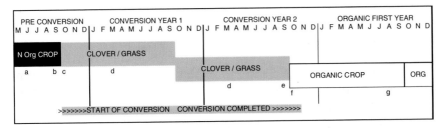

Figure 38 Autumn organic conversion commencement

autumn. A number of options are available, and these are set out below; they have different implications with regard to when a crop can be sold as organic, and whether any in-conversion cropping is planned.

Option 1: Autumn organic conversion commencement

(a) Last non-organic crop harvested in the summer of year 1.
(b) Agrochemicals applied to land at harvest or post-harvest in year 1.
(c) Organic conversion commences 1 October year 1.
(d) Twenty-four months organic conversion as red clover in years 1 and 2.
(e) Land achieves organic status 30 September in year 2.
(f) Organic crop established in October year 2.
(g) Crop harvested and sold as organic in August year 3.

Option 2: Spring organic conversion commencement

(a) Last non-organic crop harvested in the summer of year 1.
(b) Agrochemicals applied to land at harvest or post-harvest, year 1.
(c) Organic conversion commences 1 February, year 1.
(d) Twenty-four months organic conversion as red clover in years 1 and 2.
(e) Land achieves organic status on 30 March, year 3.
(f) Organic crop is established in March, year 3.
(g) Crop is harvested and sold as organic in August, year 3.

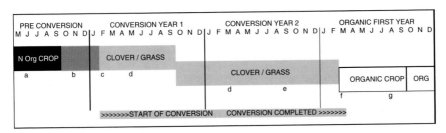

Figure 39 Spring organic conversion commencement

In-conversion Cropping

For some farms and estates, especially those undertaking a whole farm, fast-track conversion, there may be justification in planning some in-conversion cropping. This can help facilitate the establishment of a rotation and a patchwork of crops and land use on the farm. It may also assist in reducing the severity of the change on cropping, workload and cash flow for an entire farm, and provide an opportunity for the farmer and staff to learn new skills and operations during the conversion period.

As previously stated, organic crops can only be sown into land that has achieved full organic status (that is, has undergone twenty-four months organic conversion – *see* 'Conversion of land' above). Crops *harvested* a minimum of twelve months after conversion commencement and the date of the last prohibited input, can be sold as 'in-conversion crops'. There are no in-conversion markets for human consumption crops, but there are markets for in-conversion crops used in livestock feed rations. In-conversion crops are sold at discounted values compared to crops with full organic status.

It should be noted, however, that under organic standards, land is still required to undertake a component of fertility building as part of the rotation, and this is not a shortcut system to organic production, rather it is a mechanism to assist with conversion: as such, careful planning is required. Two options for in-conversion production are illustrated below.

Option 3: Autumn organic conversion commencement, with in-conversion crop in year 2

(a) Last non-organic crop harvested in the summer of year 1.
(b) Agrochemicals applied to land at harvest or post-harvest in year 1.
(c) Organic conversion commences 1 October year 1.
(d) Twenty-four months organic conversion starts, with year 1 as red clover.
(e) In-conversion cereal or pulse crop planted in October year 2.
(f) Second year as in-conversion crop.
(g) In-conversion crop harvested after land has undergone a minimum of twelve months of conversion.
(h) Land achieves organic status on 30 September year 2, and clover is re-established for year 3.
(i) Second crop planted in the late summer, August year 3.

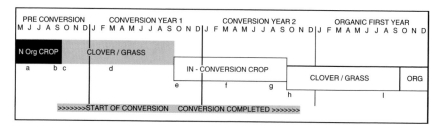

Figure 40 Autumn organic conversion commencement, with in-conversion crop in year 2

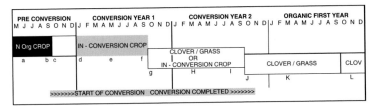

Figure 41 Autumn organic conversion commencement with in-conversion crop in years 1 and 2, which is then placed under fertility-building in years 4 and 5

Option 4: Autumn organic conversion commencement with in-conversion crop in years 1 and 2, which is then placed under fertility building in years 4 and 5

(a) Last non-organic crop harvested in the summer year 1.
(b) Agrochemicals applied to land at harvest or post-harvest year 1.
(c) Twenty-four months organic conversion commences 1 October year 1.
(d) In-conversion crop planted of beans in the spring of year 1.
(e) Year 1 in-conversion pulse crop production.
(f) Year 1 in-conversion pulse crop harvested after land has undergone a minimum of twelve months of conversion.
(g) Second in-conversion crop of cereal planted in October of year 2.
(h) Second in-conversion crop of cereal in production.
(i) Second in-conversion crop of cereal harvested after land has under gone a minimum of twelve months of conversion.
(j) Land achieves organic status on 30 September year 2; clover re-estab lished for years 3 and 4 through K and L.

The options outlined above represent a few examples only, and actual conversion strategies will need to be developed for individual farms in relation to land type, infrastructure, historic cropping and inputs and planned production. In-conversion cropping is only advisable for those farms and estates undertaking a whole farm, fast-track conversion route, who are on good soils with inherent fertility and where weeds do not pose a significant problem. In all cases it is recommended that the conversion programme is checked with the chosen organic certification body, to ensure that it is acceptable in terms of rotations and sustainability, and that at the conversion commencement and finish dates they allow in-conver- sion and organic crops to be planted and harvested. Careful planning and financial planning is required whatever route is chosen.

Agri-environment schemes

A major driver for most conversion strategies is the ability to integrate organic conversion with other agri-environment schemes. These bring a number of opportunities to the farm:

• The ability to plan and implement conservation measures on the farm.
• Access to funding for conservation measures.

- Access to funding for capital measures in some agreements (fencing, water, hedges).
- Access to organic support funding.

Different regions of the UK operate different agri-environment programmes:

England

Organic Entry Level Scheme (OELS) The aim of OELS is to encourage a large number of organic farmers across a wide area of farmland to deliver simple yet effective environmental management, recognizing the environmental benefit that organic farming systems deliver. The land to be entered into the scheme must be farmed organically and registered with an approved organic inspection body before an application can be made. Agreements run for five years. There is a payment of £60 per hectare, per year on land registered with an organic inspection body. Aid for converting conventionally farmed improved land and established top fruit orchards (planted with pears, plums, cherries and apples, excluding cider apples) is also available as a top up to OELS payments. Payment rates are £175 per hectare per year for two years for improved land, and £600 per hectare per year for three years for established top fruit orchards. OELS is administered by Natural England from regional centres.

    Higher Level Stewardship (HLS) The aim is to deliver significant environmental benefits in high priority situations and areas, with more complex environmental management where land managers need advice and support. The scheme is usually combined with OELS options. A wide range of management options are available, with some targeted to support key characteristics of the different areas of the English countryside. Payments relate to the options chosen. Capital funding is available for such items as hedgerow restoration. Unlike OELS, entry into the scheme is discretionary, with applications subject to an assessment process which takes into account how the application meets the environmental priorities identified in the local area. Agreements are for ten years. HLS is administered by Natural England from regional centres.

Scotland

Support for converting producers is available under the Organic Aid Scheme (OAS) administered by the Scottish Executive Environment and Rural Affairs Department (SEERAD). Scheme agreements are discretionary, and ranked on what changes conversion will bring to existing land management practices. A different threshold is determined each year, depending on funds available. Financial support for capital items associated with conversion to organic production is available through the rural stewardship scheme. Phased conversions are acceptable, with conversion periods starting over a maximum of five years.

Wales

Support for converting producers is available under the Organic Farming Scheme (OFS) for a five-year period, consisting of front-loaded annual payments per hectare, plus additional funds for training, advice and certification; after that, maintenance payments are available. The OFS can be combined with other agri-environment schemes such as the Tir Cynnal scheme (similar to the English OELS) and the Tif Gofal scheme (similar to the English HLS). These schemes are being reviewed in 2007.

Northern Ireland

An Organic Farming Scheme (OFS) is available in Northern Ireland to support farmers converting to organic production, consisting of front-loaded annual payments per hectare, plus additional funds for training, advice and certification. The current structure is similar to the Welsh scheme. However, the current scheme is under review by the Department of Agriculture and Rural Development (DARD), and it is expected that it will be replaced with a successor some time in 2007.

Republic of Ireland

The Republic of Ireland Rural Environment Protection Scheme (REPS) supports both existing and converting organic farmers in combination with a basic environmental scheme. The scheme is a five-year agreement to manage the farm in accordance with an agri-environment plan. Conversion payments under REPS consist of front-loaded annual payments per hectare.

Planning Thresholds

As with any major change in a farming business, it is always advisable to plan well ahead, and organic farming is no different in this respect. Because of the long nature of the farming calendar it is always advisable to start planning a conversion well in advance; combined with the requirements of the various agri-environment schemes, and it is a long lead time. Experience suggests that most arable farmers considering conversion should start planning eight to twelve months ahead of any conversion start as a bare minimum. Trying to rush through conversion commencement is likely to result in mistakes, which in the short and longer term could prove very costly. The farms that tend to perform better as organic enterprises, during conversion and thereafter, are often the farms that plan carefully.

ORGANIC ARABLE CONVERSION STRATEGIES AND OPTIONS

The continuing reluctance of farmers to convert all-arable farms hinges on the limited range of information on appropriate rotations and nutrient management strategies for different farming systems and resource circumstances. Previous research from Europe has identified that the proportion of all-arable farms in Germany varied between 20 and 50 per cent. Similarly, the all-arable system is becoming increasingly important in organic farming systems in, respectively, France and the UK.

In the recent past, the development of organic farming has been made possible by livestock production developing in a closed mixed system. As a result, most organic farming systems in Europe are based on a rotation with a large proportion of fodder crops, in combination with animal production.

However, for the all-arable farm with no livestock, it may not be economically viable for an all-arable rotation to include a long ley phase to provide a balance between fertility building and exploitative arable crops. This, then, limits the productive output of such farms, and the normal reaction is to have very short ley phases as part of the rotation, or in some extreme cases none at all.

The shorter length of, or complete absence of, a ley phase may also lead to greater problems with weeds, pests, diseases, soil structure, organic matter and fertility. Instead, short-term leguminous green manures may be used to accumulate nitrogen for the subsequent arable phases of the rotation. Trials on the duration, species composition and management of leguminous green manures demonstrated that red clover, cut and mulched, could supply a considerable amount of nitrogen.

Replicated plot trials have demonstrated that all-arable arable systems were agronomically viable in the early 1990s. Field-scale trials have also demonstrated the viability of such systems. Previously these options were highly dependent on the use of set-aside to provide income during the fertility-building phases of the rotation. However, under the new structural regimes of support in the UK from 2005 onwards, namely the Single Farm Payment Scheme (SPS), income can be generated from fertility-building green manure and ley phases of the rotation, with the additional benefit of greater flexibility in the use of this phase.

Most conversion strategies for arable-only farms have relied heavily on income from 'set-aside' during the fertility-building phase, but changes in the CAP reforms of 2005 have changed the role, and in some cases removed the requirement for, set-aside on organic farms, and hence income for this phase of the rotation as we have previously known it. Different income-generating strategies may now be considered, though these must still be used within the confines of organic standards. These could include the development of livestock enterprises, and the introduction of livestock during a ley phase, and the production of forage crops for linked farms or units.

Alternatively, where livestock enterprises are not an option, there is a need to examine systems that are not reliant on cyclic ley:arable systems. This requires the further development of continuous organic crop production systems using bi-crop or intercropping strategies, whereby legumes provide the fertility for the simultaneous production of cash crops. Research is currently underway to evaluate such systems, in which grain cereals and legumes are repeatedly planted into permanent clover green manure covers using adapted direct drilling techniques. For these systems to operate successfully, innovative cultivation and planting techniques are required, careful management of the green manure is critical, and selection of appropriate crop types and varieties is essential.

UK organic production systems that are predominately based on arable crop production are struggling to produce sufficient quantities of high quality food within the constraints of EU regulation, EEC No 2092/91, *and* meet market demands, becoming more reliant on imports. Barriers to conversion to organic arable production include EU production subsidies that continue to support non-organic arable system profitability, the lack of livestock management expertise, infrastructure or a willingness to re-introduce livestock enterprises, the lack of access to information on appropriate conversion strategies, and the current rotational systems employed. Hopefully this book goes some way to addressing at least some of these issues.

Fertility Building

Red or white clover with or without grass are the most common leguminous green manures used to build soil fertility and provide soil nitrogen for a following sequence of organic grain crops (*see* Chapter 3). However, more diverse use of different leguminous green manures to build fertility is required within the rotation than is currently practised (*see* Chapter 3 'Green Manures' page 82). This will allow more varied weed management strategies to be used, and will reduce the risks of pest build-up such as stem nematode (*Ditylenchus dipsaci*) infection, which can result in enforced changes to rotation design.

A large number of leguminous green manures can be used in organic all-arable systems to fix nitrogen, many of which have differential benefits in terms of below-ground biomass production, soil aeration and improving soil structure (*see* Chapter 3). There is the potential to improve the protein content of organic bread-making cereals by the manipulation of different green manures, the timeliness of mechanical weeding operations, or the use of intercropping or bi-cropping systems. More research is needed to validate these theories (*see* chapters 4 and 6).

All-Arable (Stockless) Conversion Strategies

Many non-organic farms have specialized in arable crop production only, and view the option of converting to organic production as too challenging. Many view the system as unworkable, through lack of understanding; probably the biggest single barrier is that farms who perceive production to be their primary function, view the fertility-building phase as a non-output, non-income-earning component of the system.

These stockless arable farms are also somewhat different in nature to mixed farms. Most no longer have any form of infrastructure associated with livestock production (fences, water, buildings, equipment), they have no livestock management skills, and in some cases no desire to operate livestock enterprises. Converting these farms to organic production requires a different approach, and planning the conversion carefully is critical to establish a viable system and maintain farm income, especially during the two-year conversion period.

Recent research at the University of Nottingham focused on this critical conversion period, under an all-arable 'stockless' system. Typically a two-year red-clover ryegrass green manure is sown, and cut and mulched regularly to improve soil structure and nitrogen status.

Alternatives to the two-year red clover-rye grass ley conversion were tested in a fully replicated experiment at Nottingham University, on sandy loam soils. Experiments were carried out between 1999 and 2002 on land previously conventionally managed with a winter wheat, second wheat, barley, oilseed rape and set-aside rotation. The study aimed to explore the dynamics between maximizing profits and building optimum soil nitrogen, by testing a range of conversion cropping strategies that stretch the economic and agronomic limits of the conversion (Table 65).

Factors considered in the design of the various strategies included the potential to increase soil nitrogen through nitrogen fixation; the potential to create revenue through the sale of crops; and the result of different management demands on the variable and fixed costs, involving altering the proportions and species of green manuring and cash cropping. For each conversion strategy, plant growth and development and soil nutrient flows were monitored, along with the impact of the conversion strategies on profit, risk, return-on-investment and cash flow.

*Table 65 The conversion strategies are two-year cropping sequences, all followed by wheat in the third year (u/s = undersown with red clover, * cut and mulched over growing season)*

Strategy	Year 1 crop	Year 2 crop	Expected fertility building	Variable costs	Fixed costs	Potential return in a free market
1	red clover/ ryegrass as green manure*	red clover/ ryegrass as green manure*	high	low	low	low
2	vetch as green manure	vetch/rye as green manure	high	high	high	low
3	red clover (seed)	red clover/ ryegrass as green manure *	medium	low	medium	high
4	spring wheat (u/s)	red clover as green manure *	medium	medium	medium	medium
5	spring wheat	winter beans	low	medium	high	high
6	spring oats	winter beans	low	medium	high	high
7	spring wheat (u/s)	spring barley/ spring pea	low	high	high	high

In the third year of the experiment (2002) winter wheat was grown across the entire experimental area as a test crop, to assess the effect of the different strategies on yield and quality of the first organic harvest. As the wheat crop will be the first eligible for organic premiums, the yield will influence the overall profitability of each strategy.

Results demonstrated that all-arable and predominantly arable organic systems are agronomically and economically viable, but that the conversion strategies were highly dependent on the use of annual green manures to provide sufficient soil nitrogen during the fertility-building phases of the rotation.

The conversion strategies used were also shown to be dependent on subsidy or agri-environment scheme conversion payments to be economically viable. The availability and level of support payments is under the discretion of the national government and the European Commission, and the continuation of such payments should not be relied upon to sustain the profitability of organic farming. Where income is not available during these phases, the economic viability of such organic systems is threatened.

Research examining alternatives to the common two-year clover-ryegrass ley conversion strategy highlighted the importance of the balance between soil nitrogen fixation, crop sale revenue, and variable and fixed costs, and in particular the key influence of soil nitrogen on yield and the importance of organic premiums to maintain a viable income.

Alternative conversion strategies and subsequent organic rotations are highly influenced by the farm enterprise balance, resource constraints and expertise, which can change during and after conversion and vary at a farm level, or even at a field or part-field level. This emphasizes the importance of designing conversion strategies and rotations that are site-specific rather than generic.

The sequence of crops in the rotation has an important influence on soil nitrogen use and for weed management, as does the use of green manures to retain and release soil nitrogen over different time scales. Grain crops that develop rapidly, tiller vigorously, and have a tall stature, are better suited to organic arable systems. This limits crop and variety choice, with many modern varieties (particularly wheat and peas) being poorly suited to organic conditions.

Current rotation design is often sub-optimal with regard to soil nitrogen, and phosphorus and potassium capture and utilization, with many commonly used rotations largely ignoring below-ground biomass production and diversity. The use of a greater diversity of fertility-building green manures within the rotation is a logical and simple next step. Recent changes to the use of set-aside also provide further utilization options.

However, alternative strategies need to be considered which optimize soil nitrogen use, minimize disturbance to the soil microbial biomass, and break the reliance on set-aside subsidies to ensure the economic viability for arable-only systems that result in a highly cyclic production system. Where livestock enterprises are not an option, choices are more limited, especially when the system requires continuous organic crop production. The use of bi-crop or intercropping strategies, in which legumes provide the fertility for the simultaneous production of cash crops, may offer an appropriate alternative. For these systems to operate successfully, innovative cultivation and planting techniques are required, careful management of the green manure is critical, and selection of appropriate crop types and varieties is essential. The current availability of a wide range of crop varieties is a limiting factor for the success of this option, with a need for crops and varieties to be bred which are better suited to organic production conditions. The expansion of the small but emerging European organic plant-breeding initiative will play an important role in developing these options further.

Site Selection

The feasibility of growing any crop, including cereals or pulses, should be investigated in a systematic way. The farmer should consider if the holding has suitable soils and climate to produce organic arable crops, and the productive capacity of the soils. There are many factors that affect the ability to successfully produce a single crop or a range of crops within a rotation. These are discussed within other chapters of this book.

Climate

Climate is an important factor to consider. The British Isles has a temperate maritime climate, meaning that the maximum and minimum temperatures are not extreme. Rainfall can vary significantly in different parts of Britain. For example, the thirty-year average for East Anglia is 605mm or 23.8in, as compared to the West Country at 1,247mm or 49in. These differences are reflected in the type of agriculture traditionally found in these areas – for example, dairy farms in the west and arable farms in the east, although this isn't always the case. In West Wales, Trawscoed's thirty-year average rainfall is 1,213.9mm or approximately 48in, whilst in Scotland rainfall can vary from 984.4mm or 39in in Auchincruive in West Scotland, to 624mm or 24.5in in Kinloss in north-east Scotland. This is also reflected in the type of agriculture found in these parts, with the Black Isle and north-east Scotland being arable, while the south-west of the country is mainly dairy with some horticultural producers.

Organic cereals and pulse crops generally fare well in the UK climate, with stable year-on-year yields when compared to countries such as Australia, where yield differences of 300 per cent year-on-year variation have been reported, making forward planning financially and practically more difficult.

Topography

Another consideration when planning conversions is the topography of the land, and also the accessibility for farm machinery. Modern farm machinery often requires wide gateways and access tracks or lanes, and many farms that have increased in size in recent years have had to invest in capital works in order that machinery can get around and be used efficiently. The topography of the land should also be considered in the context of cultivation and the risk of erosion; thus steeply sloping fields should not be cultivated where there is a risk of soil erosion.

ORGANIC ARABLE FARM MANAGEMENT OPTIONS

For all-arable farms, using legume-based annual green manure covers to build fertility is an established practice. In addition to fixing nitrogen for subsequent crop growth, the cutting and mulching of these green manures allows weed control, and soils benefit from structural and microbial enhancement from the deep-rooting characteristics of the green manures commonly used. So how does this change affect the all-arable farm?

Without the presence of livestock on the farm, there is very little change to the mode of operation or economic balance of many arable farms. However, where infrastructure (fencing, water and so on) allows, there is the potential to graze cattle (or any other livestock) on the green manure at any time of the year. Previous rules limiting the use of set-aside land no longer apply. Organic farmers who are wholly organic (including in-conversion land) do not have to set aside land and are therefore free to plant and manage it in whatever way they wish (provided it meets cross-compliance regulations). Where a farm is part organic, part non-organic, the farm is still required to have set-aside at the appropriate rate for the particular year. As these regulations change from time to time it is important that the farmer checks with DEFRA to clarify land use regulations, and how they impact upon organic conversion and land management.

If livestock perform the 'cutting and grazing' operation instead of mechanical topping, a cost saving of up to £60–£80/ha/yr may be made where the green manure is cut three to four times per year (Table 66). Some nitrogen will be lost from the system in the grazed forage, but provided livestock manures are returned to balance this out there should be no detrimental effect to the nutrient balance. However, remember that over-grazing can damage the green manure (by bruising the stolon of the clover), and grazing at inappropriate times on wet soils can cause soil damage by poaching.

Table 66 *Economic comparison of grazed and non-grazed organic green manure land*

	SPS income £/ha	*Variable costs £/ha	**Livestock gross margin £/ha	Enterprise gross margin £/ha
Arable green manure cut and mulched	242	195	0	47
Arable + livestock Share ownership	242	135	529[1]	636

*Variable cost assumed for establishment of red clover/grass ley, weed control and cutting/mulching four times per year without livestock, with no mulching assumed where livestock present.

**Livestock income assumes spring born lowland beef stores stocked at 3/ha grass finished at 18 months.

[1] income derived assuming stocking at 3 stores/ha including forage costs with the retention of 50% of gross margin, by buying stores at 9 months @ £350 into winter housing, turning out to grass/clover at 12 months and selling at 18 months from grass @ £690/head with a gross margin of £1059/ha, assuming 0.84kg daily live weight gain valued at £1.28/kg. Excludes livestock management, capital & capital borrowing interest costs.

A second option may be to establish a 'share-ownership' cattle enterprise off site, and conserve forage from the set-aside land which could be used by the part-owned cattle. The economics of this option will require a higher capital investment over a longer period of time.

With all of these options, the development of closer operational and working relationships between an arable and a livestock farm to create a 'linked unit' where livestock, grazing, conserved forage, straw and grain are exchanged, is critical for success.

As green manures such as red clover and grass establish slowly during the first year, grazing and/or forage conservation is best scheduled for the second or third year of this type of annual green manure, with cut and mulch set-aside utilized during the first year of establishment. Establishing the green manure by undersowing into a previous cereal or arable silage crop can allow grazing or forage conservation in the following year.

The cut and mulch option needs to be used with care. Frequent mulching is recommended if the crop is to be grazed subsequently, such that the crop is finely chopped and 'blown' back into the bottom of the sward. If the crop is too long when cut and is left lying on top of the remaining sward, a degree of kill will result making the crop less productive in the following year and allowing the ingress of weeds. There is also evidence that the forage crop is less palatable to grazing animals after mulching as a result of the decaying material in the sward.

MIXED ORGANIC ARABLE AND LIVESTOCK FARM OPTIONS

For organic dairy or beef farms, longer-term three to five-year grass/clover leys will be the preferred option. When establishing grass/clover leys, the use of arable silage crops containing cereals, grain legumes or even maize, which can be undersown with the following ley to provide conserved forage with different feeding values, allows the use of different rotational options (Table 67).

Table 67 Livestock farm rotational use of organic green manure land

Year A	Year B	Year C	Year D	Year E
Arable silage	Grass/clover Grazed/forage	Grass/clover Grazed/forage	Grass/clover Grazed/forage	Arable cereal
Grass/ clover	Grass/clover	Grass/clover	Grass/clover	Grass/clover
Forage harvest	Grazed/forage	Grazed/forage	Grazed/forage	Grazed/forage
Grass/ clover Forage harvest	Grass/clover Grazed/forage	Arable cereal Or arable silage	Arable cereal Or arable silage	Grass/clover Grazed/forage

With the need to produce home-grown protein, there is potential to grow arable silage crops with a predominant balance of legume (52 per cent forage pea, 48 per cent cereal) whilst retaining set-aside income on the same ground. Novel protein crops such as lupin and soya could also be considered.

Weed Management Options

In addition to being able to graze and forage land under fertility-building leys, altering the timing of establishment also offers some innovative opportunities for summer weed management, with cultivations on land used for rotational arable/ley production. Farmers could opt to grow a predominantly legume-based arable silage crop over the winter, with an early harvest in the following spring, followed by weed management or grazing (if undersown). Options are detailed in Table 68.

Table 68 Weeding opportunity management options in the rotation

Year 1	Year 2	Year 3	Year 4	Year 5
Arable crop	Winter sown vetch 52% & oats 48% Forage harvest in spring	Summer weed control cultivations from May 1st	Grass/clover Grazed	Grass/clover Grazed/forage
Arable crop	Winter sown vetch 52% & oats 48% Forage harvest in spring	Summer weed control cultivations from May 1st	Arable crop undersown	Grass/clover Grazed
Grass/arable silage + summer weed cultivations	Winter sown vetch 52% & oats 48% Forage harvest in spring	Summer weed control cultivations From May 1st	Grass/clover Grazed	Grass/clover Grazed/forage

CROP PERFORMANCE

The performance and yield of any particular crop will be determined by a wide range of resource and management factors. These include:

- **Soil** The fertility of the soil and its ability to provide adequate crop nutrition will determine yields (*see* Chapter 2).
- **Rotation** Crops closer to the fertility-building phase are likely to perform better, with better soil fertility and less weed competition (*see* Chapter 5).
- **Variety** Different varieties have different yield potential (*see* Chapter 8).
- **Weeds** Competition from weeds for light, water and nutrients will impact on yields (*see* Chapter 10).
- **Pest and disease pressure** Competition from pests or diseases will impact on yields (*see* Chapter 11).
- **Management** Including planting, harvesting, weeding.
- **Weather** The provision of adequate moisture and sunshine at the correct times of the season.

With a wide range of factors affecting crop performance, some of which the farmer can influence, some not, yields are understandably very varied in organic systems. Typical yields and ranges of yields for organic cereal and pulse crops are shown in Table 69.

Table 69 Typical yields of organic cereal and pulse crops

Organic crop	*Typical yield t/ha	Typical yield range t/ha	Crop value (£/t) 2007 harvest
Winter Wheat	4.5	2.5 – 8.0	220
Spring Wheat	3.5	2.5 – 8.0	220
Winter Barley	4.0	3.0 – 5.0	210
Spring Barley	4.0	3.0 – 5.0	210
Winter Oats	4.0	3.0 – 5.0	205
Spring Oats	3.5	2.5 – 8.0	205
Winter Triticale	4.5	2.5 – 65	210
Spring Triticale	4.0	3.0 – 5.0	210
Spelt	3.5	2.5 – 4.5	280
Rye	3.8	2.5 – 4.5	250
Winter Beans	3.0	2.5 – 4.5	260
Spring Beans	3.0	2.5 – 4.5	260
Spring Peas (dry)	3.5	2.0 – 3.8	260
Lupins	2.5	1.8 – 3.0	270
Soya**	2.5	1.8 – 3.0	280

*Source: *Organic farm management handbook (OFMH) 2007 – Lampkin, Measures & Padel & **OFMH 2004 – feed prices assumed only*

Previous work has shown a wide range in the yield performance of a broad variety of organic cereal and pulse crops. Recent work as part of the Defra-funded Organic Crops Demonstration Project (OCDP) between 2003 and 2006 evaluated a range of winter and spring wheat varieties and pulse crops across five different sites, with all crops grown on large 1-acre (0.4ha) plots to represent real field conditions. Yield results for all the sites are shown in Figure 42 for the 2004 harvest, and Figure 43 for the 2006 harvest.

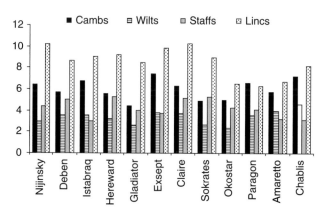

Figure 42 Wheat yields from 2004 harvest Organic Crops Demonstration Project (t/ha @ 15 per cent moisture content)

The sites used were predominately stockless arable farms. The Wiltshire site was on a thin calcareous loam over chalk, with the cereal a first crop in the rotation after clover, with much of the farm having been organic for over thirty years. The Cambridgeshire site was on heavy boulder clay, with the cereal a second wheat in the rotation, following a first wheat, preceded by clover. The Yorkshire site was on a medium sandy clay loam soil, with the cereal a first crop in the rotation after clover. The Staffordshire site was on a medium sandy loam soil, with the cereal as a first crop in the rotation after clover; and the Linconshire site was on a grade one, deep silt loam soil, with the winter cereal as a first crop in the rotation after clover and the spring cereals at the end of the rotation after crops of potatoes, onions, carrots and cereals.

Whilst there was inevitably yield variation between seasons, the greatest yield differentials were between sites as a function of soil type, fertility and rotational positions of crops.

The work demonstrates the great variability in yields depending on rotational position of crops, soil type and fertility status. In some circumstances, on very fertile soils, yields equivalent to many non-organic crops of 8–10t per ha were achievable. On soils with lower inherent fertility, organic yields of 3–5t per ha are more realistic. Second cereals can also perform well under the right rotation, with consistent yields in the region of 6t/ha achievable at the Cambridge site. Over the three years of the project 2.2t/ha was the lowest recorded yield, and 10.3t/ha the highest recorded yield.

As a general rule, on lower fertility sites spring cereals – especially those which establish rapidly, tiller strongly and are tall with planophile leaves – can achieve yields similar to winter cereals. With wheat, the spring varieties are more reliable at achieving milling quality specifications. This is reflected in the spring wheat varieties (Tybalt and Paragon) being placed in the top three economic performance ranking on the lower fertility sites (*see* Table 70).

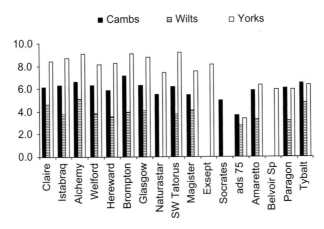

Figure 43 Wheat yields from 2006 harvest Organic Crops Demonstration Project (t / ha @ 15 per cent moisture content)

Table 70 : Economic performance rankings of different spring and winter wheat varieties from the 2006 Organic Crops Demonstration Project

	High fertility sites			Low fertility sites		
Wheat variety	Output (£/ha)	Gross Margin (£/ha)	Rank	Output (£/ha)	Gross Margin (£/ha)	Rank
Winter varieties						
Claire	1457	1326	8	826	695	2
Istabraq	1522	1391	5	673	542	6
Alchemy	1550	1419	2	874	743	1
Welford	1385	1254	7	648	517	8
Hereward	1538	1407	4	621	490	9
Brompton	1550	1419	3	675	544	5
Glasgow	1491	1360	6	684	553	4
Naturastar	1410	1279	10	n.a	n.a	n.a
SW Tatorus	1654	1523	1	669	538	7
Magister	1390	1259	11	720	589	3
Exsept	1412	1281	9	n.a	n.a	n.a
Spring varieties						
ADS75	638	507	4	536	405	4
Amaretto	1080	949	2	575	444	2
Paragon	1012	881	3	563	432	3
Tybalt	1080	949	1	833	702	1

Output – calculated as yield x crop value (£170/t feed or £185/t milling)

Gross margin – calculated as output minus variable costs (assumed at £131/ha)

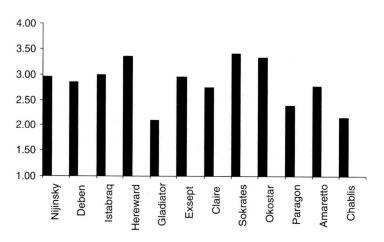

Figure 44 Wheat straw yields from 2004 harvest Organic Crops Demonstration Project (fresh weight)

On the more fertile soils, winter varieties with good all-round disease-resistance characteristics have the opportunity to utilize the higher levels of fertility and generally outperform spring cereals. This is shown in Table 70, with the winter varieties Alchemy, Brompton and SW Tatorus being ranked as the top three economic performers on the higher fertility sites in 2006, with the spring wheat crops performing less favourably with lower yields – despite some achieving milling quality and a higher price.

One of the lessons learnt from this project was that unless a milling specification can be achieved from a winter milling variety, it may be better to produce a feed variety with a lower market value but with a much improved yield, leading to better overall financial performance, to make up for the frequent yield deficit of milling varieties.

Straw yields were also recorded from the Defra-funded OCDP between 2003 and 2006, and this demonstrated that straw yield can be equally variable between varieties. There are shown in figure 44. This is an important consideration for mixed farms with livestock enterprises.

ECONOMICS AND BUDGETING

There are a number of reasons to undertake economic analyses or draw up budgets. At one level, accounts for the farm will be required for taxation or to be used in securing funding for a development or venture. At a different level they can be used to tack costs and outputs. They can also be used as comparative tools to compare different scenarios, for example growing a wheat crop versus a barley crop. Some of the different levels of economic analysis and budgets are outlined below.

Whole Farm Budgets

An analysis of the farm's overall economic performance is normally undertaken when preparing the farm accounts for the purpose of taxation, often by an accountant. These accounts often comprise a profit and loss account, trading account and balance sheet for the financial year (*see* Figure 45 below). This information can be used by the farm, or with some adjustment, can be compared with other farms. This is often termed 'benchmarking', and it can be an important planning tool.

For planning purposes the budgets can be constructed from published data such as farm management handbooks. The *Organic Farm Management Handbook* (edited by N. Lampkin, M. Measures and S. Padel), which is frequently updated, is a very useful reference book on organic farming that includes whole farm and enterprise financial data. *The Farm Management Pocket Book* by J. Nix, and the *Farm Management Handbook* published by the Scottish Agricultural College (SA), are also very useful books and published yearly.

Budgets can also be drawn up using the farm's own performance data to compare different enterprises and assist with planning and making future cropping decisions.

	2005	2006
Sales	£	£
Crops	150,000	162,000
Grazing rent	5,000	5,000
Grants and subsidies	80,000	75,000
Agri-environment income	24,000	24,000
Sundry income	750	1,200
Total sales	259,750	267,200
Costs		
Rent	40,000	40,000
Labour (wages)	40,000	42,000
Seeds and fertilizers	52,500	56,000
Machinery expenses	12,000	8,000
Water	800	800
Heat, power, light	1,800	2,100
Insurance	4,000	4,100
Office costs	2,500	2,500
Professional fees	2,000	2,000
Bank and loan interest	6,000	6,000
Depreciation	40,000	40,000
Total costs	**201,600**	**203,500**
Net profit for the year	**58,150**	**63,700**

Figure 45 Example of a typical profit and loss account

Farm accounts for the whole of the UK and each region are collected under the farm business survey (FBS). An additional study of organic farms is also undertaken by the University of Wales, Aberystwyth, Institute of Rural Sciences, both of which provide useful information on the performance of a range of farm types and systems.

Production Economics

The economic performance and profitability of most farm businesses is mainly determined by the income from marketable yield (output) minus the cost of production (inputs). Inputs on the farm include items such as land, labour, buildings, machinery and capital with its respective borrowing or opportunity costs. These inputs are normally categorized as 'fixed costs', meaning costs incurred whether a crop is grown or not, or 'variable costs', meaning costs associated with the production of crops.

Outputs are marketable yield multiplied by the sale price achieved, plus any income derived from government subsidies or grants that may be available for production or the management of agri-environment features. The factors affecting the profitability of a farm business are shown in Figure 46.

Production economics are about making the best use of the available resource to maximize the financial return, within the constraints of production such as rotation, market, and other technical and environmental requirements. While some of these factors are under the control of the farmer, such as variable costs, some are not, such as weather and markets.

Costs or inputs are normally split between fixed and variable costs, as follows:

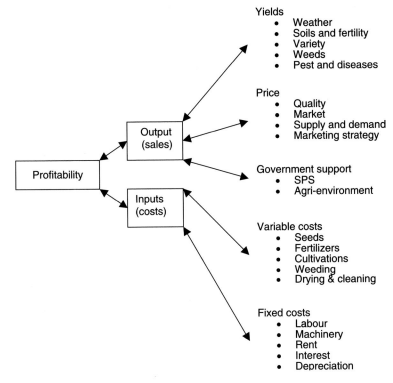

Yields
- Weather
- Soils and fertility
- Variety
- Weeds
- Pest and diseases

Price
- Quality
- Market
- Supply and demand
- Marketing strategy

Government support
- SPS
- Agri-environment

Variable costs
- Seeds
- Fertilizers
- Cultivations
- Weeding
- Drying & cleaning

Fixed costs
- Labour
- Machinery
- Rent
- Interest
- Depreciation

Figure 46 Factors affecting the profitability of farm businesses

- **Fixed costs** These remain the same, regardless of how much crop is grown or whether a crop is grown or not. These normally cover items such as rent, regular labour, machinery costs, buildings, depreciation, interest and insurance.
- **Variable costs** These normally change in relation to cropping patterns and size of enterprise. The main items are seed, fertilizer, cultivations, weeding, harvesting, crop drying or cleaning, crop storage and any casual labour associated with the crop.

Subtracting the variable costs from the financial output results in a *gross margin*. In the same way, subtracting the fixed costs from the gross margin results in a *net margin*. Gross and net margins can be calculated on a crop, field, enterprise, rotation or whole farm basis, depending on what is required. The main use of the *gross margin* is for comparison of one option against the other, whereas the main use of the net margin is to evaluate overall profitability.

Gross Margins

Gross margins are usually calculated for individual enterprises (for instance, wheat) and can be used for analysis and planning purposes. The gross margin per hectare (or per head of livestock or per crop) can be compared with standard published figures (such as those published in the *Organic Farm*

Management Handbook), or can be used to compare different proposed or completed enterprises in order to provide a comparative economic performance. A typical gross and net margin for wheat is shown in Figure 47.

Gross margins are useful for planning and the comparison of enterprises, either on the same farm or between organic holdings, or between organic and non-organic enterprises. Because they exclude fixed costs they are limited in use, and the fixed costs should be considered in the context of choosing any particular enterprise.

Certain inputs used on a rotational basis – such as fertilizers, manures, lime or composts – should have their costs spread over the entire rotation (as shown in Figure 47), and the cost should not be attributed to one particular enterprise, as the benefit is shared between different enterprises. Gross margin figures should only be compared between farms with similar production systems and resources.

Net Margins

This approach can be useful for evaluating the economics of organic systems, as it overcomes some of the limitations of gross margins. This is especially true when comparing organic with non-organic farming, where one system has very high variable costs and the other may have different fixed costs. The difficulty in using net margins is that there are no published figures for net margins, as the costs will be very farm-specific. The second limitation is that many farms do not know the true costs of operations used in fixed costs, for example machinery costs, labour, fuel. This can be overcome to some extent by using 'contractors' charges', for which there are published figures.

			£/ha	(£/ac)
Output				
Grain	4.5t/ha(1.82t/ac)	@170 £/t	765	(309)
Total output			765	(309)
Variable costs				
Seed 200kg/ha	(1.6cwt/ac)	@430£/t	86	(35)
Fertilizers (organically approved) applied on a rotational basis			40	(16)
Undersow			24	(14)
Other			10	(4)
Total variable costs			160	(69)
Gross margin			605	(240)
Cultivations			75	(30)
Planting/drilling			47	(19)
Weeding 2 pass with comb harrow @7.50/pass			15	(6)
Undersow			15	(6)
Harvesting			86	(35)
Crop drying			15	(6)
Total fixed costs			253	(102)
Total net margin			352	(138)

Figure 47 Combined net and gross margin for an organic wheat crop

Typical gross margins for a range of cereal crops are shown in Table 71, and for pulse crops in Table 72.

Table 71 Economic performance comparison of organic cereal crops

Crop	Yield t/ha	Value £/t	Output £/ha	Variable costs £/ha	Gross Margin £/ha	Rank	Market /use	Ease of marketing
Wheat W	4.5	170	765	160	605	3	H, F, AS	Easy
Wheat S	3.5	170	595	160	435	8	H, F, AS	Easy
Triticale	4.5	165	742.5	155	587.5	2	F, AS	Moderate
Oats W	4.5	170	765	150	615	1	H,F,AS	Moderate
Oats S	3.5	170	595	150	445	5	H,F,AS	Moderate
Barley W	4.0	160	640	155	485	6	S, F, AS	Easy
Barley S	4.0	160	640	155	485	9	S, F, AS	Easy
Rye	3.8	190	722	170	552	7	S	Specialist
Spelt	3.5	220	770	210	560	4	S	Specialist

H (Human), F (Livestock Feed), AS (A Silage/Home feeding), S (Specialist)

Table 72 Economic performance comparison of organic pulse crops

Crop	Yield t/ha	Value £/t	Output £/ha	Variable costs £/ha	Gross Margin £/ha	Rank	Market /use	Ease of marketing
Beans W	3.0	180	540	165	375	1	F, AS	Easy
Beans S	3.0	180	540	185	355	4	F, AS	Easy
Peas S	3.5	190	665	230	435	3	F, AS	Easy
Lupins S	2.5	220	550	415	135	7	F, AS	Easy
Soya S	2.5	20	600	250	350	5	F	Limited

H (Human), F (Livestock Feed), AS (A Silage/ Home feeding), S (Specialist)

Rotation Margins

When undertaking an economic analysis, it is often difficult to allocate costs and outputs between different enterprises at the rotation or farm level, as there are factors that impact upon these comparisons. For example, in organic farming systems it is especially important to consider the economics of the whole rotation, and interactions between crop enterprises and fertility-building components.

At a first glance it may seem that it is only the cash crops that incur costs and provide an output each year. Fertility-building components such as clover leys and green manures also incur costs in the form of seed, establishment and management (mulching or cutting), and also provide outputs, such as weed suppression or nitrogen fixation. However, the inputs and outputs associated with fertility-building components may be more difficult to quantify and cost.

The simplest method of attributing these costs and outputs for many farms, is to allocate them equally or proportionally to yield, across the cash-cropping components of the rotation. In doing so, the nitrogen fixation by a clover is charged as a cost to the cash-crop cereal in the rotation. This is more difficult when pulses are included in the rotation, as they also fix nitrogen. It is simplest to treat the nitrogen component of the pulse as net neutral – pulses fixing the nitrogen they require for their own production only.

An example of a gross and farm margin calculation for a 250ha farm operating a five block rotation is shown in Table 73. In this case the average gross margin per year over five years is £43,150 divided by five years, which equates to £8,630 per year, or a gross margin per hectare per year of £172.60.

Rotation margin analysis (gross, net or farm margin) allows fuller analysis of the economic viability of a farming system, including cropping and fertility-building phases, and therefore helps planning and optimizing rotations in the most agronomic and economically advantageous way.

Table 73 Example of a gross and farm margin calculation for a 250ha farm operating a five-block rotation

Year	Crop	Area (ha)	Performance /activity	Output (£ per ha)	Output Tot (£)	Costs (£ per ha)	Costs Tot (£)	Gross Margin (£ per ha)	Gross Margin Tot (£)
1	Clover year 1	50	Clover estab costs @ £125/ha	£0	£0	£125	£6,250	£125	−£6,250
2	Clover year 2	50	topping @ £30/ha per year	£0	£0	£30	£1,500	−£30	−£1,500
3	Wheat	50	5.0t/ha @ 160/t	£800	£40,000	£334	£16,700	£466	£23,300
4	Oats	50	4.0t/ha @ 150/t	£600	£30,000	£334	£16,700	£266	£13,300
5	Barley	50	4.0t/ha @ 155/t	£620	£31,000	£334	£16,700	£286	£14,300
			TOTAL	£2,020	£101,000	£1,157	£57,850	£863	£43,150

Whatever economic analysis of budgeting system is used, it is important for any agricultural business to operate an on-going process of planning, recording, analysing and re-planning to assist with refining and improving the profitability of the system. This may involve planning and implementing cropping, rotational and management changes as discussed in other chapters of this book.

Potential for Profit

Previous research by a number of research institutes and universities in the UK and internationally, has shown that mixed organic farming systems can compare well in economic terms with integrated and conventional systems. Where combined arable and livestock systems have been examined, it is the type of livestock enterprise that has a marked effect on overall profitability, with arable and dairy systems outperforming arable and beef, or arable and sheep systems. The economics of an all-arable system, whilst outperforming non-organic systems during the cash-cropping phase of the rotation, are reduced by the income foregone in the fertility-building phase of the rotation. Work on economic benchmarking of organic farms at the University of Wales, Aberystwyth, Institute of Rural Sciences has reported that farm incomes from all-arable organic farms are at least comparable to, or are more profitable than, their non-organic all-arable system equivalents.

However, it is apparent that whilst subsidies such as the Single Farm Payment Scheme (SPS) (and previously Arable Area Payments under the IACS system) are the same or similar for organic or non-organic systems, profitability is highly dependent on the increased premium achieved by organic systems to support the lower yields achieved with organic grain cash crops such as wheat. This is less so for other organic grain cash crops such as beans, triticale or oats, where yields are more comparable with non-organic systems, especially where mixtures of varieties are sown rather than monocrop stands.

However, if organic premiums are reduced, the profitability of some crops will decline due to the reduced yield potential under organic systems. Hence yield remains a key component in profitability, and strategies are required to improve the yields of organic crops that have the greatest yield differential to that of the non-organic equivalent, in particular wheat. The single biggest influencing factor on the yield potential of grain crops is soil nitrogen availability during key developmental periods. Hence the retention of soil nitrogen, and the manipulation of its release at appropriate stages of crop development and at different stages of the rotation, is critical for success (*see* chapters 2, 3 and 4 for nitrogen management).

BIODIVERSITY ON ORGANIC FARMS

Biodiversity is simply a term for the variety of life, and the natural processes of which all living things are a part. It represents the ways that life is organized and interacts on the planet, and is usually measured as the number of species, together with the variety of interactions between them, occurring in a given area. A wide range of research broadly agrees that the intensification and expansion of modern agriculture are amongst the greatest current threats to worldwide biodiversity. Over the last quarter of the twentieth century, dramatic declines in both range and abundance of many species associated with farmland have been reported in Europe, leading to growing concern over the sustainability of current intensive farming practices.

Biodiversity underpins all our production systems. It provides a wide range of 'ecological services' such as nutrient recycling, microclimate regulation, water regulation, pathogen suppression and detoxification of noxious chemicals. So, directly or indirectly, we rely on biodiversity for the production of our food. Generally, modern societies are putting high pressure on ecological systems and reducing biodiversity through the intensive use of natural resources. The consequences of this are potentially devastating for our food supply, among other things. For this reason, the need to care for the environment around us is becoming more and more urgent.

The government is beginning to recognize this through the various programmes that promote environmental stewardship and sustainable use of natural resources. Agriculture is no exception, and in future, systems that can demonstrate a positive environmental impact are likely to be better rewarded than those that don't. So apart from the intrinsic value of weeds in a landscape, those farming systems that at least allow weeds to survive are increasingly likely to receive financial rewards to do so in the future. Sustainable farming systems such as organic farming are now seen by many, including politicians, as a potential solution to this continuing loss of biodiversity, and receive substantial support in the form of subsidy payments through EU and national government legislation.

In 2005, the RSPB and English Nature (now Natural England) undertook a comprehensive study to see if organic farming benefited biodiversity. It assessed the impacts on biodiversity of organic farming relative to conventional agriculture, through a review of comparative studies of the two systems. It identified a wide range of taxa, including birds and mammals, invertebrates and arable flora, that benefit from organic management through increases in abundance and/or species richness. It also highlighted three broad management practices – prohibition/reduced use of chemical pesticides and inorganic fertilizers; sympathetic management of non-cropped habitats; and preservation of mixed farming – that are largely intrinsic (but not exclusive) to organic farming, that are particularly beneficial for farmland wildlife.

Also in 2005, scientists from Oxford University, the British Trust for Ornithology, and the Centre for Ecology and Hydrology undertook a five-year study on organic farming covering 180 farms from Cornwall to Cumbria. It was reported as being the largest ever wildlife survey of organic farming. The study has found very large benefits from organic farming right across the species spectrum, and its conclusions so far are that organic farms are better for wildlife than those run conventionally. Organic farms were found to contain 85 per cent more plant species, 33 per cent more bats, 17 per cent more spiders and 5 per cent more birds.

The study has found that the exclusion of synthetic pesticides and fertilizers from organic farming was a fundamental difference between the systems. Other key differences found on the organic farms included smaller fields, more grasslands, and hedges that are taller, thicker and on average 71 per cent longer. The fact that organic arable farms were more likely to have livestock on them also made them richer habitats for wildlife.

There would therefore seem to be a strong body of evidence to suggest that organic farming systems and practices benefit wildlife and biodiversity. In earlier chapters of this book, reference has been made to the importance of biodiversity in relation to the management of weeds, pests and diseases, together with the harnessing of natural systems and above- and below-ground predators to help manage problems. Greater biodiversity does not necessarily mean lower yields. On many organic farms increased biodiversity goes hand in hand with improved crop performance and profitability. It is just a question of management.

So how can organic farmers ensure that their farming systems harness biodiversity for benefit, maximize their environmental potential, and deliver public goods simultaneously? They must be sure to observe certain crucial but simple provisos:

- Participate in an agri-environment scheme (see above, page 371).
- Integrate conservation activities with farming practice.
- Manage the farmed environment in a way that is sympathetic to biodiversity.
- Encourage the establishment and management of natural habitats on the farm; this will also benefit production because of the presence of natural predators.
- Establish links and wildlife corridors between different habitats.
- Establish a patchwork of different crops and land use – avoid too much block cropping.
- No farm is an island – take note of neighbouring habits and forage linkages for greater benefit.
- Encourage neighbours to become organic farmers.

Lastly, when we think of biodiversity, we are all too often pre-occupied with above-ground production, with below-ground biomass production and biodiversity taking second place, or largely ignored. A mixture of diverse crop types below ground is as important, if not more important, than above ground. Different rooting characteristics influence soil structure, aeration, mineral availability and mobility, and also microbial activity. Combining cash crops with different rooting systems in the rotation is a first step. This can be further enhanced by integrating different annual and overwinter green manures to add to the diversity of below-ground production. Don't ignore the soil. (*See* chapters 2, 3 and 11.)

ENERGY USE ON ORGANIC FARMS

Organic farming is often criticized as being less energy efficient, or requiring 'twice the area of land for the same production as non-organic farming', or as a system of agriculture that 'will never feed the world'. Whilst each of these debates could be the subject of a book in their own right, it is important to consider what organic farming can offer in terms of energy *efficiencies*: hence this section of the book.

One of the potential benefits of organic methods is a lower energy consumption. However, measuring energy consumption and comparing it with conventional production, and with alternatives within organic systems, is complex. For instance, indirect energy inputs in machinery, fertilizers and pesticides need to be considered as well as direct inputs such as in fuels. Furthermore, energy input *per se* is not the end of the story: it should be related to the energy output of the system, that is, the efficiency of energy use. Transport costs also need to be considered, and this is especially relevant to the 70 per cent-plus of fresh produce that is imported at present.

Dr Bill Cormack and Phil Metcalfe, researchers with ADAS, produced an important piece of research looking at energy use in organic farming in 2001. The model for energy use that was used compared direct and indirect energy use in simple individual crop and livestock enterprise models; these were then combined to give whole-system models covering dairy, beef and arable farming, with three basic components:

- Basic information on energy inputs (fuels, machinery manufacture and maintenance, fertilizers, pesticides and transport).
- Energy inputs were compared to energy outputs for individual crop and livestock enterprises, and the metabolizable energy (ME) content of the output calculated. The energy ratio 'E1' (ME output/input energy) was calculated for each.
- Energy inputs and outputs were applied to model farms with sizes typical for each system, to give overall system energy inputs, outputs and ratios.

Research findings determined that the dominant energy inputs in conventional agriculture were indirect energy for the manufacture and transport of fertilizers, particularly nitrogen, and indirect energy for the manufacture and transport of pesticides. These together account for around 50 per cent of the total energy input to a potato or winter wheat crop, and as much as 80 per cent of the energy input into some vegetable crops, as shown in Figure 48.

Organically grown crops require around only half of the energy input per unit area than conventional crops, largely because of lower, or zero, fertilizer and pesticide energy inputs. However, the generally lower yields of organic crop and vegetable systems reduce the advantage to organic when energy input is calculated on a unit output basis (Figure 49).

In stockless arable crop rotations, the inclusion of fertility-building crops and winter cover crops, that have energy inputs but no direct outputs, can result in a lower whole-rotation energy efficiency from organic methods (Table 74). In livestock systems, where the fall in output may be less than in arable, and there are no dedicated fertility-building crops, overall energy efficiency is greater in organic than in comparable conventional systems. The lower energy efficiency of livestock systems is noteworthy, particularly conventional dairy where less than half of the total energy input is recovered in saleable product.

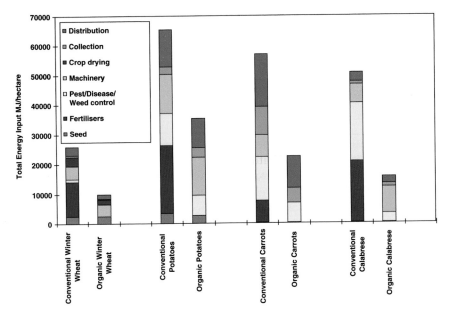

Figure 48 Energy input per hectare

Source: B Cormack & P Metcalfe, ADAS

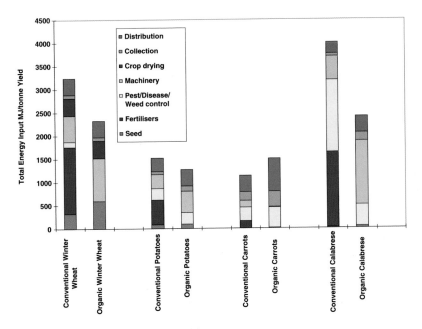

Figure 49 Energy input per tonne of yield

Table 74 Ratio of energy output to input for contrasting farming systems

	Conventional	Organic
Stockless arable	5.18	4.37
Field vegetables	0.81 – 4.80	5.31
Upland livestock	1.1	2.47
Dairy	0.43	1.67

These conclusions were made entering average yield data in the model, and need to be interpreted with caution. On more fertile soil, where the yield difference with conventional arable production is smaller, organic systems would have a relatively higher energy efficiency. The converse could occur on poorer soils. Also, in practice, energy inputs for cultivations and weed control will vary with soil type, weather, weed spectrum and population.

Using field vegetables as an example, energy costs for transport from farm to retailer distribution centre were considered for a range of scenarios, as shown in Table 75. Compared to large-scale conventional vegetable production, the modelling suggests that there is scope to reduce transport energy costs by around 40 per cent by transport to a nearby packer or by local sale, where feasible. Importing from northern Europe to the English Midlands added 44 per cent to transport energy costs, and from southern Europe it added 352 per cent to costs. Transport from further afield would be expected to have an even greater cost, particularly if it involved aircraft. Air transport was not considered as part of this study.

Table 75 Transport energy for vegetables taken from the farm to a retail distribution centre, modelled for a number of scenarios.

	Scenario	Energy MJ/tonne	Difference to scenario No. 1
1	Direct from a large production unit to the distribution centre.	600	–
2	Taken to a co-operative shipping point and then a packing centre.	628	+5%
3	Transported direct to a nearby packing centre.	376	-37%
4	Imported from northern Europe.	862	+44%
5	Imported from southern Europe.	2712	+352%
6	To a local wholesaler for local shops.	347	-42%

Next Steps

In a mainstream world agriculture that is literally hooked on fertilizers and agrochemicals, energy is often ignored, especially when it is plentiful and available. However, this reliance on energy comes at a cost, and is likely to be unsustainable in terms of food security in the long term. What is also conveniently ignored by governments, politicians, consumers and many farmers, is that world agriculture, in the main, is totally reliant on the petrochemical industry to produce energy and raw materials for fertilizers and agrochemicals. With the raw materials increasingly at the edge of the grasp of the main agricultural producers, the question should not be how do we produce food with less energy, but rather, how do we produce food when either the fertilizers and agrochemicals run out, or are too expensive and energy-consumptive to produce?

Published figures for ammonium nitrate fertilizers in 1998 were £100–£125 per tonne. These have remained fairly static for nearly a decade, but are now starting to increase in cost, rapidly fuelled by an increasing shortage of petrochemical products and higher production costs. In early 2007 ammonium nitrate fertilizers cost £160–£170 per tonne, with the potential for future cost increases.

As this section on energy efficiencies shows, organic farming is likely to have an increasingly important place in the production of food, in a world with a completely different energy climate, especially mixed organic systems where livestock can contribute an output from fertility-building phases.

ORGANIC CEREAL AND PULSE MARKETING

Understanding the Market

Marketing is an area of great importance, since crop value can be changed dramatically with the correct marketing strategy. The first objective must therefore be to gain some understanding of the market in which you will be aiming to sell into.

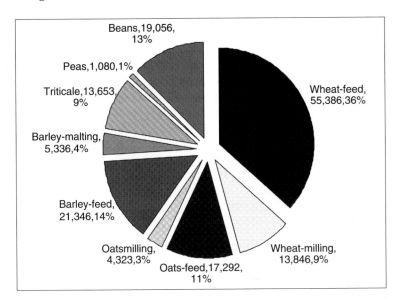

Figure 50 Tonnage of organic combinable crops produced in the UK in 2006

The organic cereal and pulse market is very small in the UK. With only 151,000 (approximately) tonnes of organic grain currently being produced per year, compared to 23,000 thousand non-organic tonnes of grain (wheat, barley, oats, OSR, linseed, peas and beans), UK organic grain production equates to approximately 0.7 per cent of all grain grown in the UK. The tonnage and percentage of organic combinable crops produced in the UK in 2006 is shown in Figure 50 (this represents full organic status crops only, and does not include any in-conversion crops).

The UK organic market is approximately 50–60 per cent self-sufficient in organic cereals (this has fluctuated between 45 per cent and 70 per cent over the last decade). The deficit of organic grain has to be imported. Some of this cannot be produced in the UK, for example high protein milling cereals or high grade protein crops. Nevertheless, imported grain plays a major part in shaping the UK domestic organic grain market.

The major difference from non-organic production, is that the organic cereal and pulse market is not so distorted by the CAP. With no export market and no intervention market for organic crops, a genuine supply and demand market exists, with crop values reflecting this situation. Crop demand and thus process is set by the value that any particular domestic buyer is prepared to pay. The main influence upon this is the availability and price of imported organic grain from Europe and further afield (typically Russia, Canada and Australia). With the advent of the Single Farm Payment Scheme (SPS) and the removal of direct crop production subsidies, there is also far more freedom to grow crops to meet market demand, rather than to secure subsidy income. This major policy change fits organic production very well indeed.

The farmer growing a commodity crop of cereal or pulse is normally one step removed from the end user of the grain (a miller or feed compounder): in between the farm (seller) and the end user (buyer) there is normally at least one intermediary in the form of a 'grain trader' – indeed there may be several intermediaries, including selling groups, and trade-to-trade sales before grain reaches an end-use destination. There is no centrally published data on organic crop values on a week-to-week or monthly basis as in the non-organic sector. Market intelligence is therefore very difficult to access and accumulate, especially for the individual farmer. This results in the farmer (seller) having an imperfect knowledge of how the market is operating and developing at any one point in the season and over the longer term.

This 'one step removed' knowledge is compounded by what happens on the continent, and knowledge of how imported grain impacts on domestic markets, as buyers cover their position in the market. For instance, most organic end users need year-round supplies of organic crop. If at any point in the season a predicted volume of supply does not arrive in the UK from the continent, the end user will contact the intermediary buyer who will then seek to cover this shortfall. However, this will be unknown to most farmers and can cause great volatility in the market place.

Marketing Strategies

All of the above impacts upon the manner in which the market functions, and the strength or weakness of the market into which the organic farm will need to sell its crops. The points raised should be remembered when considering how to market organic crops for sale. The main options available to organic cereal and pulse producers are wholesale trade, direct trade, farm to farm, collaborative ventures and specialist markets.

Wholesale Trade

This is categorized by selling cereals and grains to intermediaries; these mainly consist of grain merchants and grain co-operatives. Sales can be based on spot market values, minimum-maximum contracts over a period of time (min-max), tracker contracts or fixed forward price contracts.

Table 76 Benefits and constraints of wholesale trade markets

Benefits	Constraints
Flexible marketing arrangements	Lack of market control
Removes marketing burden from the farm	Lack of on-farm market intelligence
Can shift large volumes at any time of season	Seller at end of supply chain
Secure trade and payment terms	Subject to price volatility
Efficient transport systems	
Low risk	

Direct Trade

This is categorized by direct contracts between farmers and end users such as millers, maltsters, feed-compounding companies. Sales are normally based on fixed forward price contracts, or tracker contracts, or minimum-maximum contracts over a period of time (min-max).

Table 77 Benefits and constraints of direct trade markets

Benefits	Constraints
Shortens supply chain	Required marketing activity on-farm
Can be long term relationship	Small volumes required on a regular basis
Strong local markets are ethically good	Long term on-farm storage required
Markets generally secure	Trade and payment terms
	Crop specifications usually high
Low to medium risk	Relies on what market will pay & subject to price volatility

Farm to Farm

This is categorized by direct sales of grain, mainly for livestock feed, from arable farms to livestock farms. Small volumes can be sold to on-farm millers. Sales are normally based on an agreed price to supply, which can contain a known increase over a period of time.

Table 78 Benefits and constraints of farm to farm marketing

Benefits	Constraints
Shortens supply chain	Required marketing activity on-farm
Can be long term relationship	Small volumes required on a regular basis
Strong local markets are ethically good	Long term on-farm storage required
Unaffected by market volatility	Trade and payment terms
Crop specifications may be more flexible	Personal relationships can some times cause friction
Medium to high risk	

Collaborative Ventures

These are categorized by joint or collaborative arrangements between two or more farms such as share farming, linked arable and livestock farms, or pooled marketing activities. Sales are normally negotiated for the joint parties using any of the above arrangements.

Table 79 Benefits and constraints of collaborative venture marketing

Benefits	Constraints
Increased market intelligence	Required marketing activity on-farm
Increased volumes gives greater market presence	Requires agreement between parties
Can make use of joint infrastructure and labour	Long term on-farm storage may be required
Improved resource utilization	Trade and payment terms may be more difficult
May allow different parties to specialize with one person	
Undertaking marketing on behalf of others	
Low to medium risk	

Specialist Markets

These are categorized by direct contracts between farmers and end users such as seed merchants or suppliers, pharmaceutical or human consumption markets. Sales are normally based on fixed forward price contracts, or tracker contracts over a period of time (min-max); for example, for seed, a fixed sum over and above wholesale trade values.

Table 80 Benefits and constraints of specialist markets

Benefits	Constraints
Shortens supply chain	Required marketing activity on-farm
Can be long term relationship	Small markets
Strong local markets are ethically good	Small volumes required on a regular basis
	Long term on-farm storage required
	Trade and payment terms
	Crop specifications usually very high
	Markets generally less secure
Medium to high risk	Relies on what market will pay and subject to price volatility

Many farmers, for simplicity, use one of the above routes to market only. Others use a combination of marketing approaches in order to spread risk. Given the very small market size and relative isolation of organic farmers, who are prone to much larger buyers with better market intelligence, it is prudent to join together with other organic farmers in collaborative ventures and co-operative marketing groups in order to provide a stronger market position.

Marketing Advice and Grants

Marketing is an area of great importance, but one which is complex and time consuming if carried out correctly. It is also subject to influence from government policy initiatives, which can distort the market. There is a range of state-provided advice and funding programmes to assist farmers and rural businesses in meeting these policy initiatives, however, these are subject to change, and there is greater emphasis on regional support initiatives in the UK. A detailed description of these is beyond the scope of this book.

Government Support

Until recently, farmers in the various regions of the UK have been able to gain support from various rural development programmes such as the England Rural Development Programme (ERDP) project-based schemes. These included rural support for enterprise development, processing and marketing, vocational training, the production of energy crops and support of producer groups, amongst other things. Many organic farmers have made use of these programmes.

However, at the time of writing this book, the structure that has been in place for these schemes from 2000 to 2006 has now ceased to operate. The closure of these schemes forms part of the UK government's strategy to simplify funding for rural areas, and to better target it to the needs of rural people and businesses in each region – and provide better value for money.

Regional departments in England, Scotland, Wales and Northern Ireland are reducing the range of programmes towards a more targeted focus on environmental land management and natural resource protection; sustainable rural communities; and sustainable food and farming. These new programmes are being implemented progressively, with the final stage to be completed with the rollout of the next rural development programme between 2007 and 2013. The emphasis will be much more regional, rather than national: in England, for instance, the structure and delivery of each of these programmes varies greatly between the eight government regions. Interested farmers and groups should keep a keen eye on the development of these programmes, and obtain details where appropriate.

Local Support

Local development agencies increasingly have sources of advice and financial assistance programmes to support business development and diversification. They can often provide advice on topics such as new business start-up, business planning, business performance benchmarking, and so on. Funding is often provided on a match-funding basis, with a grant for a percentage of costs that has to be matched by the farmer. Some local authorities provide discretionary grants and low-interest loans to rural businesses, especially where there is environmental or social gain.

FUTURE MARKET PROSPECTS AND OPPORTUNITIES

The UK organic grain market continues to show a significant under-supply of feed and milling cereals, and pulse crops, especially those with higher quality proteins. As the organic cereal market develops further there is an on-going and increasing demand for wheat for human consumption and livestock feed. The rapid increase in ruminant and monogastric livestock feeds over recent years has seen the demand for crops for use by feed compounders becoming a dominant market sector. With the changes in the livestock feeding regulations in August 2005 there has been a significant increase in the demand for not only UK-produced wheat, but also protein crops, especially those that produce high grade protein with low levels of starch. There is therefore a tremendous opportunity for UK organic arable production to fulfil these demands over the coming years. If UK organic farmers fail to respond to this demand and do not 'grow for the market', the growth of the livestock sector may be impaired and imports will be used to fill the gap – and a fantastic market opportunity may be lost.

TYPICAL MARKETING SPECIFICATIONS FOR ORGANIC CEREALS AND PULSES

Wheat

Milling Wheat Specifications
There are four groups that encompass the varieties of wheat grown in the UK. The groups are numbered one to four, and 'nabim' group them according to the following;

nabim group 1 Hard suitable for most UK grists
nabim group 2 Milling potential
nabim group 3 Biscuit, starch blending
nabim group 4 Feed and minor grists

The nabim group should be referred to when choosing a variety. Invariably if a variety doesn't make the grade it will then be sold as feed. The typical market requirements for milling wheat for bread making are shown below. Producers should note that individual buyer specifications may vary. It is also important to remember that yields of milling wheat and malting barley will tend to yield less than feed varieties.

Grain sampling   Before any grain is bought off the farm, a sample of it will need to be taken in order that the buyer knows what they're purchasing. A sample will be sent to potential buyers, or alternatively some buyers may send a sampler to the farm and use a grain spear to sample at depth. Samples should be taken from the grain in a 'W' pattern in order to get a representative sample. Once the grain has been sampled it will be subjected to a number of laboratory tests in order to assess its quality.

Protein   The protein content affects the strength of the dough, and premiums are paid for a high protein content. The availability of nitrogen will affect the protein content, and more fertile sites will produce wheat with a higher protein content. Protein levels should be above a minimum of 11.5 per cent, and more ideally 13 per cent in order to be accepted for milling.

Hagberg falling number   This is a measure of the alpha amylase content of the grain. A high alpha amylase content leads to bread having a sticky structure. The hagberg falling number (hfn) is a measure of how long it takes a plunger to pass through a prepared sample of water and flour mixed to a slurry, and is measured in seconds. If the plunger passes through slowly, then the enzyme alpha amalyse is low and has not reduced the viscosity, and will therefore result in a high hagberg number. Most millers require a minimum hagberg falling number of 250.

Specific weight This measures the grain density, and is a measure of the individual grains and their packing density, expressed as kilograms per hectolitre. It is calculated by filling a 1ltr container with grain, which is then weighed. A specific weight of 76 indicates good flour extraction, and in simple terms the buyer is getting more grain for a given volume than a sample with a low specific weight. Samples should be measured at the contractual moisture content. The specific weight is affected by variety, husbandry and the weather, and if there is a lack of moisture available to the crop.

Moisture It is important that the crop is stored at 14 per cent moisture to ensure it will reach the buyer at no more than 15 per cent moisture content. At this moisture content the likelihood of infestation by insects is greatly reduced, unless the grain is very warm, and it can be safely stored for long periods. Mould is also less likely to occur at this moisture content. Buyers prefer grain at this content as they are buying more grain, rather than water, for a given tonnage, than grain at a high moisture content.

Sample purity This test basically measures the presence in the batch of broken grains, and non-grains such as straw, dust, other cereal grains, weed seeds, ergot pieces, insects and pests. The result is a measure of a proportion by weight of the usable grain. Sieving machinery is used. The sample is sieved for half a minute, and the admixture is weighed and is then expressed as a percentage of the sample weight. This should not exceed 2 per cent for milling.

Sample appearance This is a subjective test and may be undertaken by an experienced grain sampler. The grain is viewed in front of a clear background and assessed for colour, contamination, chaff, and shrivelled and broken grains. The producer can control some of these factors: chaff in the grain, broken grains and contamination are affected by settings on the combine harvester. Pinched grain can be influenced by the grower to some extent through the choice of variety, seed rate and husbandry (for example, the control of weeds, so more moisture is available to the crop), drilling date, and the level of crop cleaning carried out after harvest. There are also factors that cannot be controlled by the grower, such as rainfall and the availability of moisture. Colour is not so easy to influence, as sooty moulds occurring at harvest time cannot be controlled where rainfall has been high at harvest time.

Feed Wheat Specification
The criteria for the feed wheat market are more easily satisfied. The sample should still be at no more than 15 per cent moisture content and have no more than 2 per cent impurities by weight. Typical minimum specific weights for the feed market are 72kg/hl.

Barley

Although 80 per cent goes into the animal feed market, growing for the brewing or malting markets can significantly increase the value of the crop.

Malting Barley Specifications

It should be remembered that buyers of malting barley all have their own specifications, and the following should be taken as a guide only. Buyers of malting grain usually have a preference for specific varieties that are known to perform well in malting conditions.

Germination   Germination is a vital part of the malting process, therefore it is important that the sample germinates at as close to 100 per cent as possible. If the sample germinates at less than 96 to 98 per cent, then it is likely to be rejected for malting. This test is carried out in the laboratory where the grains are split in two and then soaked in tetrazolium for 10 minutes at 55°C; after this if the germ has turned red, it is deemed viable.

Contamination   As with wheat, the presence of anything other than barley is undesirable in the sample as it may affect the malting process. Buyers may vary in their specification, however; it is usually 2 per cent by weight.

Nitrogen   Barley containing a low nitrogen content usually indicates a high carbohydrate content: this will govern the production of fermentable sugars, and therefore alcohol production. Specifications vary for the nitrogen content, but premiums are usually paid for barley that is under 1.65 per cent nitrogen. Barley too high in nitrogen may cause the beer to go cloudy during the brewing process; however, there are some buyers who may purchase barley up to 2 per cent nitrogen content.

Screenings   This test measures the amount of screenings that pass through a 2.2mm sieve. Normally maltsters specify that 95 per cent of grain should pass over the sieve and be retained; however, in some exceptional years, both good and bad, these specifications may change. Maltsters prefer bold grain, as it gives a higher yield of malt extract.

Feed Barley specification

The criteria for the feed market is far more easily satisfied. The sample should still be at no more than 15 per cent moisture on content on arrival, and have no more than 2 per cent impurities by weight. Typical minimum specific weights for the feed market are 63kg/hl.

Oats

The general requirements for oats are that they are free from mould, moisture (anything above 15 per cent) and infestations. The oats should also be clean and bright, with a minimum of 50kg/HL specific weight, and no more than 2 per cent impurities.

Triticale

The typical specification for triticale is usually just a maximum of 15 per cent moisture, free from mould and infestations, and no more than 2 per cent impurities.

Rye and Spelt

Where grown for human consumption, rye and spelt have to meet stringent standards, and the grower should determine these prior to growing and agree a contact for this, as the market is extremely small.

Varietal Mixtures or Blends or Cereals

There is increasing interest in growing varietal mixtures/blends with an aim of reducing the disease risks of growing single varieties whilst maintaining the quality required by the customer. This technique has generally shown that performance can match the best variety in the blend, but rarely exceed it; it also has the benefit of reducing performance variability from year to year.

However, the technique may compromise the ability to market the grain depending on customers' requirements, and many buyers will not handle blends of varieties in the sample, and potential buyers and markets should be sourced first before drilling such mixtures. This is particularly true for milling grain, less so for feed grains (if the grain is to be home fed, then this would not be an issue).

Pulse Crop Market Requirements

Peas, field beans and lupins invariably enter the animal feed market, though there are markets for human consumption where specific varieties are used. The basic standards for the feed markets are a maximum moisture content of 15 per cent, as well as the sample being free from mould, infestations and below a maximum impurity level of 2 per cent.

Market specifications for human consumption beans can vary widely, and buyers' specifications should be obtained before drilling. Typically buyers look for sound, hard beans with a good pale colour and skin finish. The beans are passed over a 9mm sieve to remove undersize beans.

The specification for human consumption peas can vary, though the peas should be clean and free from staining and taint, as well as having 15 per cent moisture and no more than 2 per cent impurities. Growers should ensure that a potential buyer has been sought prior to drilling.

Chapter 14

Sources of Information and Advice

THE NEED FOR INFORMATION

Converting to organic farming is a daunting prospect to many, and the need for careful planning is of paramount importance. As part of this process there is a great need to gather information, expertise and experience in order to assist with the process. Even farmers who have been farming organically for a number of years, often decades, find that knowledge develops, crops and markets change, and regulations alter. They too are constantly seeking to be kept up to date with the latest information. As one farmer commented to the author, 'Converting the farm to organic was the easy bit, converting the way that I think and plan was the challenge.'

The sections below outline a few options where information, expertise and experience can be gathered. The refreshing aspect for many is that most organic farmers are more than willing to share their experiences in an open and honest way, with a high level of participative, and often vocal contribution at events.

SOURCES OF INFORMATION

The Farming Press

The farming press is generally a good source of information on current events, either on a daily, weekly or monthly basis. General press articles can provide outline information on subjects which can be followed up by personal contact for more in-depth information. Useful sources include press aimed at farmers in general, such as the BBC Radio 4 *Farming Today* programme, weekly journals such as *Farmers Weekly* and *Farmers Guardian*, monthly journals such as *Arable Farming* and *Crops*, and specialist organic journals such as the quarterly *Organic Farming* magazine produced by the Soil Association. Many research organizations such as EFRC and HDRA and NIAB also produce regular bulletins or other publications; all of these tend to be on a subscription basis. However, many are now available on the organizations' respective web sites, which are usually fairly up-to-date and don't incur subscription fees to all areas.

National Conferences and Shows

There is a range of national conferences and shows which can be useful to organic farmers and those considering organic conversion, which often have trade representation for organizations with an interest in organic production. The Soil Association run a food and farming conference each year, normally in January at a regional location somewhere in the UK. This is an excellent event and a good place to get a broad understanding of the organic sector, although more recently there has been less technical information for farmers at these events, with a wider focus on human health, markets and energy.

The national cereals' event held in July each year is one of the best events in Europe to review machinery and talk to trade representatives, although there tends to be little specialist organic presence. Regional events such as LAMA (Lincolnshire Agricultural Machinery ssociation), the Royal Show, the Royal Welsh Show, the Highland Show and the Bath and West Show, all provide an opportunity to gather the latest information and discuss options with dealers, buyers and the trade.

Trade

The seed trade and grain-buying organizations are generally a good source of technical information on varieties and the market for crops. Many provide technical bulletins and market reports. Some arrange open days and farm walks.

Open Days and Farm Walks

The Soil Association, EFRC and HDRA all run a programme of farm walks and demonstration farms. These can be invaluable to look at a range of different farms and learn different experiences from farmers.

Organic Centres

There are a number of government and EU-supported organic centres in the UK. Current centres are supporting the South West, North West, North East and Yorkshire and Humberside regions of England and Wales from the University of Wales, Institute of Rural Sciences at Aberystwyth. (*See* contact details below).

Farmer Groups

Most regions of the UK have established organic farmer groups. These can be structured around topics (livestock management, horticulture, marketing) or farm types. These are often self-run, or they are based around local colleges or other institutions. Farmers' groups often run a programme of farm walks around members' farms, to which most people are welcome.

On-farm organic farmer event.

Other Organic Farmers

Most organic farmers are more than willing to share their experiences in an open and honest way, and many are good advocates of organic farming. One of the easiest ways to learn about organic farming is to find out who are the organic farmers local to your location, contact them and arrange a visit. Most will be willing to show you round and discuss the opportunities, constraints and challenges of organic farming. Most organic certification bodies can provide a list of names of local organic farmers.

Advisory Services

Advisory services generally offer tailored person-to-person advice on aspects of organic farming. This is normally on a contract basis between the farmer and the consultant. It can also be between a group of farmers and a service provider. Advice may be varied, and normally covers most aspects of the farm business: technical agronomic issues, organic certification and compliance, agricultural legislation, business management, conservation and assistance with obtaining relevant farming payments and agri-environment schemes. This may be delivered on-farm, by telephone or by e-mail, or by a combination of all of these. In addition to on-farm services, many consultants can access specialist services such as pest and disease identification, soil analysis, and so on.

Consultants and advisers should be expected to keep themselves up-to-date with the latest farming policies, schemes, research and best practice. The best consultants are able to demonstrate this. Farmers would normally expect to pay consultants sufficiently for their services to enable them to do this as well as the direct advisory work, although this cost would naturally be spread over all the farmers to which a particular consultant is providing services.

There are independent consultants and those employed by organizations. These may also offer training workshops or provide learning opportunities (such as training seminars, open days or farm walks). Most reputable consultants and advisers hold professional qualifications and operate professional indemnity insurance.

Advice concerning conversion to organic farming can be obtained from private consultancy organizations, or individually and from the Organic Conversion Information Services (OCIS), which is a government-funded initiative in England and Wales.

There are many organizations that claim to be able to deliver organic consultancy. Many try and deliver non-organic and organic advice. Whilst it *is feasible* to have sufficient knowledge and experience of both systems, the main challenge for those trying to do both is in being able to think organically and strategically rather than always taking a very conventional 'fix it' approach. There are a few organizations that specialize only in organic farming consultancy, for example Abacus Organic Associates (For a list of organic advisory services, *see* the useful addresses given below).

Organic Certification Bodies

There are a number of approved organic certification and inspection bodies in the UK (*see* useful addresses below). The various bodies provide organic certification services to ensure farmers are complying with organic production, processing and retailing standards and legislation. Certification bodies generally charge for this service. Some certification bodies can provide a level of support, usually by telephone. This is especially useful for the interpretation of organic standards. However, organic certification bodies are prohibited from advising farmers whom they certify.

Memberships

Various organizations provide membership facilities for a wide range of information. This is provided as printed material, web based and e-mail based. Some organizations run training events on a regular basis. The Soil Association producer services membership is to be recommended for the wide range of technical support it offers. Abacus Organic Associates offer a membership for farmer groups and telephone and e-mail support services. Elm Farm Research Centre offers a membership for farmer groups, and a regular news bulletin. The Institute of Organic Trainers and Advisers offers professional membership for advisers and consultants in the organic sector. (For a list *see* useful addresses below.)

Research Institutes

There are a number of government and private research institutes in the UK. These organizations can exclusively carry out research into organic farming (such as HDRA and EFRC) or carry out research into both organic and non-organic farming (such as NIAB, ADAS, SAC etc).

Research is either privately funded or more often funded through EU or government research programmes. Organic agricultural research covers a wide range of topics, from agronomic and technical issues through to marketing, policy and environmental management. Most research projects have some obligation to disseminate their findings to farmers, and may do this through various channels.

ADAS operate two research farms where arable crop enterprises are managed: ADAS High Mowthorpe (Yorkshire) and ADAS Terrington (Norfolk). These are worth visiting on the various open days that are held.

Agricultural Colleges and Universities

A number of universities and agricultural colleges offer part- and full-time courses on organic farming. These can result in a formal qualification where appropriate (*see* education and training section below).

Government

Defra and its equivalent in Wales, Scotland and Northern Ireland are responsible for policy on organic farming in their respective regions of the UK. All of these provide information on organic farming in some way or another. The most accessible and up-to-date way of obtaining information from these bodies is to access their web sites, or visit their regional offices. (For a list of organic government offices and web sites, *see* useful addresses below.)

Agricultural Levy Bodies

The levy bodies for cereals and pulses are HGCA and PGRO. They have both undertaken a limited amount of research and development work for organic production. Some information leaflets and research results are available via printed matter off web sites. Farmers, who all pay a levy on all organic crop grown and sold, should encourage the levy bodies to undertake more organic research and development work.

Leaflets and technical guides

Many of the institutes and organizations mentioned above also produce a range of technical bulletins, fact sheets and leaflets for farmers. These aim to provide practical information to farmers to cover most aspects of production, marketing and policy. The web sites of many of the institutes will provide free access to some of this information.

The internet

The internet is a very powerful tool for gathering information on organic production from across the globe, with information becoming ever more accessible. Of notable local interest for cereal and pulse growers are the Soil Association SA daily news bulleting, and the monthly NIAB bulletins by e-mail.

ORGANIZATIONS HOLDING NATIONAL AND REGIONAL EVENTS FOR PRODUCERS

A number of organizations hold regular events for organic and interested conventional farmers in various regions. The following are useful contacts for events:

The Soil Association produces a list of events throughout England, Wales and Scotland. E-Mail: ps@soilassociation.org, Internet: www.soilassociation.org

The Yorkshire Organic Centre for events in Yorkshire and the Humber region E-Mail: info@yorkshireorganiccentre.org, Internet: www.yorkshireorganiccentre.org

Organic Centre Wales organizes a demonstration farm network, and events and training courses regularly at venues throughout Wales. Internet: www.organic.aber.ac.uk

The Organic Demonstration Farm Network of Elm Farm Organic Research Centre hosts events (seminars, farm walks and training days) in England. E-Mail: education@efrc.com, Internet: www.efrc.com/

The Organic Advisory Service holds regular meetings and farm walks in England for converting and established organic farmers, covering dairy, beef and sheep, arable and horticulture. Tel. (01488) 658 279

Garden Organic (HDRA) organizes events mainly for horticultural producers and gardeners E-Mail: events@hdra.org.uk Internet: www.gardenorganic.org.uk/events/index.php

Norfolk and Suffolk Organic Farmer Group (NOFG) meets regularly throughout the year with farm visits in the summer, speakers from the industry in the winter. Contacts: Bill Starling Tel: (01379) 674100 E-Mail: bstarling@soilassociation.org or Stephen Briggs Tel: (07855) 341309 E-Mail: stephen.briggs@abacusorganic.co.uk

East Midlands Organic Farmer Group (EMOFG) for producers in Cambs, Beds, Northants, Leics, Rutland and Lincs meets regularly throughout the year with farm visits in the summer, speakers from the industry in the winter. Contacts: Rebecca Rayner Tel: (01487) 773282 or Stephen Briggs Tel: (07855) 341309 E-Mail: stephen.briggs@abacusorganic.co.uk

Organic South West promotes organic farming in Cornwall and Devon. Organic South West provides producer information and training and organizes sector groups. Contact: Organic Southwest, Kyle Coberparc, Stoke Climsland, Callington, Cornwall, PL17 8PH. Tel: (01579) 371147 E-Mail: osw@soilassociation.org.

Graig Farm Organics, meat marketing, farm walks, discussion group and publications. Contact: Bob Kennard, Graig Farm, Dolau, Llandrindod Wells, Powys, LD1 5TL , Tel: (01597) 851655 E-Mail: rwk@graigfarm.co.uk

The North-West Organic Co-Op Society Ltd promotes organic farming, supports marketing and provides some advice and training in the North West of Republic of Ireland/Northern Ireland. E-Mail: info@nworganic.com Internet: www.nworganic.com

Irish Organic Farmers and Growers Association (IOFGA) have a number of monitoring farms where farm walks and demonstrations are held on a regular basis. Internet: www.iofga.org

Highlands & Islands Organic Association Discussion group, farm walks and publications. Marketing horticultural crops and meat. Contact: Catherine Wares, Glenorrin,

Great North Road, Muir of Ord, Ross-shire, IB6 7XR, Tel. (01463) 870360 E-Mail: glendusky@btopenworld.com

PUBLICATIONS

Periodicals

Clover Quarterly magazine of the Organic Trust, Dublin.

Elm Farm Organic Research Centre Bulletin Research updates and regular technical, management and financial advice for producers. Organic Advisory Service/Elm Farm Organic Research Centre.

Farmers Weekly Weekly press and news, tel 0845 0777744

Farmers Guardian Weekly press and news, tel 01858 43883

Living Earth Quarterly magazine on issues linking agriculture, food, health and the environment. Soil Association.

Organic Farming A technical quarterly magazine for producers. Soil Association.

Organic Matters Bi-monthly newsletter on organic farming in Ireland. Irish Organic Farmers and Growers Association.

Organic Today A quarterly journal for producers. Organic Farmers and Growers Ltd.

Soil Association *Certification News* Bi-monthly newsletter for all Soil Association certified producers. Regular updates on production standards issues. Market contacts service.

Star and Furrow Twice yearly journal for the Biodynamic Agricultural Association.

The Organic Way A quarterly newsletter on organic gardening. HDRA.

Technical Bulletin A quarterly update from the Organic Studies Centre, Duchy College, Cornwall

Useful Books and Leaflets

Cranfield University Silsoe *A Guide to Better Soil Structure* National Soil Resources Institute, 2002.

Lampkin N., Measures M., and Padel, S. *Organic Farm Management Handbook* (seventh edition 2007) updated annually.

Lampkin, N. *Organic Farming* Old Pond Publishing, 2002.

HDRA *Organic Weeds Project See* www.organicweeds.org

Home Grown Cereals Authority, *Recommended list 2006/07 for Cereals and Oilseeds* Home Grown Cereals Authority, 2006

Organic Cereals and Pulses *Weed Control in Organic Cereals and Pulses* D.H.K. Davies and J.P.Welsh pp77

OPICO Ltd, Advanta and Elm Farm Research Centre *Mechanical Weeding in Organic Production Systems: How, Why, When*, OPICO, 2000

Processors and Growers Research Organisation (PGRO) *Pulse Agronomy Guide: Advice on agronomy and varieties of peas, field beans and lupins*, PGRO, 2006 PGRO, The Research Station Thornhaugh, Peterborough, PE8 6HJ

Schering *Grain Quality Guide* Green Science, 1986

Schering *Agriculture, Cereal Disease Guide* Green Science, 1988

Schering Agriculture, *Weed Guide* Green Science, 1990

Soil Association *Storage of Organic Combinable Crops*, technical guide range, Soil Association, 2002

Soil Association *Organic Grain Storage Guideline Sheet*, laminated advisory sheet, Soil Association, 2005

Soil Association *Organic Market Report 2006*, Soil Association, 2006

Soya UK *The Soya UK Guide to the Production and Utilisation of Lupins in the UK* Soya UK, 2005

Steel in the Field A farmer's guide to weed management tools, edited by Greg Bowman from the *Sustainable Agriculture Network Handbook Series* (no 2). A book incorporating a large amount of farmer experience with mechanical weeding tools, also containing case studies although some are more suited to the US than the UK. Available here.

Weeds and Weed Management on Arable Land. An Ecological Approach. by Sigurd Hakansson. A comprehensive overview of the ecological approach to managing weeds in arable crops. Published by CABI (274 pages)

Younie D., Taylor B., Welsh J. and Wilkinson J. (eds) *Organic Cereals and Pulses* Chalcombe Press, 2001

Biological Agriculture and Horticulture. A scientific journal on every aspect of organic farming. AB Academic Publishers, PO Box 97, Berkhamsted, Herts HP4 2PX.

Ecology and Farming English language bulletin of international news and research reports on organic farming. IFOAM.

Altieri, M. (1995) *Agroecology – the scientific basis for alternative agriculture* 2nd edition. Intermediate Technology Publications, London.

Blake, F. (1994) *Organic Farming and Growing* Crowood Press, Swindon.

OCW/COR (2002) *UK Organic Research 2002.* Proceedings of the Conference 26–28th March, Aberystwyth. see www.organic.aber.ac.uk

Lampkin, N. H. (1990) *Organic Farming.* Old Pond Ipswich

Technical guides from the Soil Association: Bristol

Newton, J. (1995) *Profitable Organic Farming* Blackwell Science, Oxford.

Younie, D., B.R. Taylor, J.P. Welsh and J.M. Wilkinson (2002) *Organic Cereals and Pulses* Chalcombe Publications, Lincoln

USEFUL WEB SITES

Abacus Organic Associates: www.abacusorganic.co.uk

British Society of Plant Breeders : www.bspb.co.uk

CABI Organic Research: www.organic-research.com

COSI seed database: www.cosi.org.uk

Defra Organic Farming Pages: www.defra.gov.uk/farm/organic

European Association for Grain Legume Research: www.grainlegumes.com

Elm Farm Organic Research Centre: www.organicresearchcentre.com

EU Organic Farming Pages

www. europa.eu.int/comm/agriculture/qual/organic/index_en.htm

Marketing : www.bepa.co.uk Information on quality requirements and markets

FAO Organic Agriculture: www.fao.org/organicag

FiBL Organic Research: www.fibl.org/english/index.php

IFOAM International Federation of Organic Agriculture Movements: www.ifoam.org

HDRA Garden Organic: www.gardenorganic.org.uk/

Lupins : www.lupins.iger.bbsrc.ac.uk Information on lupins

OMIaRD Organic Marketing Initiatives and Rural Development: www.irs.aber.ac.uk/omiard

Organic Centre Wales: www.organic.aber.ac.uk

Organic Europe: www.organic-europe.net

Organic X Seeds: www.organicxseeds.com/

Organic Soil Fertility Project: www.organicsoilfertility.co.uk

Organic production information, Northern Ireland: www.ruralni.gov.uk/bussys/organic/business_management/organic_food/

Pulse Crop Genetic Improvement Network: www.pcgin.org

SAFO Sustaining Animal Health and Food Safety in Organic Farming: www.safonet-work.org/index.html

Scottish Executive's organic farming pages: www.scotland.gov.uk/Topics/Agriculture/Agricultural-Policy/15869/3748

Soil Association: www.soilassociation.org

Scottish College of Agriculture: www.sac.ac.uk/consultancy/organic/

USDA Alternative Farming Systems Information Centre Agriculture: www.nal.usda.gov/afsic/

Weed control in organic agriculture from HDRA: www.organicweeds.org.uk

Research Centre websites:

ADAS: www.adas.co.uk
IGER: www.iger.ac.uk
HDRA: www.hdra.org.uk
Elm Farm: www.efrc.com
SAC: www.sac.ac.uk/consultancy/organic/
Newcastle University: www.ncl.ac.uk/tcoa/producers/

ADVICE – ORGANIC CONVERSION

Abacus Organic Associates
Rowan House, 9 Pinfold Close, South Luffenham, Rutland, LE15 8NE Tel/Fax: 01780 721019 Internet: www.abacusorganic.co.uk
E-Mail: enquiry@abacusorganic.co.uk
The UK's leading group of independent organic consultants.

Organic Conversion Information Service (OCIS):
In all regions of the UK. Advice and information concerning organic farming can be obtained from the Organic Conversion Information Service (OCIS) helpline.

In England, farmers and growers can arrange for a free half-day visit and report with a follow-up full day visit and expanded report by an adviser experienced in organic production and marketing who will provide impartial advice relevant to the business. The visits are provided by advisers with experience of organic farming.
OCIS Helpline number (England) – 0117 922 7707

Organic Centre Wales (OCW)
In Wales, an information pack and up to two free advisory visits are provided plus detailed conversion planning linked to the Farming Connect Farm Business Development Plans.

Scottish Agricultural College
In Scotland a free telephone advice and information to converting and existing organic farmers from a network of local advisory offices. Funding is available towards the cost of advisory help in preparing conversion plans. Face to face consultations are available on a charged basis.

In Northern Ireland, the Department of
Agriculture and Rural Development (DARD)
DARD has an advisory team based at Greenmount College and offers OCIS.

England Tel: (0117) 922 7707

Wales Tel: (01970) 622100

Scotland Tel: (01224) 711072

Northern Ireland Tel: (028) 9442 6765

UK ORGANIC CERTIFICATION BODIES

Organic Farmers and Growers Ltd. (UK2)
The Elim Centre, Lancaster Road,
Shrewsbury, Shropshire SY1 3LE
Tel: 01743 440512
Fax: 01743 461441
E-Mail: info@organicfarmers.uk.com
Website: www.organicfarmers.uk.com

Scottish Organic Producers
Association (UK3)
Scottish Organic Centre, 10th Avenue,
Royal Highland Centre, Ingliston,
Edinburgh EH28 8NF
Support & Development:
Tel: 0131 333 0940
Fax: 0131 333 2290
Certification:
Tel: 0131 335 6606
Fax: 0131 335 6607
E-Mail: sopa@sfqc.co.uk
Website: www.sopa.org.uk

Organic Food Federation *(UK4)*
31 Turbine Way, Eco Tech Business Park,
Swaffham, Norfolk PE37 7XD
Tel: 01760 720444
Fax: 01760 720790
E-Mail: info@orgfoodfed.com
Website: www.orgfoodfed.com

Soil Association Certification Ltd (UK5)
South Plaza, Marlborough Street,
Bristol BS1 3NX
Farmers and Growers: Tel: 0117 914 2406
Processors: Tel: 0117 914 2407
Fax: 0117 925 2504
E-Mail: prod.cert@soilassociation.org
Website: www.soilassociation.org

Bio-Dynamic Agricultural
Association (UK6)
The Painswick Inn Project, Gloucester
Street, Stroud GL5 1QG
Tel: 01453 759501
Fax: 01453 759501
E-Mail: bdaa@biodynamic.freeserve.co.uk
Website: www.biodynamic.org.uk

Irish Organic Farmers and Growers
Association (UK7)
Harbour Building, Harbour Road,
Kilbeggan, Co Westmeath,
Republic of Ireland
Tel: 00 353 506 32563
Fax: 00 353 506 32063
E-Mail: iofga@eircom.ne
Website: www.irishorganic.ie

Organic Trust Limited (UK9)
Vernon House, 2 Vernon Avenue,
Clontarf, Dublin 3,
Republic of Ireland

Tel: 00 353 185 30271
Fax: 00 353 185 30271
E-Mail: organic@iol.ie
Website: www.organic-trust.org

Quality Welsh Food Certification
Ltd (UK13)
Gorseland, North Road, Aberystwyth,
Ceredigion SY23 2WB
Tel: 01970 636688
Fax: 01970 624049
E-Mail: mossj@wfsagri.net

Ascisco Ltd (UK15)
Bristol House, 40-56 Victoria Street,
Bristol BS1 6BY
Farmers and growers: Tel: 0117 914 2406
Processors: Tel: 0117 914 2407
Fax: 0117 925 2504
E-Mail: DPeace@soilassociation.org

ORGANIC MEMBERSHIP ORGANIZATIONS

Biodynamic Agricultural
Association (BDAA)
The Secretary, Painswick Inn, Stroud,
GL5 1QG Tel/Fax: 01453 759501
Internet: www.biodynamic.org.uk
E-Mail: office@biodynamic.org.uk

Organic Gardens (HDRA), Ryton
Organic Gardens, Coventry, CV8 3LG
Tel: 024 7630 3517 Fax: 024 7663 9229
Internet: www.gardenorganic.org.uk
E-Mail: enquiry@hdra.org.uk

International Federation of Organic
Agriculture Movements (IFOAM)
Head Office, Charles-de-Gaulle-Str. 5,
53113 Bonn - Germany
Tel: +49 (0) 228 926 50-10
Fax: +49 (0) 228 926 50-99
Internet: www.ifoam.org
E-Mail: headoffice@ifoam.org

Irish Organic Farmers and Growers
Association (IFOGA)
Main Street, Newtownforbes, Co.
Longford, Republic of Ireland
Tel: +353 (0) 43 42495
Fax: +353 (0) 43 42496
Internet: www.iofga.com
E-Mail: iofga@eircom.net

Scottish Organic Producers Association Ltd
10th Avenue, Royal Highland Centre,
Ingliston, Edinburgh, EH28 8NF
Tel: 01313 333 0940 Fax: 01313 333 2290
Internet: www.sopa.org.uk E-Mail:
info@sopa.org

Soil Association
South Plaza, Marlborough Street,
Bristol BS1 3NX
Tel: 0117 314 5000 Fax: 0117 3145001
Internet: www.soilassociation.org
E-Mail: info@soilassociation.org

Soil Association Scotland
18 Liberton Brae, Tower Mains,
Edinburgh, EH16 6AE
Tel: 0131 666 2474
Fax: 0131 666 1684
E-Mail: contact@sascotland.org

The Institute of Organic Training and
Advice (IOTA)
Cow Hall, Newcastle, Craven Arms,
Shropshire SY7 8PG
Tel: (01588) 6640118.
E-Mail: iota@organicadvice.org.uk
Internet: www.organicadvice.org.uk.
Provides information, training and
support to specialist organic advisers
and trainers.

World Wide Opportunities on Organic
Farms WWOOF
PO Box 2675, Lewes, East Sussex,
BN7 1RB
Tel: 01273 476 286 Fax: 01273 476 286
Internet: www.wwoof.org
E-Mail: hello@wwoof.org.uk
Tel: 0117 314 5000
Fax: 0117 314 5001
E-Mail: info@soilassociation.org

ADVICE/CONSULTANCY/ RESEARCH

Abacus Organic Associates
Rowan House, 9 Pinfold Close,
South Luffenham, Rutland,
LE15 8NE
Tel/Fax: 01780 721019
Internet: www.abacusorganic.co.uk
E-Mail: enquiry@abacusorganic.co.uk

The UK's leading group of independent organic consultants
ADAS Woodthorne, Wergs Road, Wolverhampton, WV6 8TQ
National help desk: 0845 766 0085
Internet: www.adas.co.uk/
E-Mail: use "contact us" page on website

ADAS-Pwllpeiran Cwmystwyth, Aberystwyth, SY23 4AB
Tel: 01974 282229 Fax: 01974 282302
ADAS-High Mowthorpe Duggleby, Malton, North Yorkshire YO17 8BP
Tel: 01944 738646 Fax: 01944 738434
ADAS-Terrington Terrington St.Clement, King's Lynn, Norfolk, PE34 4PW
Tel: 01553 828621 Fax: 01553 827229
E-Mail: Bill.Cormack@adas.co.uk

Central Science Laboratory (CSL)
Sand Hutton, York YO41 1LZ
Tel: 01904 462000 Fax: 01904 462111
E-Mail: science@csl.gov.uk
Website: www.csl.gov.uk

Elm Farm Research Centre
Hamstead Marshall, Newbury, Berkshire, RG20 0HR
Tel: 01488 658298 Fax: 01488 658503
Internet: www.efrc.com
E-Mail: elmfarm@efrc.com

Farming and Wildlife Advisory Group FWAG
Internet: www.fwag.org.uk (See website for other regional offices)
English Head Office: FWAG, National Agricultural Centre, Stoneleigh, Kenilworth, Warwickshire, CV8 2RX
Tel: 02476 696 699 Fax: 02476 696 760
E-Mail: info@fwag.org.uk
Northern Ireland: FWAG, 46b Rainey Street,
Magherafelt, Co Derry, BT45 5AH
Tel: 028 79 300606
E-Mail: n.ireland@fwag.org.uk
Scottish Head Office: FWAG Scotland, Algo Business Centre, Glenearn Road, Perth, PH2 0NJ Tel: 01738 450500 Fax: 01738 450495
E-Mail: steven.hunt@fwag.org.uk
Wales Head Office: FWAG Cymru, Ffordd

Arran, Dolgellau, Gwynedd, LL40 1LW
Tel: 01341 421456 Fax: 01341 422757
E-Mail: cymru@fwag.org.uk

Food Standards Agency
Aviation House, 125 Kingsway, London, WC2B 6NH
Tel: 020 7276 8000
Website: www.food.gov.uk

Garden Organic (HDRA)
Ryton Organic Gardens, Coventry, CV8 3LG
Tel: 024 7630 3517 Fax: 024 7663 9229
Internet: www.gardenorganic.org.uk
E-Mail: enquiry@hdra.org.uk

Greenmount Campus, College of Agriculture, Food and Rural Enterprise (CAFRE)
Antrim, Northern Ireland, BT41 4PU
Tel: 028 9442 6765   Fax: 028 9442 6606
Internet: www.ruralni.gov.uk/
bussys/organic E-Mail: adrian.saunders@dardni.gov.uk (See website for more information and contacts for organic production in Northern Ireland)

Growing Seed Company Ltd
15 Victoria Road, Stamford, Lincolnshire, PE9 1HB,
Tel 07875 239608 Fax 01780 482423.
Organic seed supply specialists –
A partnership between European organic farmers and UK supply specialists ensuring competitive process and ethical returns to growers.

HDRA see Garden Organic
Horticultural Research International Association (HRI-A)
Wellesbourne (Headquarters), Warwick CF35 9EF
Tel: 01789 470382 Fax: 01789 470552
Internet: www.hri.ac.uk
E-Mail: hri.association@hri.ac.uk

Institute of Grassland and Environmental Research (IGER)
Heather McCalman, IGER, Plas Gogerddan, Aberystwyth, SY23 3EB
Tel: 01970 823026
Internet: www.iger.bbsrc.ac.uk
E-mail: heather.mccalman@bbsrc.ac.uk

Institute of Organic Training and Advice (IOTA)
Cow Hall, Newcastle-on-Clun, Craven Arms, Shropshire, SY7 8PG
Tel: 01547 528546
Internet: www.organicadvice.org.uk
E-Mail: iota@newinvention.plus.com

Institute of Rural Sciences (IRS)
Organic Research Group, University of Wales, Aberystwyth, Ceredigion, SY23 3AL.
Tel: 01970 622248   Fax: 01970 622238.
Internet: www.irs.aber.ac.uk/
E-Mail: organic@aber.ac.uk

Nafferton Ecological Farming Group
Nafferton Farm, Stocksfield, Northumberland, NE43 7XD
Tel: 01661 830222   Fax: 01661 831006
Internet: www.ncl.ac.uk/tcoa/producers
E-Mail: tcoa@ncl.ac.uk

National Institute of Agricultural Botany
Huntingdon Road, Cambridge, Cambs CB3 0LE
Tel: 01223 342200 Fax: 01223 277602
E-Mail: info@niab.com
Website: www.niab.com
Internet: www.niab.com
E-Mail: info@niab.com

National Association of Farmers' Markets (FARMA)
P.O. Box 575, Southampton, Hampshire, SO15 7BZ
Tel. (0845) 458 8420
Internet: www.farmersmarkets.net
E-Mail: nafm@farmersmarkets.net

North-West Organic Co-Op Society Ltd. (NWOPG)
Northern Ireland Office, 2 Foreglen Road, Kilaloo, Derry/Londonderry, BT45 3TP
Tel: 028 7133 7950
Internet: www.northwestorganic.org
E-Mail: info@nworganic.com

Northwest Organic Centre:
Rural Business Centre,
Myerscough College, Myerscough Hall, Bilsborrow, Preston PR3 0RY

Tel: 01995 642206
Fax: 01995 642107
E-Mail: enquiries@nworganiccentre.org
Website: www.nworganiccentre.org

North East Organic Programme
PO Box 321, Newcastle upon Tyne NE3 2YP
Tel/fax: 0845 121 7645
E-Mail: neop@northeastorganic.org
Website: www.northeastorganic.org

Organic Advisory Service
Hamstead Marshall, Newbury, Berkshire, RG20 0HR
Tel: 01488 658279 Fax: 01488 658503
Internet: www.efrc.com
E-Mail: gillian.w@efrc.com

Organic Centre Wales
University of Wales, Aberystwyth, Ceredigion, SY23 3AL
Tel: 01970 622248 Fax: 01970 622238
Internet: www.organic.aber.ac.uk
E-Mail: organic@aber.ac.uk

ORA, Organic Resource Agency Ltd
Malvern Hills Science Park, Geraldine Road, Malvern, WR14 3SZ
Tel: 01684 585423 Fax: 01684 585422
Internet: www.o-r-a.co.uk
E-Mail: info@o-r-a.co.uk

Organic Seed Producers Ltd (OSP) – contact: Roger Wyatt
Tel. (01359) 270410.

Organic Studies Centre
Duchy College, Rosewarne, Camborne, Cornwall TR14 0AB
Tel: 01209 722155   Fax: 01209 722156
Internet: www.organicstudiescornwall.co.uk
E-Mail: j.burke@cornwall.ac.uk

Organic South West
Kyl Cober Parc, Stoke Climsland, Callington, Cornwall PL17 8PH
Tel: 01579 371147
Fax: 01579 371148
E-Mail: osw@soilassociation.org
Website: www.organicsouthwest.org

Pesticides Action Network PAN UK
Development House, 56-64 Leonard Street, London EC2A 4JX

Tel: 020 7065 0905 Fax: 020 7065 0907
Internet: www.pan-uk.org
E-Mail: admin@pan-uk.org

Scottish Agricultural College (SAC)
Organic Farming
Ferguson Building, Craibstone Estate,
Bucksburn, Aberdeen, AB21 9YA
Tel: 01224 711072 Fax: 01224 711293
Internet: www.sac.ac.uk/consultancy/
organic E-Mail: David.Younie@sac.co.uk

Sustain: Alliance for Better Food and
Farming
94 White Lion Street, London, N1 9PF
Tel: 020 7837 1228 Fax: 020 7837 1141
Internet: www.sustainweb.org
E-Mail: sustain@sustainweb.org

The Organic Centre (Ireland)
Rossinver, Co. Leitrim, Ireland
Tel: +353 71-98-54338
Fax: +353 71-98-54343
Internet: www.theorganiccentre.ie
E-Mail: organiccentre@eircom.net
Veterinary Epidemiology and Economics
Research Unit (VEERU)

Soil Association Producer Services
South Plaza, Marlborough Street, Bristol
BS1 3NX Tel: 0117 914 2400
Fax: 0117 925 2504
E-Mail: ps@soilassociation.org
Website: www.soilassociation.org/ps

Soil Association Scotland
Tower Mains, 18 Liberton Brae,
Edinburgh EH16 6AE
Tel: 0131 666 2474
Fax: 0131 666 1684
E-Mail: contact@sascotland.org
Website: www.soilassociationscotland.org

University of Reading, Earley Gate
Reading, RG6 6AR
Tel: 0118 378 8478 Fax: 0118 926 2431
Internet: www.veeru.reading.ac.uk
E-Mail: veeru@reading.ac.uk
Yorkshire Organic Centre

Yorkshire Organic Centre
Skipton Auction Market, Gargrave Road,
Skipton,
North Yorkshire BD23 1UD.

Tel. (01756) 796222, Fax: (01756) 796333
Internet:
www.yorkshireorganiccentre.org
E-Mail: info@yorkshireorganiccentre.org

EDUCATION AND TRAINING

Abacus Organic Associates
Rowan House, 9 Pinfold Close, South
Luffenham, Rutland, LE15 8NE
Tel/Fax: 01780 721019
Internet: www.abacusorganic.co.uk
E-Mail:
Stephen.briggs@abacusorganic.co.uk
Services: Seminars, courses, workshops
and discussion groups for farmers.

ACS Distance Education
P.O. Box 4717, Stourbridge, DY8 2WZ
Tel: 0800 328 4723 (9am – 4pm) Fax: 0207
681 2702
Internet: www.acsedu.co.uk E-Mail:
admin@acsedu.co.uk
Course titles: Organic Farming (corre-
spondence)
Organic Plant Culture (100 hours corre-
spondence)

Centre for Alternative Technology
Macchynlleth, Powys, SY20 9AZ
Tel: 01654 705981
Fax: 01654 703605
E-Mail: info@cat.org.uk
Website: www.cat.org.uk

City of Bristol College
Diploma in Organic Enterprise
(f/t 2 years)
Short courses including workshops in
organic food and farming
Bedminster Centre, Marksbury Road,
Bristol, BS3 5JL
Tel: 0117 312 5000 Fax: 0117 312 5050
Internet: www.cityofbristol.ac.uk
E-Mail: enquiries@cityofbristol.ac.uk

Derby College
Broomfield Hall, Morley, Ilkeston, Derby,
DE7 6DN
Tel: 01332 836600
Fax: 01332 836601
E-Mail: enquiries@derby-college.ac.uk
Website: www.derby-college.ac.uk

Dromcollogher Organic College
An t-Ionad Glas, Dromcollogher, Co.
Limerick, Ireland.
E-Mail: ionadglas.ias@eircom.net
Web: www.organiccollege.com/

Dundee College
Kingsway Campus, Old Glamis Road,
Dundee, DD3 8LE, UK.
Tel: 01382 834834
Fax: 01382 858117
E-Mail: enquiry@dundeecoll.ac.uk

Duchy College, Stoke Climsland,
Callington, Cornwall PL17 8PB
Tel: 01579 372233
Internet: www.cornwall.ac.uk/duchy
E-Mail: stoke.enquiries@duchy.ac.uk
Course title: FdSc Organic Farm Business
Development (f/t or p/t)
FdSc Horticulture (Organic Horticultural
Production) (f/t or p/t)
Horticulture HNC (Organic Horticultural
Production) (f/t or p/t)

Easton College
Easton, Norwich, NR9 5DX
Tel: 01603 731200 Fax: 01603 741438
Internet: www.easton-college.ac.uk
E-Mail: info@easton-college.ac.uk
Course title: National Award Organic
Horticulture (p/t 1 year)

Elm Farm Organic Research Centre
Hamstead Marshall, Newbury, Berkshire,
RG20 0HR
Tel: 01488 658298 Fax: 01488 658503
Internet: www.efrc.com E-Mail: elm-
farm@efrc.com
Services: Specialist short courses and
Organic Demonstration
Farm Network coordination.

Emerson College
Forest Row, East Sussex, RH18 5JX
Tel: (01342) 822238 Fax: (01342) 826055
Internet: www.emerson.org.uk E-Mail:
info@emerson.org.uk
Course titles: Diploma (EU level 4)
Biodynamic Organic Agriculture
(f/t 3 years)

Summer course (July): Biodynamic
Agriculture
Weekend course (June): Biodynamics for
the Backyard

Greenmount Campus, College of
Agriculture,
Food and Rural Enterprise (CAFRE)
Antrim, Northern Ireland, BT41 4PU
Tel: (028) 9442 6765 Fax: 028 9442 6606
E-Mail: adrian.saunders@dardni.gov.uk
Internet: www.ruralni.gov.uk/bussys/
organic/courses
www.greenmount.ac.uk/studying_at_
greenmount/courses
Course titles: National Diploma in
Agriculture (Organic options)
(f/t 3 years)
Higher National Diploma in Agriculture
(Organic options) (f/t 3 years)
Introductory courses in organic
production – subject to demand.

Hadlow College
Hadlow, Tonbridge, Kent, TN11 0AL
Tel: (Freephone) 0500 551434
Internet: www.hadlow.ac.uk
E-Mail: enquiries@hadlow.ac.uk
Course title: National Award in
Horticulture (Design or Organic options)
(f/t 1 year or p/t)

Horticultural Correspondence College
Freepost Notton, Chippenham, Wilts,
SN15 2BR
Tel: 01249 730326   Fax: 01249 730326
Internet: www.hccollege.co.uk
E-Mail: info@hccollege.co.uk
Course titles: RHS Advanced Certificate
in Horticulture (organic part module)
(correspondence)
Certificate in Organic Gardening
(correspondence)
Certificate in Organic Arable Farming
(correspondence)
Certificate in Organic Livestock Farming
(correspondence)

Holme Lacy College
Holme Lacy, Hereford, HR2 6LL
Tel: 01432 870316 Fax: 01432 870566

E-Mail: holmelacy@pershore.ac.uk
Website: www.projectcarrot.org

Institute of Rural Sciences
University of Wales, Aberystwyth,
Ceredigion, SY23 3AL; Tel: 01970 621614
Fax: 01970 611264
Internet: www.irs.aber.ac.uk/
brochure/organic.shtml
E-Mail: irs-enquiries@aber.ac.uk
Course titles: BSc(Hons) Organic
Agriculture (f/t)
BSc(Hons) Agriculture (organic
options) (f/t)
BSc(Hons) Sustainable Rural Development
(organic options) (f/t)
Higher National Diploma in Agriculture
(organic option) (f/t)
Postgraduate Certificate in Organic
Agriculture (f/t 1 semester)
Postgraduate Diploma in Organic
Agriculture (f/t 2 semesters)

Irish Organic Farmers and Growers
Association
Organic Farm Centre, Harbour Road,
Kilbeggan, Co. Westmeath, Ireland
Tel: (+353) (01) 506 32563
Fax: (+353) (01) 506 32063
Internet: www.irishorganic.ie/services
E-Mail: info@irishorganic.ie
Services: Placement list and other
web information on education and
training.

Kingston Maurward College
Kingston Maurward, Dorchester, DT2 8PY
Tel: 01305 215000 Fax: 01305 215001
Internet: www.kmc.ac.uk
E-Mail: administration@kmc.ac.uk
Course title: Sustainable Land Use
(organic module)
(4 days per week for 9 weeks)
BTEC ND Unit Principles and Practice of
Organic Agriculture (p/t)
Certificate of Higher Education in
Horticulture (organic unit) (p/t 20 weeks)
National Diploma in Horticulture
(includes organic module) (f/t 2 years)

London South Bank University
Faculty of Engineering, Science and Built
Environment, 103 Borough Road,
London SE1 0AA Tel: 020 7815 7815
Internet: www.lsbu.ac.uk/esbe/courses
E-Mail: enquiry@lsbu.ac.uk
Course title: BSc (Hons) Organic
Food Studies (f/t 3 years or sandwich
4 years)

LANTRA – the Sector Skills Council for
the Environmental
and Land-based Sector
Lantra House, Stoneleigh Park, Nr
Coventry, Warwickshire CV8 2LG
Tel: 024 7669 6996 Fax: 024 7669 6732
E-Mail: connect@lantra.co.uk
Website: www.lantra.co.uk

North-West Organic Co-Op Society Ltd.
Northern Ireland Office, 2 Foreglen
Road, Kilaloo, Derry/Londonderry,
BT47 3TP
Tel: 028 71 337 950 Fax: 028 71 337 146
Internet: www.northwestorganic.org
E-Mail: info@nworganic.com
Courses: Training services
Open Learning Centre International
24 King Street, Carmarthen, SA31 1BS
Tel: 0800 393 743 Fax: 01267 238 179
Internet: www.olci.info
E-mail: info@olci.info

Otley College of Agriculture and
Horticulture
Otley, Ipswich, Suffolk IP6 9EY Tel/fax:
01473 785543
E-Mail: course_enquiries@
otleycollege.ac.uk
Website: www.otleycollege.ac.uk

Permaculture courses (correspondence)
An t-Ionad Glas
Organic College, Dromcollogher, Co.
Limerick, Ireland.
Internet: www.organiccollege.com
E-Mail: oifig@organiccollege.com
Course titles: Certificate Course Organic
Growing and Sustainable Living Skills
with options including agriculture,

horticulture and sustainable development
(f/t 1 year and p/t options available)

Pershore College Pershore,
Worcestershire WR10 3JP
Tel: 01386 552443
Fax: 01386 556528 E-Mail: pershore@
pershore.ac.uk

Royal Agricultural College
Cirencester, Gloucestershire, GL7 6JS
Tel: (01285) 652531
Fax: (01285) 650219
Internet: www.rac.ac.uk E-Mail:
steve.chadd@rac.ac.uk
Course title: BSc (Hons) Agriculture
(organic agriculture)
MSc Organic Agricultural Systems

Scottish Agricultural College
Ferguson Building, Craibstone,
Bucksburn, Aberdeen, AB21 9YA
Tel: 0800 269453 Internet: www.sac.ac.uk
E-Mail: recruitment@sac.ac.uk
Course titles: PgC/PgD/MSc Organic
Farming (f/t or p/t by distance learning)
HND in Agriculture (Organic options) (f/t)
Training services for farmers.

University of Newcastle Upon Tyne
Newcastle Upon Tyne NE1 7RU
Tel: 0191 222 5594 Fax: 0191 222 8685
Internet: www.ncl.ac.uk
E-Mail: enquiries@ncl.ac.uk
Course titles: BSc (Hons) Organic Food
Production (honours option in BSc
Agriculture) (f/t) Pg/D & MSc
Sustainable Land Management and Rural
Development (organic option)

Welsh College of Horticulture
Northop, Mold, Flintshire CH7 6AA
Tel: (01352) 841000 Fax: (01352) 841031
Internet: www.wcoh.ac.uk
E-Mail: enquiries@wcoh.ac.uk
Course titles: HNC/HND in Horticulture
(organic options) (f/t)
(http://www.wcoh.ac.uk/sa_hiht.htm#org)

Wiltshire College, Lackham
Lacock, Chippenham, Wiltshire SN15 2NY

Tel: 01249 466800 Fax: 01249 444474
E-Mail: *info@wiltscoll.ac.uk*
Website: www.wiltscoll.ac.uk

GOVERNMENT/GRANT-AWARDING AGENCIES

Crofters Commission
Castle Wynd, Inverness, IV2 3EQ
Tel: 01463 663408   Fax: 01463 711820
Internet: www.crofterscommission.org.uk
E-Mail: info@crofterscommission.org.uk
Countryside Agency (Landscape, access
& recreation division will transfer to
Natural England)
John Dower House, Crescent Place,
Cheltenham, GL50 3RA
Tel: (01242) 521381 Fax: (01242) 584270
Internet: www.countryside.gov.uk
E-Mail: info@countryside.gov.uk

Countryside Council for Wales (CCW)
Maes y Ffynnon, Penrhosgarnedd,
Bangor, Gwynedd, LL57 2DW
Tel: 01248 385500   Enquiry line:
0845 1306229 Fax: 01248 355782
Internet: www.ccw.gov.uk
E-Mail: enquiries@ccw.gov.uk
See website for regional area office
contact details N.B. Tir Gofal and
Tir Cymen will be administered by
WAG-DEPC from 2007

Department for Environment, Food and
Rural Affairs (DEFRA)
Nobel House, 17 Smith Square,
London SW1P 3JR
DEFRA Help line: 08459 335577
Fax: 020 270 8419
Internet: www.defra.gov.uk/farm/index
E-Mail: helpline@defra.gsi.gov.uk

DEFRA Organic Strategy Branch
Area 4D, Nobel House, 17 Smith Square,
London SW1P 3JR
Tel: 020 7238 5605   Fax: 020 7238 6148
Internet: www.defra.gov.uk/
farm/organic
E-Mail: organic.standards@
defra.gsi.gov.uk

Department of Agriculture and Food (Republic of Ireland)
Organic Unit, Johnstown Castle Estate, Co. Wexford, Ireland
Tel: +353 (0) 53 63400
Fax: +353 (0) 53 43965
Internet: www.agriculture.gov.ie/
E-Mail: organics@agriculture.gov.ie

Department of Agriculture and Rural Development (DARD-NI)
Dundonald House, Upper Newtownards Road, Belfast, BT4 3SB
Tel: 028 9052 4999 Fax: 028 9052 5546;
E-Mail: library@dardni.gov.uk
Internet: www.dardni.gov.uk
E-Mail: dardhelpline@dardni.gov.uk

Environment Agency for England and Wales
Local offices throughout England and Wales on website
Agricultural Waste Registration
Tel: 0845 603 3113
Internet: www.environment-agency.gov.uk;
E-Mail enquiries via main website

English Nature (EN) – (Will transfer to Natural England in 2007)
Northminster House, Northminster Road, Peterborough, PE1 1UA
Tel: 01733 455000 Fax: 01733 568834
Internet: www.english-nature.org.uk
E-Mail: enquiries@english-nature.org.uk
Food from Britain
4th Floor, Manning House, 22 Carlisle Place, London, SW1P 1JA
Tel: 020 7233 5111   Fax: 020 7233 9515
Internet: www.foodfrombritain.com
E-Mail: info@foodfrombritain.com
Forestry Commission
Forestry Commission Head Office, 231 Corstorphine Road, Edinburgh, EH12 7AT
Tel: 0131 334 0303 or 0845 367 3787
Fax: 0131 334 3047
Internet: www.forestry.gov.uk
E-Mail: enquiries@forestry.gsi.gov.uk

Forest Service (Northern Ireland)
Customer Services Manager, Dundonald House, Upper Newtownards Road, BELFAST, BT4 3SB; Tel: 028 9052 4480
Fax: 028 9052 4570
Internet: www.forestserviceni.gov.uk
E-Mail: customer.forestservice@dardni.gov.uk

Forest Service – Republic of Ireland
Department of Agriculture and Food, Johnstown Castle Estate, Co. Wexford, Republic of Ireland Tel: +353 (0) 53 60200
Fax: +353 (0) 53 43834 Internet: www.agriculture.gov.ie/index.jsp?file=forestry/pages/index.xml

Highlands and Island Enterprise
Cowan House, Inverness Retail and Business Park, Inverness, IV2 7GF
Tel: 01463 234171 Fax: 01463 244469
Internet: www.hie.co.uk
E-Mail: hie.general@hient.co.uk

Natural Rural Resource Tourism Initiative
Rural Development North Branch, Ecos Centre, Kernoghans Lane, Broughshane Road, Ballymena, BT43 7QA
Tel: 028 2563 3812
Internet: www.rdpni.gov.uk/nrrt
E-Mail: una.morgan@dardni.gov.uk

REGIONAL DEVELOPMENT AGENCIES (RDA)

National Secretariat, Broadway House, Tothill Street, London, SW1H 9NQ
Tel: 020 7222 8180 Fax: 020 7222 8182
Internet: www.englandsrdas.com See homepage for contact details and web addresses for the local and regional RDA offices

Rural Development Service (RDS)-Agri-environment
scheme admin transfers to Natural England, in 2007 Nobel House, 17 Smith Square, London, SW1P 3JR
Tel: 02072 385432 Fax: 02072 385372
Internet: www.defra.gov.uk/rds

Rural Payments Agency (RPA)
Reading HQ, Kings House, 33 Kings Road, Reading, RG1 3BU

Tel: 0118 958 3626 Fax: 0118 959 7736
Internet: www.rpa.gov.uk
E-Mail: enquiries@rpa.gsi.gov.uk
Helpline numbers: Single Payment
Scheme and bovine
schemes 0845 603 7777
Dairy schemes inc.
Milk Quotas 01392 266 466
External Trade 0191 226 5050
Slaughter Schemes 0118 968 7333
Scottish Natural Heritage
12 Hope Terrace, Edinburgh, EH9 2AS
Tel: 0131 447 4784, For grants
Tel: 01738 458677 Fax: 0131 446 2277
Internet: www.snh.org.uk E-Mail:
enquiries@snh.gov.uk

Scottish Executive Environmental and
Rural Affairs Department (SEERAD)
Pentland House, 47 Robb's Loan,
Edinburgh EH14 1TY
Tel: 0131 556 8400 or 08457 741741;
Organic Stakeholder Group
Tel: 0131 244 4765
Internet: www.scotland.gov.uk/
Topics/Agriculture
E-Mail: ceu@scotland.gov.uk

Welsh Assembly Government,
Department for Environment, Planning
and the Countryside (WAG-DEPC)
Crown Building, Cathays Park,
Cardiff CF10 3NQ
Tel: 02920 825111
E-Mail: agriculture@wales.gsi.gov.uk
Internet: new.wales.gov.uk/topics/
environmentcountryside
Carmarthen Divisional Office, Government
Buildings, Picton Terrace, Carmarthen,
SA31 3BT; Tel: 01267 225300
Fax: 01267 235964
E-Mail: agriculture.carmarthen@
wales.gsi.gov.uk
Caernarfon Divisional Office, Government
Buildings, Penrallt,
Caernarfon, LL55 1EP
Tel: 01286 674144 Fax: 01286 677749
E-Mail: agriculture.caernarfon@
wales.gsi.gov.uk
Llandrindod Wells Divisional Office,

Government Buildings, Spa Road East,
Llandrindod Wells, LD1 5HA
Tel: 01597 823777 Fax: 01597 828304
E-Mail: agriculture.llandrindod@
wales.gsi.gov.uk
CAP Management Division, Ffynnon
Las, TŷGlas Avenue, Llanishen, Cardiff,
CF14 5EZ; Tel: 029 2075 2222
Fax: 029 2068 1381
E-Mail: agriculture@wales.gsi.gov.uk

WAG-DEPC Food and Market
Development
Plas Glyndwr, Kingsway, Cardiff,
CF10 3AH
Tel: 08450 103300 (English) /
08450 104400 (Welsh)
Mid Wales Office, Ladywell House,
Newtown, Powys, SY16 1JB
Tel: 01686 613153 Fax: 01686 622499
Y Lanfa, Trefechan, Aberystwyth,
SY23 1AS
Tel and Fax – use main contact numbers
South East Wales Division, QED Centre,
Main Avenue, Treforest Estate,
Pontypridd, CF37 5YR
Tel: 01443 845500
Fax: 01443 845589
South West Wales Office, Llys-Y-Ddraig,
Penllergaer Business Park,
Swansea, SA4 1HL
Tel: 01792 222422 Fax: 01792 222498
Unit 7, Fford Richard Davies, St. Asaph
Business Park, St Asaph,
North Wales, LL17 0LJ Tel: 01745 586244
Fax: 01745 586259

GENERAL FARM BUSINESS AND BUSINESS ADVICE

As part of the CAP reform, all member
states will need to introduce a voluntary
farm advisory system by 2007. This is a
combination of farm audit and advice in
order to enhance the performance of all
farms with respect to environmental, animal welfare, food safety and occupa-
tional health and safety.

Organic Centre Wales, the Organic
Study Centre in Cornwall and SAC offer

some financial benchmarking services for organic producers.

ADAS
Woodthorne, Wergs Road,
Wolverhampton, WV6 8TQ
National help desk: 0845 766 0085
Internet: www.adas.co.uk/;
Many local centres throughout England and Wales. See website
for local address details.
Tel: 08457 96 97 98

Business Gateway – Scotland
Many local centres throughout Scotland (see also separate entry for Highlands and Islands) See website for local address details ("Business information" link under "Services" category)
Tel: 0845 609 6611
Internet: http://www.bgateway.com
E-Mail: (use "contact us" page on website)

Business Link-Local Centres throughout England Tel: (0845) 600 9 006
Internet: www.businesslink.gov.uk
E-Mail: (find local centre e-mails on website)

DEFRA Farm Business Benchmarking online
Internet: farmbusinessbenchmark.
defra.gov.uk/ Farm Business Advice – England regional help desks
North West Region Tel: (0870) 870 7380;
North East Region Tel: (0870) 870 7381;
Yorkshire & Humber Region Tel: (0870) 870 7382 East Midlands Region
Tel: (0870) 870 7383; East of England
Tel: (0870) 870 7384.
Internet: www.farmbusinessadvice.co.uk
E-Mail: info@farmbusinessadvice.co.uk

Farming Connect Wales
Tel: 08456 000813; Internet:
www.wales.gov.uk/farmingconnect/
Food Chain Centre
Food Chain Centre at IGD, Letchmore Heath, Watford, WD25 8GD
Tel: 01923 857141 Fax: 01923 852531
Internet: www.foodchaincentre.com
E-Mail: foodchaincentre@igd.com

Invest Northern Ireland
44-58 May Street, Belfast, BT1 4NN
Tel: 028 9023 9090 Fax: 028 9049 0490
Internet: www.investni.com E-Mail:
info@investni.com

Scottish Enterprise
5 Atlantic Quay; 150 Broomielaw;
Glasgow, G2 8LU
Tel: 0141 248 2700 Fax: 0141 221 3217
Internet: www.scottish-enterprise.com
E-Mail: network.helpline@scotent.co.uk

GOVERNMENT AGENCIES

Defra

Information Resource Centre, Lower Ground Floor, Ergon House,
c/o Nobel House, 17 Smith Square,
London SW1P 3JR
Defra helpline: 08459 33 55 77
E-Mail: helpline@defra.gsi.gov.uk
Website: www.defra.gov.uk

Defra Rural Development Service for Organic Entry Level Scheme and Higher Level Scheme
Defra (RDS East of England)
Eastbrook, Shaftesbury Road, Cambridge CB2 2DR
Tel: 01223 462727
E-Mail: enquiries.east@defra.gsi.gov.uk
Website: www.defra.gov.uk/rds/ee

Defra (RDS East Midlands)
Block 7, Chalfont Drive, Nottingham NG8 3SN
Tel: 0115 929 1191
E-Mail:
enquiries.eastmidlands@defra.gsi.gov.uk
Website: www.defra.gov.uk/rds/em

Defra (RDS North West)
Crewe Business Park, Electra Way,
Crewe, Cheshire CW1 6GJ
Tel: 01270 754000
E-Mail: enquiries.crewe@defra.gsi.gov.uk
Website: www.defra.gov.uk/rds/nw

Defra (RDS South East)
Government Buildings, Coley Park,
Reading, Berkshire RG1 6DT

Tel: 0118 958 1222
E-Mail: enquiries.southeast@
defra.gsi.gov.uk
Website: www.defra.gov.uk/rds/se

Defra (RDS South West)
Block 3, Government Buildings,
Burghill Road,
Westbury-on-Trym, Bristol BS10 6NJ
Tel: 0117 959 1000
E-Mail: enquiries.southwest@
defra.gsi.gov.uk
Website: www.defra.gov.uk/rds/sw

Defra (RDS West Midlands)
Block B, Government Buildings,
Whittington Road, Worcester WR5 2LQ
Tel: 01905 763355
E-Mail: enquiries.westmidlands@
defra.gsi.gov.uk
Website: www.defra.gov.uk/rds/wm

Defra (RDS Yorkshire and Humber)
Government Buildings, Otley Road,
Lawnswood, Leeds LS16 5QT
Tel: 0113 230 3750
E-Mail: enquiries.yorkshumber@
defra.gsi.gov.uk
Website: www.defra.gov.uk/rds/yh

Farming Connect Wales
Tel: 08456 000813
Website: www.wales.gov.uk/
farmingconnect

Rural Payments Agency
RPA Customer Service Centre –
0845 603 7777
Email: *customer.service.centre@*
rpa.gsi.gov.uk
Website: www.rpa.gov.uk

**Scottish Executive Environment and
Rural Affairs Department**
Pentland House, 47 Robb's Loan,
Edinburgh EH14 1TY
Tel: 0131 556 8400
Fax: 0131 244 6116
E-Mail: ceu@scotland.gsi.gov.uk
Website: www.scotland.gov.uk

**Welsh Assembly Government
Department for Environment, Planning
and the Countryside**

Crown Building, Cathays Park,
Cardiff CF10 3NQ
Tel: 02920 825111
E-Mail: agriculture@wales.gsi.gov.uk
Website: www.countryside.wales.gov.uk

OTHER USEFUL ADDRESSES AND CONTACTS

Abacus Organic Associates
Tel/Fax 01780 721019
E sales@abacusorganic.co.uk

Association of Independent Crop
Consultants: Agriculture House, Station
Road, LISS, Hants GU33 7AR
Tel 01730 895354

Assured Combinable Crops: Secretariat,
11 Orchard Avenue, Thames Ditton,
Surrey KT17 0BB
Tel 01993 885652

Central Science Laboratory: Sand
Hutton, York YO41 1LZ
Tel 01904 2000
E -Mail science@csl.gov.uk

Farming and Wildlife Advisory Group:
NAC, Stoneleigh, Kenilworth,
Warwickshire CV8 2RX
Tel 024 7669 6699

Home Grown Cereals Authority:
Caledonian House, 223 Pentonville Road,
London N1 9HY
Tel 020 7520 3926

Lazy Dog Tools Ltd: Hill Top Farm,
Spaunton Appleton-Le-Moors, Yorkshire
YO62 6TR
Tel 01751 417351
Fax 01751 417642

Machinery Rings Association: Wood
St Farm Cottage, Catfield, Great
Yarmouth NR29 5DF
Tel 01629 582276

National Association of Agricultural
Contractors: 8 High Street, Maldon,
Essex CM9 5PJ
Tel 01621 841675

Organic Conversion Information Service (OCIS): England 0117 922 7707 • Scotland 01224 711072 Wales 01970 622100 • Northern Ireland (crops and horticulture) T 028 9070 1115 (livestock) Tel 028 9442 6752

Soil Association food and farming department: Bristol House, 40–56 Victoria Street, Bristol BS1 6BY • Tel 0117 914 2400 • E-Mail ff@soilassociation.org

Stephen Briggs, Rowan House, 9 Pinfold Close, South Luffenham, Rutland, LE15 8NE
Tel/Fax: 01780 721019
Internet: http://www.abacusorganic. co.uk E-Mail: Stephen.briggs@ abacusorganic.co.uk

UKASTA: 3 Whitehall Court, London SW1A 2EQ • Tel 020 7930 3611 • E-Mail enquiries@ukasta.org.uk

Windrow turning machines: Westcon Equipment, Unit 2A, 27 Brook Road, Wimborne, Dorset tel 01202 880380 – Nick Rumsey – have a very good Sandberger machine circa £12K-£15K

Morawetz composting – 10 Falcon Knowl Ing, Darton, Barnsley, South Yourk, Tel 01226 388078 - Richard Newton

INDEX